Supported Reagents and Catalysts in Chemistry

Supported Reagents and Catalysts in Chemistry

Edited by

B.K. Hodnett
University of Limerick, Ireland

A.P. Kybett
The Royal Society of Chemistry, Cambridge, UK

J.H. Clark
University of York, UK

K. Smith
University of Wales, UK

THE ROYAL
SOCIETY OF
CHEMISTRY
Information
Services

The proceedings of The Royal Society of Chemistry Third International Symposium on Supported Reagents and Catalysts in Chemistry (3rd ISSRC) held at the University of Limerick on 8–11 July, 1997.

Special Publication No. 216

ISBN 0-85404-797-2

A catalogue record for this book is available from the British Library

Published by The Royal Society of Chemistry,
Thomas Graham House, Science Park, Milton Road,
Cambridge CB4 4WF, UK
For further information visit our web site at www.rsc.org

Printed by Bookcraft (Bath) Ltd

Preface

As the chemical industry comes under increasing pressure to develop production methods based on cleaner technologies, there is now a sustained effort to develop a range of catalysts and supported reagents to form the backbone of these new technologies. Towards this end there are significant advances being made in the application of microporous and mesoporous zeolitic-type materials, some with isomorphous substitution for framework aluminium ions, as catalysts (i) for selective oxidation reactions with hydrogen peroxide and other clean oxidizing agents and (ii) for a wide range of acid-base reactions, where these materials may eliminate waste associated with the disposal of spent mineral acid solutions. In addition, the field of enantioselective hydrogenation, mediated by heterogeneous catalysts is developing close to the point of commercial realisation. Other areas are also being addressed, including the challenge of finding a catalyst to treat exhaust from lean-burn petrol and diesel engines.

These areas were addressed at the 3rd International Symposium on Supported Reagents and Catalysts in Chemistry held at the University of Limerick, 8-11 July 1997. The 12 keynote and invited lectures and the 24 other contributions present the latest developments in these areas.

Cover illustration. The Pol na mBrón (Pool of the sorrows) dolmen in County Clare, Ireland. This megalithic structure dates from 2500BC and is one of the oldest known man-made supported structures.

Contents

KL:Keynote Lecture
IL : Invited Lecture
CL: Contributed Lecture
PL: Postgraduate Lecture

Dedication

Russell Drago (5 Nov 1928–5 Dec 1997)

Professor Russell Drago, Graduate Research Professor of Chemistry at the University of Florida and the Opening Keynote Lecturer at the Third International Symposium on Supported Reagents and Catalysts in Chemistry died on the 5th of December 1997. As a mark of respect, and on behalf of all symposium participants, we dedicate the proceedings of this conference to our late colleague.

Russ Drago was one of the academic world's most talented and versatile research chemists and teachers. His contributions to the study of catalysis and acid–base phenomena including his recent seminal work on solid acids are enormous and yet represent only some of his interests and a fraction of his scientific output. The Florida Catalysis Conferences became major international events in the catalysis calendar and many of us have fond memories of balmy days at Palm Coast. The mixed academic–industrial audience and age spread of the participants at those meetings reflected Russ's appreciation of the need to bridge the pure–applied chemistry gap and his enjoyment of working with younger colleagues. He was an inspiration to us all and he will be sadly missed.

B.K. Hodnett
A.P. Kybett
J.H. Clark
K. Smith

List of Contributors

Main Authors indicated by *

KL1.   Russell S. Drago*.

Russell S. Drago*, Department of Chemistry, University of Florida, Gainesville, FL 32611 - 7200, USA.

CL1.   P. J. Kunkeler*, J.A. Elings, R. S. Downing and H. van Bekkum.

P. J. Kunkeler, Delft University of Technology, Laboratory of Organic Chemistry and Catalysis, Julianalaan 136, 2628 BL, Delft, The Netherlands.

J. A. Elings, Delft University of Technology, Laboratory of Organic Chemistry and Catalysis, Julianalaan 136, 2628 BL, Delft, The Netherlands.

R. S. Downing, Delft University of Technology, Laboratory of Organic Chemistry and Catalysis, Julianalaan 136, 2628 BL, Delft, The Netherlands.

H. van Bekkum, Delft University of Technology, Laboratory of Organic Chemistry and Catalysis, Julianalaan 136, 2628 BL, Delft, The Netherlands.

CL2.   A. J. H. Klunder*, A. C. L. M. van der Waals and B. Zwanenburg

A. J. H. Klunder*, Department of Organic Chemistry, NSR Center for Molecular Structure, Design and Synthesis, University of Nijmegen, Toernooiveld, 6525 ED Nijmegen, The Netherlands.

A. C. L. M. van der Waals, Department of Organic Chemistry, NSR Center for Molecular Structure, Design and Synthesis, University of Nijmegen, Toernooiveld, 6525 ED Nijmegen, The Netherlands.

B. Zwanenburg, Department of Organic Chemistry, NSR Center for Molecular Structure, Design and Synthesis, University of Nijmegen, Toernooiveld, 6525 ED Nijmegen, The Netherlands.

IL 1   A.Corma*,

A.Corma, Instituto de tecnologia Quimica, UPV-CSIC, Universidad Politnicia de Valencia, Avda de los Naranjos s/n, 46022 Valencia, Spain

CL 3 Tomasz Wasowicz and Larry Kevan*.

Larry Kevan*, University of Houston, Department of Chemistry, Houston, Texas 7204-5641, USA.

Tomasz Wasowicz, University of Houston, Department of Chemistry, Houston, Texas 77204-5641, USA.

CL 4. Á. Molnár*, T. Beregszászi, Á. Fudala, B. Török, M. Rózsa-Tarjáni and I. Kiricsi.

Á. Molnár*, Department of Organic Chemistry , Jozsef Attila University, Dom ter 8, Szeged, H-6720, Hungary.

T. Beregszászi, Department of Organic Chemistry , Jozsef Attila University, Dom ter 8, Szeged, H-6720, Hungary.

Á. Fudala, Applied Chemistry Department, Jozsef Attila University, Dom ter 8, Szeged, H-6720, Hungary.

B. Török, Department of Organic Chemistry and Organic Catalysis Research Group, Jozsef Attila University, Dom ter 8, Szeged, H-6720, Hungary.

M. Rózsa-Tarjáni, Department of Organic Chemistry , Jozsef Attila University, Dom ter 8, Szeged, H-6720, Hungary.

I. Kiricsi, Applied Chemistry Department, Jozsef Attila University, Dom ter 8, Szeged, H-6720, Hungary.

CL 5. Keith Smith*.

Keith Smith*, Department of Chemistry, University of Wales Swansea, Swansea SA2 8PP, UK.

KL 2. R. A. Sheldon*, I.W.C.E. Arends and H.E.B. Lempers.

R. A. Sheldon Laboratory for Organic Chemistry and Catalysis, Delft University of Technology, Julianalaan 136, 2628 BL Delft, The Netherlands.

I.W.C.E. Arends Laboratory for Organic Chemistry and Catalysis, Delft University of Technology, Julianalaan 136, 2628 BL Delft, The Netherlands

H.E.B. Lempers Laboratory for Organic Chemistry and Catalysis, Delft University of Technology, Julianalaan 136, 2628 BL Delft, The Netherlands

KL 3   Sir J M Thomas*

Sir J M Thomas, The Master's Lodge, Peterhouse, Cambridge CB2 1QY, UK

CL 6.   I. Belkhir*, A. Germain, F. Fajula and E. Fache.

I. Belkhir*, Laboratoire de Matériaux Catalytiques et Catalyse en Chimie Organique, UMR-CNRS 5618, ENSCM, 8, Rue de l'Ecole Normale, 34296 Montpellier Cedex 5, France.

A. Germain, Laboratoire de Matériaux Catalytiques et Catalyse en Chimie Organique, UMR-CNRS 5618, ENSCM, 8, Rue de l'Ecole Normale, 34296 Montpellier Cedex 5, France.

F. Fajula, Laboratoire de Matériaux Catalytiques et Catalyse en Chimie Organique, UMR-CNRS 5618, ENSCM, 8, Rue de l'Ecole Normale, 34296 Montpellier Cedex 5, France.

E. Fache, Rhône-Poulenc Industrialisation, CRIT-Carrieres, 85, Avenue des Frères Perret, BP 62, 69192 Saint-Fons Cedex, France.

CL 7   P. Sutra and D. Brunel*.

D. Brunel*, Laboratoire de Matériaux Catalytiques et Catalyse en Chimie Organique UMR 5618 ENSCM/CNRS, 8, Rue Ecole Normale, 34296 Montpellier, Cedex 5, France.

P. Sutra, Laboratoire de Matériaux Catalytiques et Catalyse en Chimie Organique UMR 5618 ENSCM/CNRS, 8, Rue Ecole Normale, 34296 Montpellier, Cedex 5, France.

CL 8.   Christopher H. Barclay and John M. Winfield*.

John M. Winfield*, Department of Chemistry, University of Glasgow, Glasgow G12 8QQ, Scotland.

Christopher H. Barclay, Department of Chemistry, University of Glasgow, Glasgow G12 8QQ, Scotland.

CL 9.   T. Mallat*, L. Seyler, M. Mir Alai and A. Baiker.

A. Baiker*, Department of Chemical Engineering and Industrial Chemistry, Swiss Federal Institute of Technology, CH-8092 Zürich, Switzerland.

T. Mallat, Department of Chemical Engineering and Industrial Chemistry, Swiss Federal Institute of Technology, CH-8092 Zürich, Switzerland.

L. Seyler, Department of Chemical Engineering and Industrial Chemistry, Swiss Federal Institute of Technology, CH-8092 Zürich, Switzerland.

M. Mir Alai, Department of Chemical Engineering and Industrial Chemistry, Swiss Federal Institute of Technology, CH-8092 Zürich, Switzerland.

IL2.   J. A. Cairns and J. Thomson*.

J. Thomson*, Department of Chemistry, University of Dundee, Dundee, DD1 4HN, Scotland, UK.

J. A. Cairns, Department of Applied Physics and Electronic and Mechanical Engineering, University of Dundee, Dundee, DD1 4HN, Scotland, UK.

CL10.   H. Ogawa*.

H. Ogawa*, Department of Chemistry, Tokyo Gakugei University, Koganei, Tokyo 184, Japan.

CL11.   Joseph Cunningham* and James. A. Sullivan.

Joseph Cunningham, Department of Chemistry, University College, Cork, Cork City, Ireland.

James. A. Sullivan, Department of Chemistry, University College, Cork, Cork City, Ireland.

IL 3   N. Essayem, V. Martin, A. Riondel, G. Coudurier, J. C. Védrine*.

J. C. Védrine*, Institut de Rechereches sur la Catalyse, associé à l'UCB Lyon I, CNRS, 2 avenue Albert Einstein, F-69626 Villeurbanne, France.

N. Essayem, Institut de Rechereches sur la Catalyse, associé à l'UCB Lyon I, CNRS, 2 avenue Albert Einstein, F-69626 Villeurbanne, France.

V. Martin, Institut de Rechereches sur la Catalyse, associé à l'UCB Lyon I, CNRS, 2 avenue Albert Einstein, F-69626 Villeurbanne, France.

A. Riondel, Centre De Recherches et de Developpement de l'Est, Elf Atochem, F-57501 Saint Avold, France.

G Coudurier, Institut de Rechereches sur la Catalyse, associé à l'UCB Lyon I, CNRS, 2 avenue Albert Einstein, F-69626 Villeurbanne, France.

IL 4   Hans-Ulrich Blaser*, Benoit Pugin.

Hans-Ulrich Blaser*, Catalysis & Synthesis Services, Novartis Services AG, R 1055.6 CH-4002 Basel, Switzerland.

Benoit Pugin, Catalysis & Synthesis Services, Novartis Services AG, R 1055.6 CH-4002 Basel, Switzerland.

CL 12   Y. V. Subba Rao*, D. E. De Vos and P. A. Jacobs.

P. A. Jacobs*, Centre for Surface Science and Catalysis, Katholieke Universiteit Leuven, Kardinaal Mercierlaan 92, 3001 Heverlee, Belgium.

Y. V. Subba Rao, Centre for Surface Science and Catalysis, Katholieke Universiteit Leuven, Kardinaal Mercierlaan 92, 3001 Heverlee, Belgium.

D. E. De Vos, Centre for Surface Science and Catalysis, Katholieke Universiteit Leuven, Kardinaal Mercierlaan 92, 3001 Heverlee, Belgium.

CL 13   A. R. Maguire* and L. L. Kelleher

A. R. Maguire*, Department of Chemistry, University College Cork, Cork City, Ireland.

L. L. Kelleher, Department of Chemistry, University College Cork, Cork City, Ireland.

CL 14   N. W. A. Geraghty* and M. J. Monaghan.

N. W. A. Geraghty, Chemistry Department, University College Galway, Ireland.

M. M. Monaghan, Chemistry Department, University College Galway, Ireland.

CL15   H. L. Holland*, J. -X. Gu, and D. Riffle, E. N. Vulfson and J. A. Khan.

H. L. Holland, Department of Chemistry Brock University, St Catherines, On L2S 3A1, Canada.

J. -X. Gu, Department of Chemistry Brock University, St Catherines, On L2S 3A1, Canada.

D. Riffle, Department of Chemistry Brock University, St Catherines, On L2S 3A1, Canada.

E. N. Vulfson, Institute of Food Research, Early Gate, Whiteknights Road, Reading RG 6 6BZ, UK.

J. A. Khan, Institute of Food Research, Early Gate, Whiteknights Road, Reading RG 6 6BZ, UK.

KL 4   Makoto Misono*, Toshio Okuhara and Sawami Shikata.

Toshio Okuhara, Graduate School of Environmental Earth Science, Hokkaido University, Sapporo 060, Japan.

Makoto Misono, Department of Applied Chemisty, Graduate School of Engineering, The University of Tokyo, Bunkyo-ku, Tokyo 113, Japan.

Sawami Shikata, Department of Applied Chemisty, Graduate School of Engineering, The University of Tokyo, Bunkyo-ku, Tokyo 113, Japan.

IL5   Alfons Baiker*

Alfons Baiker*, Department of Chemical Engineering and Industrial Chemistry, Swiss Federal Institute of Technology, ETH-Zentrum, CH-8092 Zurich, Switzerland.

CL16   A. Choplin*, F. Quignard, P. Leyrit, S. dos Santos, C. McGill, O. Graziani, D. Sinou.

A. Choplin*, Institut de Recherches sur la Catalyse, and Laboratoire de Synthèse Asymétrique, CNRS-Université Lyon-1, Villeurbanne, France.

F. Quignard, Institut de Recherches sur la Catalyse, and Laboratoire de Synthèse Asymétrique, CNRS-Université Lyon-1, Villeurbanne, France.

P. Leyrit, Institut de Recherches sur la Catalyse, and Laboratoire de Synthèse Asymétrique, CNRS-Université Lyon-1, Villeurbanne, France.

S. dos Santos, Institut de Recherches sur la Catalyse, and Laboratoire de Synthèse Asymétrique, CNRS-Université Lyon-1, Villeurbanne, France.

C. McGill, Institut de Recherches sur la Catalyse, and Laboratoire de Synthèse Asymétrique, CNRS-Université Lyon-1, Villeurbanne, France.

O. Graziani, Institut de Recherches sur la Catalyse, and Laboratoire de Synthèse Asymétrique, CNRS-Université Lyon-1, Villeurbanne, France.

D. Sinou, Institut de Recherches sur la Catalyse, and Laboratoire de Synthèse Asymétrique, CNRS-Université Lyon-1, Villeurbanne, France.

CL17   Stewart J. Tavener*, James H. Clark, Paul A. Heath, Duncan J. Macquarrie and John Rafelt.

James H. Clark, Department of Chemistry University of York, Heslington, York, UK YO1 5DD.

Stewart J. Tavener, Department of Chemistry University of York, Heslington, York, UK YO1 5DD.

Paul A. Heath, Department of Chemistry University of York, Heslington, York, UK YO1 5DD.

Duncan J. Macquarrie, Department of Chemistry University of York, Heslington, York, UK YO1 5DD.

John Rafelt, Department of Chemistry University of York, Heslington, York, UK YO1 5DD.

CL 18   N. Bellocq*, D. Brunel, M. Laspéras and P. Moreau

P. Moreau, Laboratoire de Matériaux Catalytiques et Catalyse en Chimie Organique - UMR 5618 Ecole Nationale Supérieure de Chimie de Montpellier - CNRS, 8, rue de l'Ecole Normale, 34296 Montpellier, Cedex 5, France.

N. Bellocq, Laboratoire de Matériaux Catalytiques et Catalyse en Chimie Organique - UMR 5618 Ecole Nationale Supérieure de Chimie de Montpellier - CNRS, 8, rue de l'Ecole Normale, 34296 Montpellier, Cedex 5, France.

D. Brunel, Laboratoire de Matériaux Catalytiques et Catalyse en Chimie Organique - UMR 5618 Ecole Nationale Supérieure de Chimie de Montpellier - CNRS, 8, rue de l'Ecole Normale, 34296 Montpellier, Cedex 5, France.

M. Laspéras, Laboratoire de Matériaux Catalytiques et Catalyse en Chimie Organique - UMR 5618 Ecole Nationale Supérieure de Chimie de Montpellier - CNRS, 8, rue de l'Ecole Normale, 34296 Montpellier, Cedex 5, France.

CL 19   S. O'Brien, T. R. Spalding S. E. Lawrence and M. A. Morris*

M. A. Morris*, Materials and Inorganic Sections, Department of Chemistry, University College Cork, Cork, Ireland.

S. O' Brien, Materials and Inorganic Sections, Department of Chemistry, University College Cork, Cork, Ireland.

T. R. Spalding, Materials and Inorganic Sections, Department of Chemistry, University College Cork, Cork, Ireland.

S. E. Lawrence, Materials and Inorganic Sections, Department of Chemistry, University College Cork, Cork, Ireland.

IL 6   Duncan J. Macquarrie*, James H. Clark, Dominic B. Jackson, Arnold Lambert, James E. G. Mdoe, and Andrew Priest.

Duncan J. Macquarrie*, Department of Chemistry, University of York, Heslington, York YO1 5DD, UK.

James H. Clark, Department of Chemistry, University of York, Heslington, York YO1 5DD, UK.

Dominic B. Jackson, Department of Chemistry, University of York, Heslington, York YO1 5DD, UK.

Arnold Lambert, Department of Chemistry, University of York, Heslington, York YO1 5DD, UK.

James E. G. Mdoe, Department of Chemistry, University of York, Heslington, York YO1 5DD, UK.

Andrew Priest, Department of Chemistry, University of York, Heslington, York YO1 5DD, UK.

CL 20   B Corain*, A.A. D'Archivio, L. Galantini, K. Jeřàbek, M. Králik, S. Lora, G. Palma, M. Zecca.

B. Corain*, Universita' di L'Aquila, Dipartimento di Chimica, Ingegneria Chimica e Materiali, via Vetoio, 67010 L'Aquila, Italy, also Centro per lo Studio della Stabilità e Reattività dei Cimposti di Coordinazione, C.N.R., via Marzolo 1, 35131 Padova, Italy.

A.A. D'Archivio, Universita' di L'Aquila, Dipartimento di Chimica, Ingegneria Chimica e Materiali, via Vetoio, 67010 L'Aquila, Italy.

L. Galantini, Universita' di L'Aquila, Dipartimento di Chimica, Ingegneria Chimica e Materiali, via Vetoio, 67010 L'Aquila, Italy.

K. Jeřábek, Institute of Chemical Processes Fundamentals, Academy of Sciences of the Czech Republic, Rozvojova 135, 165 02 Suchdol, Praha 6, Czech Republic.

S. Lora, Instituto F.R.A.E., C.N.R., via Romea 4, 35020 Legnaro, Italy.

G. Palma, Dipartimento di Chimica Fisica, via Loredan 2, 35131 Padova, Italy.

M Zecca, Dipartimento di Chimia Inorganica, Metallorganica e Analitica, via Marzolo 1, 35131 Padova, Italy.

CL 21   J.H. Clark, S. Evans and J.R. Lindsay Smith*.

J.R. Lindsay Smith*, Department of Chemistry, The University of York, York, YO1 5DD, UK.

J.H. Clark, Department of Chemistry, The University of York, York, YO1 5DD, UK.

S. Evans, Department of Chemistry, The University of York, York, YO1 5DD, UK.

CL 22   B. Kondratowicz, J. Sithamparapillai, A. Subramaniam and P. A. Sermon*

P. A. Sermon, Fractal Solids and Surfaces Research Group, Department of Chemistry, Brunel University, Uxbridge, Middlesex, UB8 3PH, UK.

B. Kondratowicz, Fractal Solids and Surfaces Research Group, Department of Chemistry, Brunel University, Uxbridge, Middlesex, UB8 3PH, UK.

J. Sithamparapillai, Fractal Solids and Surfaces Research Group, Department of Chemistry, Brunel University, Uxbridge, Middlesex, UB8 3PH, UK.

A. Subramaniam, Fractal Solids and Surfaces Research Group, Department of Chemistry, Brunel University, Uxbridge, Middlesex, UB8 3PH, UK.

CL 23.   J. M. Bobbitt*, Z. Ma, D. Bolz, T. Osa, Y. Kashiwagi, Y. Yanagisawa, F. Kurashima, J. Anzai, J.E. Tacorante-Morales.

J. M. Bobbitt*, Department of Chemistry, University of Connecticut, Storrs, CT. 06269-3060, USA.

Z. Ma, Department of Chemistry, University of Connecticut, Storrs, CT. 06269-3060, USA.

D. Bolz, Department of Chemistry, University of Connecticut, Storrs, CT. 06269-3060, USA.

T. Osa, Pharmaceutical Institute, Tohoku University, Ayobayama, Sendai 980-77, Japan.

Y. Kashiwagi, Pharmaceutical Institute, Tohoku University, Ayobayama, Sendai 980-77, Japan.

Y. Yanagisawa, Pharmaceutical Institute, Tohoku University, Ayobayama, Sendai 980-77, Japan.

F. Kurashima, Pharmaceutical Institute, Tohoku University, Ayobayama, Sendai 980-77, Japan.

J. Anzai, Pharmaceutical Institute, Tohoku University, Ayobayama, Sendai 980-77, Japan.

J.E. Tacorante-Morales, Fac, Quimica, Universidad de la Habana, C. Habana, Cuba 10400.

IL 7   Timothy P. Smyth* and Orla Kennedy.

Timothy P. Smyth*, Department of Chemical & Environmental Sciences, University of Limerick, National Technological Park, County Limerick, Ireland.

Orla Kennedy, Department of Chemical & Environmental Sciences, University of Limerick, National Technological Park, County Limerick, Ireland.

CL 24   E. Cariati*, E. Lucenti, D. Roberto and R. Ugo.

E. Cariati, Dipartimento di Chimica Inorganica, Metallorganica e Analitica and Centro CNR, Università di Milano, via G. Venezian 21, 20133 Milano, Italy.

E. Lucenti, Dipartimento di Chimica Inorganica, Metallorganica e Analitica and Centro CNR, Università di Milano, via G. Venezian 21, 20133 Milano, Italy.

D. Roberto, Dipartimento di Chimica Inorganica, Metallorganica e Analitica and Centro CNR, Università di Milano, via G. Venezian 21, 20133 Milano, Italy.

R. Ugo, Dipartimento di Chimica Inorganica, Metallorganica e Analitica and Centro CNR, Università di Milano, via G. Venezian 21, 20133 Milano, Italy.

KL 5   D. C. Sherrington*.

D. C. Sherrington*, Department of Pure and Applied Chemistry, University of Strathclyde, 295 Cathedral Street, Glasgow G1 1XL, Scotland.

PL 1   Peter M. Price*, James H. Clark, Keith Martin, Duncan J. Macquarrie, and Tony W. Bastock.

Peter M. Price*, Department of Chemistry, University of York, Heslington, York, YO1 5DD, UK.

James H. Clark, Department of Chemistry, University of York, Heslington, York, YO1 5DD, UK.

Keith Martin, Contract Chemicals Ltd, Knowsley Business Park, Prescot, Merseyside, L34 9HY, UK.

Duncan J. Macquarrie, Department of Chemistry, University of York, Heslington, York, YO1 5DD, UK.

Tony W. Bastock, Contract Chemicals Ltd, Knowsley Business Park, Prescot, Merseyside, L34 9HY, UK.

PL 2   Nicoletta Ravasio, Michaela Finiguerra*, Michele Gargano.

Michaela Finiguerra*, Centro C.N.R. MISO, Dipartimento di Chimica dell'Universita, via Amendola 173, 1-70126 Bari (Italy).

Nicoletta Ravasio, Centro C.N.R. MISO, Dipartimento di Chimica dell'Universita, via Amendola 173, 1-70126 Bari (Italy).

Michele Gargano, Centro C.N.R. MISO, Dipartimento di Chimica dell'Universita, via Amendola 173, 1-70126 Bari (Italy).

PL 3 Z. Wang, J. Cunningham and J. McCarthy*.

J. McCarthy*, Department of Chemistry, University College Cork, Ireland.

Z. Wang, Department of Chemistry, University College Cork, Ireland.

J. Cunningham, Department of Chemistry, University College Cork, Ireland.

PL4 D. Sutton* and J.R.H. Ross.

D. Sutton, Centre for Environmental Research, University of Limerick, Limerick, Ireland

J.R.H. Ross, Centre for Environmental Research, University of Limerick, Limerick, Ireland

PL 5 F. E. Cassidy*, C. Batiot and B. K. Hodnett.

F. E. Cassidy, Department of Chemical and Environmental Sciences, University of Limerick, Limerick, Ireland.

C. Batiot, Department of Chemical and Environmental Sciences, University of Limerick, Limerick, Ireland.

B. K. Hodnett, Department of Chemical and Environmental Sciences, University of Limerick, Limerick, Ireland.

CHARACTERIZING AND SYNTHESIZING STRONG SOLID ACIDS

Russell S. Drago

Department of Chemistry
University of Florida
Gainesville, FL 32611-7200

1 INTRODUCTION

The use of solids as reagents, catalysts, or supports introduces important, new, fundamental, challenges that do not exist in homogeneous solution reactions. Added influences which relate to the heterogeneity of the system must be considered to understand reactivity. Unfortunately, we are poorly-equipped to measure the relevant solid state properties and often employ measurements that can most charitably be described as gross approximations. It is little wonder that property - reactivity correlations often fail in heterogeneous systems. This presentation will offer new measurements for characterizing solid surface area, porosity, and acidity. The new insights that result from the measurements will be illustrated with applications to poorly understood, reported solid acids and to the characterization of a new solid acid.

Solids are conveniently divided into porous and non-porous classifications. Porous materials consist[1] of micropores (< 20 Å), mesopores (20 to 500 Å), and macropores (>500 Å). Surface areas range from 300 to 1200 m^2g^{-1}. Non-porous materials consist of macropores and have surface areas that are less than 300 m^2g^{-1}. Non-porous solids chemisorb adsorbates and require extensive surface functionality to obtain high adsorptive capacities. Porous materials have high capacities for physisorption by virtue of adsorptive dispersion interaction with the walls of the pores.[1] Maximum interaction occurs when the adsorptive dimensions match the pore dimensions.[2] Characterization of solids in terms of pore dimensions, pore capacities, surface areas, acidity and basicity is essential to understand their reactivity.

Surface areas are calculated from gas-solid isotherms (usually N_2) with the BET equation.[3] This equation is only capable of reproducing about one-third of the isotherm. In porous solids, surface areas per gram that exceed the surface area calculated for a monomolecular solid surface result because of capillary condensation. Analysis of the isotherms by empirical methods produces the porosity. Harkins-Jura t-plots[4] give the micropore composition and Barrett-Joyner-Halenda analyses[5] give the mesopore and macropore composition.

1.1 Multiple Equilibrium Adsorption Analyses

A new approach[6,7] to characterizing porous solids treats heterogeneous physisorption with an equilibrium expression much like a distribution coefficient.

$$\text{Ads}_{(g)} + \text{S}_{(\text{solid})} \Leftrightarrow \text{SAds}_{(\text{solid})} \qquad (1)$$

The resulting equilibrium constant is,

$$K = \frac{n_{\text{ads}}}{P_{\text{eq}}[n - n_{\text{ads}}]} \qquad (2)$$

where n_{ads} is the number of moles adsorbed, n is the total capacity of the solid for the adsorption process, P_{eq} is the equilibrium pressure of the gas over the solid, and K is the equilibrium constant for the process in atm^{-1}. The equilibrium constant expression can be rearranged to a form that resembles the Langmuir equation:

$$\frac{P_{\text{eq}}}{n_{\text{ads}}} = \frac{1}{nK} + \frac{P_{\text{eq}}}{n} \qquad (3)$$

Conceptually, this model is very different than the Langmuir model. The n values refer to process capacities and not to sites. The process is recognized as consisting of a distribution of pore dimensions that are close enough in size to be treated as a single process within the accuracy of the measurement.

When Equation 3 is applied to the isotherm for adsorption of gases by microporous carbons or zeolites at up to 1 atm external pressure, poor fits result. This occurs because there is more than one process involved that corresponds to adsorption into different size micropores. Expanding Equation 2 to treat multiple equilibria and rearranging it produces Equation 4. With adsorption measurements expressed as moles adsorbed vs. the equilibrium pressure, Equation 4 is in a form to fit the isotherm.

$$n_{\text{ads}} = \sum_i \frac{n_i K_i P_{\text{eq}}}{1 + K_i P_{\text{eq}}} \qquad (4)$$

For a two process system, one is fitting the isotherm to four unknowns, while six are required for a three process system. Clearly in the three process example, meaningless fits could result because of the large number of parameters. This problem is circumvented by studying isotherms at several temperatures and combining them in the data fit. Since the process capacities are expected to be nearly temperature independent, like solid densities, each new temperature only introduces new Ks as unknowns.

The type of information that can be obtained[7] from the multiple equilibrium analysis is illustrated in Table 1 for the adsorption of CO on Ambersorb® -572. These measurements are carried out above the critical temperatures of the adsorptives and at pressures up to 1 atm, so conclusions are relevant to these conditions.

First note that the enthalpies of the processes are considerably larger than the enthalpies of vaporization as a result of the dispersion interaction of the adsorbate with the solid. For a wide range of adsorptives on carbon the different process enthalpies plot up linearly with the square root of the van der Waals constant. The other significant aspect of these data involves the n-values. Using the molar volumes of the adsorbates one can convert millimoles g^{-1} to volume g^{-1} giving the volumes corresponding to the

Table 1 *Multiple Equilibrium Parameters for the Adsorption of CO on A-572*

Process Number	n-values (mmol g^{-1})	$-\Delta H$ (kcal mole)	K (atm^{-1}) (T °C)
1	0.51	5.2	204 (-93)
2	1.93	4.8	8.44 (-42)
3	5.32	3.0	0.67 (25)

$-\Delta H_{vap}$of CO 1.44

processes, which gives a measure of the distribution of various size pores. In a similar fashion for monomolecular surface coverage, molecular cross sectional areas can be used to determine the surface areas corresponding to the different processes. Determining the isotherm above the critical temperature ensures monomolecular coverage. The very different results obtained from these analyses of porous carbons and conventional surface area and porosity measures have been reported.[8]

Molar volumes, cross sectional areas and kinetic diameters have been calculated[9] from semiempirical molecular orbital calculations, ZINDO. Excellent correlations to measured molar volumes are found. This result allows the determination of these most important molecular properties for systems that have not been measured and provides a consistent quantity for molecular areas and volumes.

Multiple equilibrium analyses, MEA, have been carried out[10] on HZSM-5. The n-values, process volumes and areas are given in Table 2. The different probes all indicate that the volumes of these channels sum up to 0.11 ± 0.02 ml g^{-1}. The molecular areas can

Table 2 *Process Capacities, Volumes and Surface Areas for HZSM-5*

Adsorptive	n_1	n_2	Σ
N$_2$	1.50 mmol g^{-1} 0.038 mL g^{-1} 144 m^2 g^{-1}	1.30 mmol g^{-1} 0.033 mL g^{-1} 125 m^2 g^{-1}	0.071 mL g^{-1} 269 m^2 g^{-1}
CO (n_2, n_3)	1.22 mmol g^{-1} 0.035 mL g^{-1} 132 m^2 g^{-1}	1.70 mmol g^{-1} 0.048 mL g^{-1} 184 m^2 g^{-1}	0.083 mL g^{-1} 358 m^2 g^{-1} (inc n_1)
CH$_4$	1.99 mmol g^{-1} 0.060 mL g^{-1} 248 m^2 g^{-1}	0.920 mmol g^{-1} 0.027 mL g^{-1} 115 m^2 g^{-1}	0.087 mL g^{-1} 363 m^2 g^{-1}
C$_2$H$_6$	1.92 mmol g^{-1} 0.091 mL g^{-1} 293 m^2 g^{-1}	0.600 mmol g^{-1} 0.029 mL g^{-1} 74 m^2 g^{-1}	0.12 mL g^{-1} 367 m^2 g^{-1}
C$_3$H$_8$	0.819 mmol g^{-1} 0.053 mL g^{-1} 175 m^2 g^{-1}	0.85 mmol g^{-1} 0.055 mL g^{-1} 182 m^2 g^{-1}	0.11 mL g^{-1} 357 m^2 g^{-1}
SF$_6$	0.560 mmol g^{-1} 0.043 mL g^{-1} 119 m^2 g^{-1}	1.20 mmol g^{-1} 0.093 mL g^{-1} 256 m^2 g^{-1}	0.14 mL g^{-1} 375 m^2 g^{-1}

be used to determine the surface area and this value is 360 ± 10 m^{2}g^{-1}. Since the pore dimensions of the zeolite lead to monomolecular surface coverage in the BET measurement, the MEA results, in contrast to the case for carbons, are in fair agreement with the BET surface area of 402 m^{2}g^{-1}. The t-plot micropore volume of 0.14 ml g^{-1} is also in fair agreement with MEA but the unreasonable mesopore volume of 0.17 ml g^{-1} from B-J-H analysis is not found. For reactions involving gases at 1 atm pressure, the multiple equilibrium data are most relevant.

The equilibrium constants indicate the concentrating power that can result from porosity. This is a very important property for catalysis and reactivity. A dramatic way to illustrate this advantage uses a concept called effective pressure.[7] This quantity is the pressure that would have to be exerted on an ideal gas to attain the concentration of adsorbate contained in the pores of the solid. For methane, this value is 98 atm for A-572, 194 atm for HZSM-5 and 21 atm for silica gel.

1.2 Cal-Ad Measurement of Acidity

Our next concern is with the measurement of surface reactivity. Hammett indicators, gas-solid calorimetry, temperature programmed desorption, thermogravimetric analysis and differential scanning calorimetry have all been suggested as ways to measure the acidity of solids and all have been criticized as measures in the literature.[11] In work from this laboratory, a calorimetric titration of the solid slurried in cyclohexane is advocated[12] as a way to measure the donor–acceptor component of the interaction free of the dispersion component which is included in gas-solid measurements. The titration curve (heat evolved vs. moles of base added) is fit to an equilibrium constant, an enthalpy and quantity of sites, n for each different strength acid site. For a two site problem this corresponds to fitting the curve to six unknowns. In cal-ad the adsorption isotherm is measured in addition to the calorimetric titration and these two measurements are solved simultaneously to produce the Ks, ΔHs and ns. This dispersion free, donor-acceptor enthalpy is proposed as a temperature independent scale of acceptor strength and the ns provide information about the number of sites available. To date most studies have involved pyridine but study of more donors will lead to a more complete dual parameter characterization of the acceptor strength.

The literature contains a description of the procedure and its application to the study of Pd/C[12a] and silica gel.[12b] More recently, the acidity of the suggested superacid, sulfated zirconia has been measured.[13] With pyridine as the donor, sulfated zirconia is found to be an acid of moderate strength(31 kcal mole^{-1}), weaker than the zeolite HZSM-5 (42 kcal mole^{-1})[14] or Si OAlCl$_2$(51 kcal mole^{-1}) and comparable to the HY zeolite (34 kcal mole^{-1}). Contrary to reports,[15] the acidity is not enhanced by the addition of transition metal promoters.

Cal-ad analysis of the reaction of pyridine with HZSM-5 showed[14] the presence of two acid sites. The strong site (42 kcal mole^{-1}) was present to the extent of 0.0415 mmol g^{-1} and the weaker site (9 kcal mole^{-1}) to the extent of 0.53 mmol g^{-1}. The results are compared to those from gas-solid calorimetry[16] and it is shown that both sites are averaged into a single site at the elevated temperatures employed in the gas-solid study. Comparison of cal-ad results for titrations with 2,6-lutidine and pyridine show that the access to the strongest sites is limited. Titration with 2,6-di-t-butyl pyridine shows that none of the strong sites and very few of the weaker sites are located on the exterior surface of the zeolite.

1.3 A New Moderate Strength Solid Acid

Solid acids of different strength are required for different chemical reactions. For example, an isomerization catalyst should not be so strong that it causes cracking. Current research has as its goal the synthesis of a series of reproducible solid acids of varying strength using the above methodology to characterize them. This series will be used to determine threshold acidity ranges for acid catalyzed reactions.

A new acid of moderate strength is synthesized by reacting WCl_6 with properly conditioned silica and hydrolyzing the resulting solid with nitric acid. Four moles of HCl are evolved in the reaction of WCl_6 with silica gel. The sum of the first (32 kcal mole^{-1}) and second (16 kcal mole^{-1}) acid sites (n_1+n_2) total the moles of tungsten used in the reaction. This indicates that both sites arise from tungsten and the synthesis leads to an excellent dispersion of the tungsten. Excellent dispersion is confirmed by SEM and the ratio of the intensities of the tungsten and silicon bands in the XPS which indicate that the tungsten is 97% disperse. Comparison of this new solid acid with calorimetric titration results for WO_3 and H_2WO_4 indicate that the second site resembles H_2WO_4 and is a hydrated tungsten oxide polymer. The ratio of $n_1:n_2$ indicates that the anchored $(SiO)_{4-n}$ $W(OH)(O-)_{1+n}$ strong acid sites have on average 5.5 $WO_3 \cdot xH_2O$ units attached forming the weaker site. These conclusions are confirmed by infrared spectroscopy which show the supported material contains hydrated, polymeric tungsten oxide species, as in H_2WO_4, and does not form the Keggin, $H_4SiW_{12}O_{40}$ compound.

Cal-ad has considerable potential for characterizing solid acid strengths. The enthalpy criterion is expected to be temperature independent from room temperature up to the usual elevated temperatures used in catalysis. Spent catalysts can be studied and the loss in activity correlated to the loss of acid sites. Threshold acidities for catalytic reactions can be established. When combined with the results from multiple equilibrium studies, a greater fundamental understanding of catalyst reactivity will emerge that can be reflected in rational catalyst design.

REFERENCES

1. A. W. Adamson, "Physical Chemistry of Surfaces", 5th Ed., John Wiley & Sons, New York, 1990.
2. D. H. Everett and J. C. Powl, *J. Chem. Soc. Faraday Trans. 1*, 1976, **72**, 619.
3. S. Brunauer, P. H. Emmett and E. Teller, *J. Am. Chem. Soc.*, 1938, **60**, 309.
4. W. D. Harkins and G. Jura, *J. Am. Chem. Soc.*, 1944, **66**, 1366.
5. E. P. Barrett, L. G. Joyner and P. P. Halenda, *J. Am. Chem. Soc.*, 1951, **73**, 373.
6. R. S. Drago, D. S. Burns and T. J. Lafrenz, *J. Phys. Chem.*, 1996, **100**, 1718.
7. a). R. S. Drago, W. S. Kassel, D. S. Burns, J. M. McGilvray, T. J. Lafrenz and S. K. Showalter, submitted.
   b). W. S. Kassel and R. S. Drago, *Microporous Materials*.
8. W. S. Kassel and R. S. Drago, submitted.
9. C. E. Webster, R. S. Drago and M. Zerner, submitted.
10. R. S. Drago and J. M. McGilvray, submitted.
11. a). D. Farcasiu and A. Ghenciu, J. Q. Li, *J. Catal.*, 1996, **158**, 116.
    b). R. Srinivasan, R. A. Keogh, A. Ghenciu, D. Farcasiu and B. H. Davis, *J. Catal.*, 1996, **158**, 502.

c). K. T. Wan, C. B. Khouw and M. E. Davis, *J. Catal.*, 1996, **158**, 311.

d). W. E. Farneth, R. J. Gorte, *Chem. Rev.* 1995, **95**, 615.

12. a). Y. Y. Lim, R. S. Drago, M. W. Babich, N. Wong and P. E Doan, *J. Am. Chem. Soc.*, 1987, **109**, 169.

b). C. W. Chronister and R. S. Drago, *J. Am. Chem. Soc.*, 1993, **115**, 793.

13. R. S. Drago and N. Kob, *J. Phys. Chem. B.*, 1997, **101**, 3360.

14. R. S. Drago, S. C. Dias, M. Torrealba and L. de Lima, *J. Am. Chem. Soc.*, 1997, **119**, 4444.

15. J.Cheng, J. D'Itri and B. Gates, *J. Catal.*, 1995, **151**, 464.

16. a). N. Cardona-Martinex and J. A. Dumesic, *Adv. Catal.*, 1992, **38**, 149.

b). G. Yaluris, R. B. Jarson, J. M. Kobe, M. R. Gonzalez, K. B. Fogash and J. A. Dumesic, *J. Catal.*, 1996, **158**, 336.

c). D. T. Chen, S. B. Sharma, I. Filimonov and J. A. Dumesic, *Catal. Lett.*, 1992, **12**, 201.

d). D. J. Parrillo, C. J. Lee and R. J. Gorte, *Appl. Catal. A* 1994, **110**, 67.

e). W. E. Farneth, R. J. Gorte and D. J. Parrillo, *J. Am. Chem. Soc.* 1993, **115**, 12441.

f). A. I. Biaglow, R. J. Gorte, G. T. Kototailo and D. White, *J. Catal.* 1994, **148**, 779.

g). C. Lee, D. J. Parrillo, R. J. Gorte and W. E. Farneth, *J. Am. Chem. Soc.* 1996, **118**, 3262.

17. N. Kob and R. S. Drago, submitted.

ALLYLIC N-FUNCTIONALIZATION OF ANILINE AND CONSECUTIVE AMINO-CLAISEN REARRANGEMENT

P. J. Kunkeler, J.A. Elings, R.S. Downing and H. van Bekkum
Delft University of Technology, Laboratory of Organic Chemistry and Catalysis,
Julianalaan 136, 2628 BL, Delft, The Netherlands

1 INTRODUCTION

The [3,3]-sigmatropic rearrangement of vinyl and allyl ethers, known as the Claisen rearrangement, is a useful method for regiospecific carbon-carbon bond formation. The reaction has been extensively studied both from mechanistic and synthetic points of view[1]. By contrast, the nitrogen analogue of the Claisen-rearrangement, the amino-Claisen rearrangement or 3-aza-Cope rearrangement, has received much less attention due to several limitations including the need of temperatures as high as 250-280°C. Under these conditions, N-allylamines tend to undergo thermal decomposition accompanied by tar formation, as was reported for N-allylaniline[2], the simplest aromatic N-allylamine. The first recorded example of an amino-Claisen rearrangement was the thermally induced rearrangement of N-allyl-1-naphthylamine into 2-allyl-1-naphthylamine[3].

Catalysts often permit the occurrence of an aromatic amino-Claisen rearrangement (AACR) that is thermally inaccessible. Lewis acids, like anhydrous $ZnCl_2$, were applied in stoichiometric amounts with varying success[1,4]. The most efficient and generally applicable Lewis acid catalyst reported to date is BF_3-etherate[1,5]; e.g. N-allylaniline rearranged to 2-allylaniline in 73% yield with BF_3-OEt_2 in refluxing xylene, compared with 40% using $ZnCl_2$. Brønsted acids have also been used extensively to promote the AACR[1]; the rearrangement is then often followed by a cyclization to yield indolines and indoles[6,7]. In Scheme 1 the AACR of N-allylaniline followed by ring closure is shown as an example. The asterisk indicates the inversion of the allyl group.

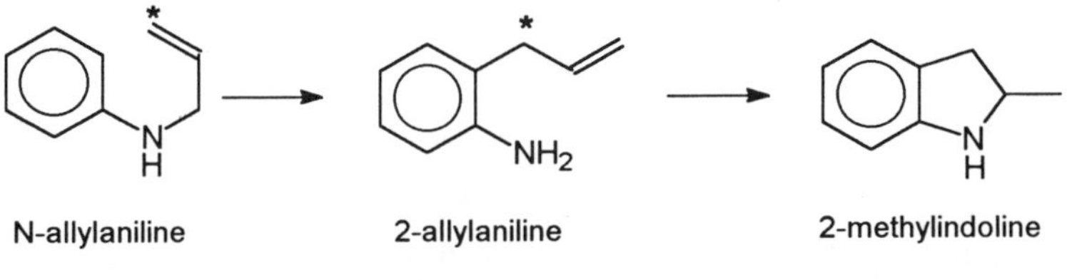

N-allylaniline   2-allylaniline   2-methylindoline

Scheme 1   Amino-Claisen rearrangement of N-allylaniline followed by ring closure

Coordination of the nitrogen to the Lewis acid or formation of the ammonium ion induces a positive charge which accelerates the reaction[1,4]. Electron-withdrawing groups within the molecule itself also facilitate the rearrangement[5]. A detailed mechanism is given in reference 1, which also accounts for the usually observed decomposition to anilines and concomittant intermolecular allyl transfers.

A disadvantage of the above mentioned catalysts is that they are not recyclable. Recently, zeolites H-Beta and H-Mordenite were reported as heterogeneous catalysts for the Claisen rearrangement of allyl phenyl ethers using benzene as the solvent[8] and in the photo-Claisen rearrangement of allyl phenyl ether with H-ZSM-5 [9]. Besides easy separation of the catalyst from the reaction mixture, these catalysts can be regenerated by a simple burn-off in air. In an attempt to use these zeolites under the same conditions as catalysts for the AACR, we only obtained a trace amount of 2-allylaniline. Application of higher boiling solvents (xylene) did not result in an increased conversion of N-allylaniline; probably the amine interacts too strongly with the acid sites thus making desorption difficult. However, the use of higher temperatures under gas-phase conditions proved to be a new and interesting method for performing the AACR of N-allylanilines.

Concerning selective N-alkylation of aniline with (m)ethanol by acidic and basic zeolites, reports indicate that low temperatures (~250°C) favour N-alkylation, whereas high temperatures favour ring alkylation either by direct alkylation or by rearrangement of N-alkylated anilines[10-12]. Also medium strength acids or bases are preferred for N-alkylation while strong acids will give predominantly ring alkylation. The allylation of ammonia with allyl alcohol is reported very recently by Göbölös *et al.*[13]

In the present work, attempts have been made to synthesize N-allylaniline by direct allylic N-functionalization of aniline with allyl alcohol in the gas-phase, see Scheme 2.

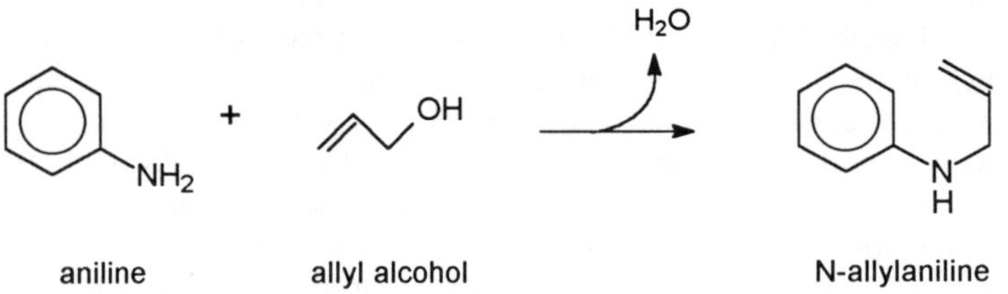

Scheme 2 Direct synthesis of N-allylaniline from aniline and allyl alcohol

The application of zeolites in the synthesis of 2-allylaniline and 2-methylindoline from aniline and allyl alcohol *via* N-alkylation and subsequent AACR was investigated.

2   EXPERIMENTAL

2.1 *Reactants and products*

N-allylaniline (NAA) (95% purity) was obtained from Acros. Allyl alcohol (AA) was supplied by Fluka. Aniline (A) was purified by distillation prior to use. 2-Allylaniline (2AA) and N-propylaniline (NPA) were obtained by column chromatography (silica gel, eluent: hexane/ethyl acetate = 9:1) from the combined condensates of several reaction runs and analysed by ¹H-NMR. 2-Methylindoline (2MIE) and 2-methylindole (2MIL) were purchased from Aldrich.

2.2 *Gas-phase reactor*

The fixed-bed continuous down-flow reactor consisted of a vertically mounted borosilicate glass tube (7 mm I.D.) heated by a fluidized bed oven. To be able to run the reactor under plug flow conditions, the catalysts were pelletized and subsequently crushed and sieved to collect the particles in the range of 0.7-1.0 mm. One gram of catalyst was used to obtain a sufficiently high catalyst bed. To exclude influences of metallic components (tubing, connectors, etc.), all auxiliary reactor parts were made of glass and heated externally to ca. 200°C to prevent condensation of reactants and products. The reactant mixtures were pumped into a stream of preheated carrier gas (nitrogen) by means of a motor-driven syringe pump. All reactions were run under atmospheric pressure. Usually 1 vol.% of NAA or A and AA was used in a total gas stream of 50 ml/min (1.22 mmol/h per $g_{cat.}$). Products were collected in a condensation vessel at room temperature and analyzed by GC-MS. Samples of the reactor effluent were taken regularly by an on-line autosampler and analyzed by a GC equipped with a CP SIL-19 CB column (50 m, 0.53 mm i.d., N_2 carrier gas) and FID. (selectivity = mol product formed per mol reactant converted)

3   RESULTS AND DISCUSSION

Performing the reaction under continuous gas-phase conditions has several advantages such as easy recovery of products and catalyst and the possibility to influence the amount of consecutive products by changing the contact times. A consequence of working under gas-phase conditions is the limitation to relatively volatile components.

Improvements in selectivity might be expected by the use of zeolites as the catalysts as a result of the confined space of their micropores. An overview of the applied catalysts with their compositions and activation temperatures is given in Table 1.

Table 1 Catalyst compositions as determined by flame AAS and activation temperatures

catalyst	Si/Al	Na/Al	K/Al or Zn /2Al	activation (°C, min.)	reference / source / remarks
H-Beta	11.6	0.06		400, 30	ref. 14, air calcined
Na-Beta-A	11.6	0.96		200, 30	ref. 15, air-calcined
Na-Beta-B	11.6	1.02		200, 30	ref. 15, NH_3-calcined
K-Beta	11.6		0.89	200, 30	ref. 15, air-calcined
Na-Y	2.4	0.94		200, 30	
Na-X	1.3	1.00		200, 30	
H-Na-Y	2.7	0.20		400, 30	
US-Y	20			200, 30	*, steamed Y
Zn-Na-Y	2.4	0.35	0.72	200, 30	0.2 M Zn^{2+} solution
Zn-Na-X	1.3	0.30	0.76	200, 30	0.2 M Zn^{2+} solution
Na-Mor	6.8	0.90		200, 30	PQ-zeolites
H-Mor	6.8	0.02		400, 30	PQ-zeolites
3-DDM	37.8	0.02		200, 30	ref. 16, acid leached
HA-HPV-707	3.4			#	AKZO-Nobel•

* kindly donated by Shell R&T Centre, Amsterdam
# 200°C, 30 min. or 480°C, 120 min., see text.
• ASA (Amorphous Silica-Alumina, 25wt% Al_2O_3)

3.1 *Amino-Claisen rearrangement of N-allylaniline*

To test whether the Lewis or Brønsted sites are the most important sites for the AACR, the catalysts were pretreated to influence the amount of these sites; see Table 1. Zeolite Beta was calcined in air or in ammonia to influence the amount of Lewis sites[15]. Also differently sized cations (Na or K) were introduced. Zeolites NaX and NaY were ion-exchanged with Zn^{2+} to introduce additional Lewis acidity. Zeolite Mordenite was dealuminated to 'open up' the (uni-dimensional) structure to improve access to the micropores. The (mesoporous) ASA was activated at different temperatures to vary the amount of Lewis sites; it also served as a comparison with the zeolites to test whether the micropores actually offer an advantage.

In Table 2 the selectivities are listed for the AACR of NAA at 4h on stream. The catalysts can be divided by the type of sites present: i) weak Brønsted (acidic hydroxyl groups) and Lewis sites and ii) strong Brønsted and Lewis sites. Overall, the strong Brønsted catalysts are the most active with zeolite H-Beta being the most stable catalyst as can be seen from Figure 1. It is clear that the selectivity depends strongly on the time on stream. For the weak Brønsted acidic catalysts NaMor shows the highest selectivity to 2AA. The Zn^{2+} exchanged Faujasites appear to deactivate fast combined with a low selectivity to 2AA, especially for the ZnNaX sample. The introduced Zn^{2+} actually decreases the activity compared to the parent sodium form. The activity and selectivity to 2AA for the ASA depend very much on the pretreatment; a higher activation temperature (480°C, 120 min. *vs.* 200°C, 30 min.) causes an increased conversion as well as an increased production of the ring closed product 2MIE. It was reported that a higher activation temperature results in partial dehydroxylation of the surface, thereby converting Brønsted sites into Lewis sites[18].

All the catalysts undergo deactivation and the spent catalysts are found to be slightly coloured. TGA of the deactivated ZnNaX sample showed it to contain 30 wt% of deposited material.

Table 2 AACR of N-allylaniline (NAA) over various catalysts
Selectivities (%) to the main products (4h T.O.S.) and conversions (%). Conditions: 1 % NAA in 50 ml/min, 200°C. (bar - indicates no trace detected)

catalyst	2AA	2MIE	2MIL	A	conv. 2h T.O.S.	conv. 6h T.O.S.
H-Beta	30	55	2	2	100	99
Na-Beta-A	62	3	-	7	75	24
Na-Beta-B	65	7	-	8	75	24
K-Beta	63	7	-	9	70	24
Na-Y	54	-	-	7	63	48
Na-X	68	5	-	5	93	82
H-Na-Y	65	15	<1	1	100	50
US-Y	53	9	1	5	80	52
Zn-Na-Y	50	0	-	<1	30	<1
Zn-Na-X	25	8	-	10	90	<1
Na-Mor	74	-	-	2	58	30
H-Mor	66	4	-	2	93	50
3-DDM	60	7	-	4	40	<1
HA-HPV 200°C	44	-	-	11	57	31
HA-HPV 480°C	57	15	2	5	96	84

The strength of adsorption depends on both the compound and the catalyst; thus a basic reactant will adsorb strongly on an acidic catalyst. The relatively low basicities of aniline (pKa = 4.63) and N-allylaniline (pKa = 4.17) are therefore an advantage. However, heavier (e.g. transalkylated) products adsorb more strongly and the catalyst may become deactivated. H-Beta shows an initial conversion of 100% (see Figure 1) which after 6h declines slowly. The very high external surface of zeolite Beta will contribute to the high activity[17]. The decrease in selectivity of 2MIE and concomitant increase in 2AA selectivity up to 86% is explained by catalyst deactivation: the catalyst bed is poisoned from the top down thus leaving few active sites available for the consecutive cyclization of 2AA.

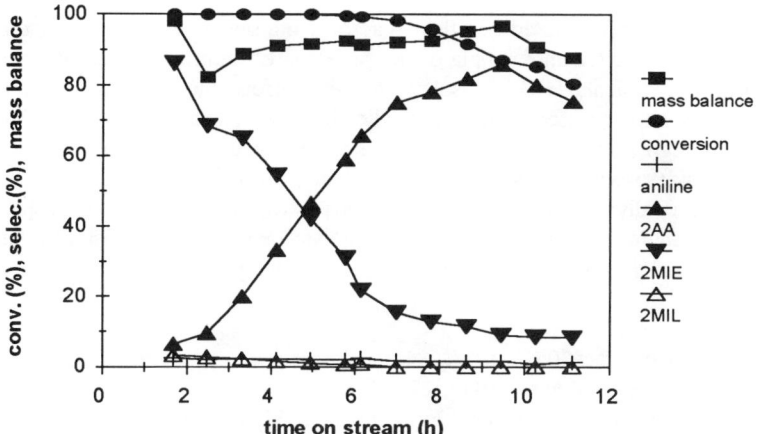

Figure 1 Catalytic behaviour of zeolite H-Beta for the AACR of NAA in time

3.2 *Direct route to N-allylaniline from allyl alcohol and aniline*

Compared with (m)ethanol or propanol, allyl alcohol is more reactive due to the resonance stabilization of the allyl cation. On the other hand, allyl alcohol contains two functional groups thus making it a difficult reagent for selective alkylation. Other side reactions that may occur are self-condensation of allyl alcohol towards diallyl ether and dehydrogenation of allyl alcohol forming acrolein and hydrogen[19]. The catalyst should preferably not catalyze the formation of acrolein since then the aniline present will, *via* a Michael-type condensation, react with acrolein resulting in non-desorbing polymers which will deactivate the catalyst. According to the results of Hutchings and Lee[19], basic faujasites (LiNaY, NaY, KNaY and Cs-NaY) are a possibility: they show low acrolein formation and are active in the self-condensation of allyl alcohol towards diallyl ether.

Preliminary results on the condensation of aniline and allyl alcohol showed that much lower temperatures than required for the AACR (140 *vs.* 200°C) are necessary to avoid formation of by-products. Desorption becomes difficult at this temperature and the activity drops accordingly. From the catalysts examined, only ASA and zeolite H-Beta show a significant activity; the basic faujasites mentioned above are inactive or deactivate at a high rate. When using HA-HPV as the catalyst, the main side-product is N-propylaniline, which becomes the major product at increasing time on stream. Probably, NAA formed is reduced *in situ* by allyl alcohol (possibly by a hydrogen transfer reaction) giving NPA and acrolein; this might also explain the observed catalyst deactivation. This is further supported by the observation that ASA catalyzes the formation of acrolein with increasing T.O.S. when allyl alcohol is the only reactant; this was also reported by Hutchings and Lee[19] for the Brønsted acidic catalysts. On the other hand, when feeding a mixture of NAA and AA, the NAA was not reduced to NPA when using ASA as the catalyst. Since the acrolein selectivity is a function of the T.O.S. it can be assumed that the sites active for hydrogen-transfer have to be created by materials deposited or other sites need to be poisoned first before products from (de)hydrogenation reactions can evolve.

CONCLUSIONS

The combination of a gas-phase fixed-bed flow reactor and a sensitive reaction like the amino-Claisen rearrangement offers an example of the potential of heterogeneous catalysts for the production of fine chemicals. H-Beta displays the best performance for this reaction. Amorphous silica alumina shows potential for the allylic N-functionalization of aniline due to the mild acidity, although side reactions still cause a rapid deactivation of the catalyst. Interestingly, the main side-product is N-propylaniline which presumably stems from the reduction of N-allylaniline. Allyl alcohol probably functions as the reductor giving acrolein which polymerizes with aniline causing the observed catalyst deactivation. Most zeolites examined suffer from severe deactivation, probably caused by pore-plugging.

References
1. R.P Lutz, *Chem. Rev.*, 1984, **84**, 205.
2. C.D. Hurd and. Jenkins, *J. Org. Chem.*, 1957, **22**, 1418.
3. S. Marcinkiewicz, J. Green and P. Mamalis, *Tetrahedron*, 1961, **14**, 208.
4. N.S. Barta, G.R. Cook, M.S. Landis and J.R. Stille, *J. Org. Chem.*, 1992, **57**, 7188.
5. W.K. Anderson and G. Lai, *Synthesis*, 1995, 1287.
6. J.E. Hyre and A.R. Bader, *J. Am. Chem. Soc.*, 1958, 80, 437.
7. A.R. Bader, R.J. Bridgwater and P.R. Freeman, *J. Am. Chem. Soc.*, 1961, **83**, 3319.
8. J.A. Elings, R.S. Downing and R.A. Sheldon, *Stud. Surf. Sci. Catal.* 1995, **94**, 487.
9. K. Pitchumani, M. Warrier and V. Ramamurthy, *J. Am. Chem. Soc.*, 1996, **118**, 3311.
10. B. Lian Su and D. Barthomeuf, *Appl. Catal. A*, 1995, **124**, 73-80 and 81.
11. S. Narayanan, V. Durga Kumari and A. Sudhakar Rao, *Appl. Catal. A*, 1994, **111**, 133.
12. S. Yuvaraj and M. Palanichamy, *React. Kinet. Catal. Lett.* 1996, **57,** 159.
13. S. Göbölös, E. Tálas, M. Hegedüs, I. Bertóti and J.L. Margitfalvi, *Appl. Catal. A*, 1997, **152**, 63.
14. R.L Wadlinger, G.T. Kerr and E.J. Rosinski, *US Patent* 3,308,069 (1967).
15. E.J. Creyghton, S.D. Ganeshie, R.S. Downing and H. van Bekkum, *J. Mol. Catal. A*, 1997, **115**, 457.
16. J.G. Lee, J.M. Garcés, J.J. Maj and S.C. Rocke, *Eur. Pat.* 0,433,932 (1990)
17. P.J. Kunkeler, D. Moeskops and H. van Bekkum, *Microporous Mat.*, in press.
18. Th. M. Wortel, W.H. Esser, G. van Minnen-Pathuis, R. Taal, D.P. Roelofsen and H. van Bekkum, *Recl. Trav. Chim. Pays-Bas*, 1977, **96**, 44.
19. G.J. Hutchings and D.F. Lee, *J. Chem. Soc., Chem. Commun.*, 1994, 2503.

Catalytic Flash Vacuum Thermolysis using Solid Acids[1]

A.J.H. Klunder[*], A.C.L.M. van der Waals and B. Zwanenburg

*Department of Organic Chemistry,
NSR Center for Molecular Structure, Design and Synthesis,
University of Nijmegen, Toernooiveld, 6525 ED Nijmegen, The Netherlands*

1 INTRODUCTION

The term "Flash Vacuum Thermolysis" (FVT) designates a dynamic gas phase process in which a substrate is passed through a hot quartz tube at low pressure ($1\text{-}1.10^{-3}$ mbar) and whereby the products are collected in a cold trap[2]. The applied low pressure prevents intermolecular reactions and reduces the residence time in the hot zone, allowing the isolation of kinetically labile compounds which are not attainable using static thermolysis techniques[3]. The aim of our current research is to investigate the application of modern and environmentally benign mineral solid acid catalysts, such as amorphous (silica-) aluminas, clays and zeolites to effect acid catalyzed reactions under Flash Vacuum Thermolysis (FVT) conditions in order to achieve thermolysis at lower temperatures and obtain higher product selectivities than under non-catalyzed FVT conditions. For this purpose, a so-called "Catalytic flash Vacuum Thermolysis" set-up was designed and applied (Fig. 1).

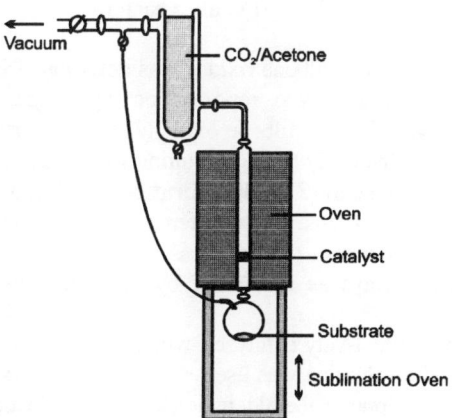

Figure 1 *Catalytic flash Vacuum Thermolysis set-up*

2   CATALYTIC FLASH VACUUM THERMOLYSIS OF *CIS-* AND *TRANS-* STILBENE OXIDE

Epoxides are versatile intermediates in organic synthesis. Reactions generally involve heterolytic fission of the C-O bond, whereas homolytic cleavage of the C-O or the C-C bond is less frequently encountered. The acid-catalyzed rearrangement of epoxides to carbonyl compounds has been extensively studied and is synthetically probably the most important transformation of epoxides (Scheme 1). A large variety of protic and Lewis

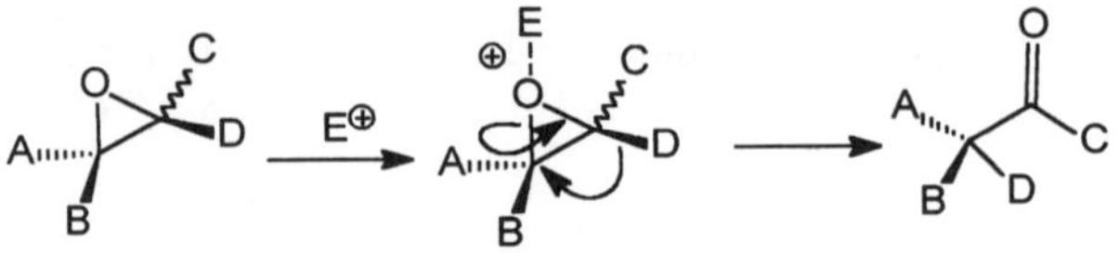

Scheme 1

acids have been successfully applied to effect this reaction. Besides conventional acids, such as BF_3, also environmentally more benign solid catalysts such as silica gel[4], alumina[5], alumino-silicates[6], clays[6] and zeolites[4a,7] have been used to accomplish this isomerization under mild conditions. However, all these studies were carried out in the liquid-phase.

In order to establish whether such an acid-catalyzed rearrangement of epoxides to carbonyl compounds is feasible in the gas phase using mineral solid acids, we studied the isomerization of *cis-* and *trans-*stilbene oxide (**1**) under catalytic flash vacuum thermolysis conditions and applying amorphous silica-alumina HA-SHPV as the solid acid catalyst[8]. This catalyst has an open mesoporous surface structure with a total pore volume of 0.75 ml/g with an average pore diameter (ϕ) of about 7.5 nm. Qualitative and (semi-) quantitative infrared (DRIFT) analysis using pyridine as a probe species showed that the HA-SHPV catalyst possesses predominantly Lewis acidity.

Isomerization reactions were performed using the catalytic-FVT apparatus described above. Experiments were carried out at 0.05 mbar with 0.15 ml (*i.e.* ~58 mg) or 0.75 ml (*i.e.* 290 mg) of a fractured (150-425 µm) catalyst. Stilbene oxide (50 mg) was vaporized in about 45 min at 100°C. Thus, a series of runs was performed using the same catalyst while the temperature was increased by 50°C after each run.

The thermal reactivity of the stilbene oxides was determined in a few control runs. In contrast to Oyewale and Aitken[9] who reported no significant reaction below 600°C, almost complete conversion of *trans-*stilbene oxide (**1**) was observed already at 300°C, to give mainly diphenylacetaldehyde (**2**) (85%) resulting from a 1,2-phenyl shift (Scheme 2). Other products were deoxybenzoin (**3**) (14%) formed by a formal 1,2-H shift and a small amount of benzophenone (**4**). Increasing the temperature to 600°C alters the reaction selectivity in favor of deoxybenzoin (**3**), whereas at 700°C poor selectivity results from complex fragmentation pathways as illustrated by the formation of diphenylmethane **5** (11%).

Epoxide-opening was effectively catalyzed when 0.15 ml (*i.e.* 1.04 g/h.g) of the HA-SHPV type amorphous silica-alumina was used. Complete conversion was reached already at 100°C whereby 88% of diphenylacetaldehyde (**2**) was produced together with 11% of deoxybenzoin (**3**). Increasing the reaction temperature to 350°C shifted the product composition toward **3**, analogous to the non-catalyzed case. Further increase of the temperature to 400°C gave a new product which was identified as diphenylacetylene (**6**).

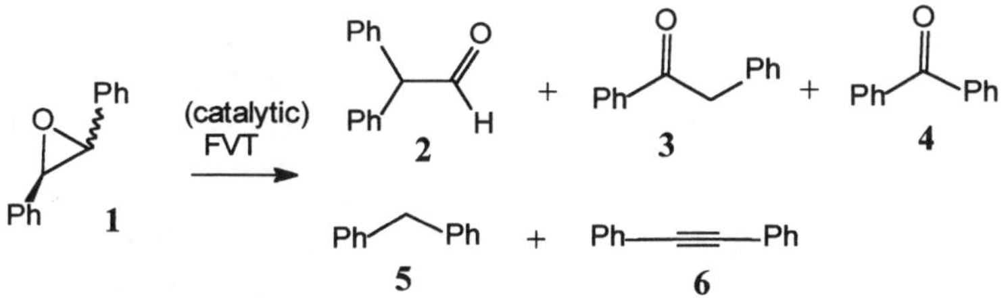

Scheme 2

This remarkable dehydration reaction became more pronounced at 500°C when the pyrolysate contained 26% of this acetylene. It turned out that diphenylacetylene could be obtained as the exclusive product when the reaction was performed at about 500°C using 0.75 ml of catalyst (*i.e.* 0.21 g/h.g).

Although dehydration of alcohols and hydration of acetylene over solid acid catalysts are well-known processes, dehydration of epoxides is unknown whereas dehydration of α-hydrogen containing carbonyl compounds to substituted acetylenes has only a few precedents[10], all employing exotic catalysts and at least stoichiometric amounts of bases like triethylamine and/or giving low yields.

Interesting mechanistic information of the acetylene formation was obtained when *cis-* or *trans*-stilbene oxide (**1**), diphenylacetaldehyde (**2**) and deoxybenzoin (**3**) were subjected to catalytic-FVT conditions applying temperatures between 300 and 550°C. No significant stereoisomeric effect was observed since both *cis-* and *trans*-stilbene oxide displayed a similar isomerization pattern, although a somewhat higher tolane (**6**)/deoxybenzoin (**3**) ratio was observed for *cis*-stilbene oxide. More interestingly, aldehyde **2** predominantly rearranged to ketone **3** at lower temperatures but yielded acetylene **6** at 550°C, whereas deoxybenzoin (**3**) did not produce any aldehyde but instead dehydrated to diphenylacetylene (**6**) very cleanly in the entire temperature range. It can be concluded that the isomerization pathways are temperature dependent and that at higher temperature dehydration becomes dominant.

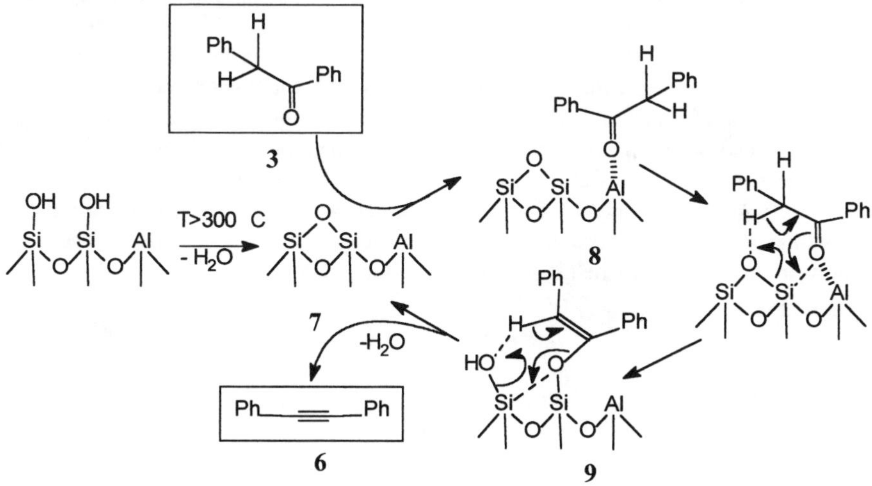

Scheme 3

The above results can be explained in analogy with previously postulated[11] working models for silica-alumina type surfaces. It is well established that pure silica is gradually dehydroxylated upon heating between 200-400°C, thereby generating reactive siloxane sites like **7** (Scheme 3). When an epoxide or carbonyl compound co-ordinates to a Lewis-acidic center (**8**) a grafting reaction creates an enol-type surface species (**9**). The free silanol group generated can act as a base forming the acetylenic bond with concomitant regeneration of the active site (**7**). More extensive surface dehydroxylation at higher temperatures then accounts for the increased yield of diphenylacetylene (**6**) at 500-550°C compared with 300°C.

3 CATALYTIC FVT OF DIBENZOYLMETHANE AND BENZOYLACETONE

The mechanism proposed above for dehydration of *cis*- and *trans*-stilbene oxide (**1**) *via* diphenylacetaldehyde (**2**) and deoxybenzoin (**3**) implies that other enolizable carbonyl compounds should, in principle, undergo this novel reaction. To substantiate this hypothesis dibenzoylmethane (**10**) and benzoylacetone (**14**) were thermolyzed under catalytic conditions.

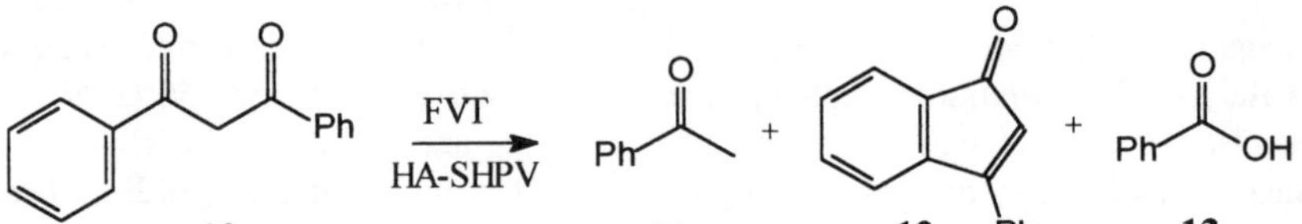

Scheme 4

A number of control runs prior to the catalytic experiments showed that dibenzoylmethane (**10**) was quite inert since at 600°C a mere 4% conversion was observed With 0.75 ml of the HA-SHPV type amorphous silica-alumina, however, 75% conversion was achieved already at 300°C yielding mainly acetophenone (**11**) (73%) together with 19% of a dehydration product that was identified as 3-phenyl-1-indenone (**12**) (Scheme 4). At 300°C and with a fresh catalyst the most active sites produced predominantly **11**, whereas at higher temperatures indenone **12** became the major reaction product. At temperatures ≥450°C nearly complete conversion was achieved with maximum indenone selectivity (50%) but a reduced production of acetophenone (**11**). Also some benzoic acid had now been produced, however no alkynes were formed at all.

Formation of indenone **12** is most conveniently rationalized by an intramolecular electrophilic aromatic substitution initiated by acid activation of one of the carbonyl functions in dibenzoylmethane (**10**) as shown in Scheme 5.

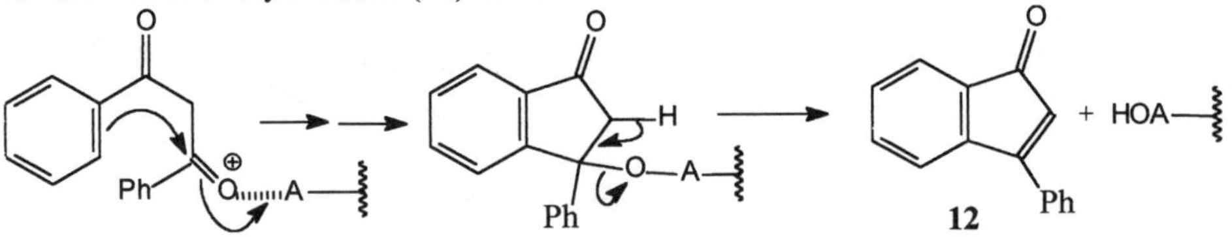

Scheme 5

Since concurrent formation of benzoic acid (**13**) was observed at high temperatures, both fragmentation products **11** and **13** probably arise from α-cleavage of dibenzoylmethane (**10**), as shown in Scheme 6.

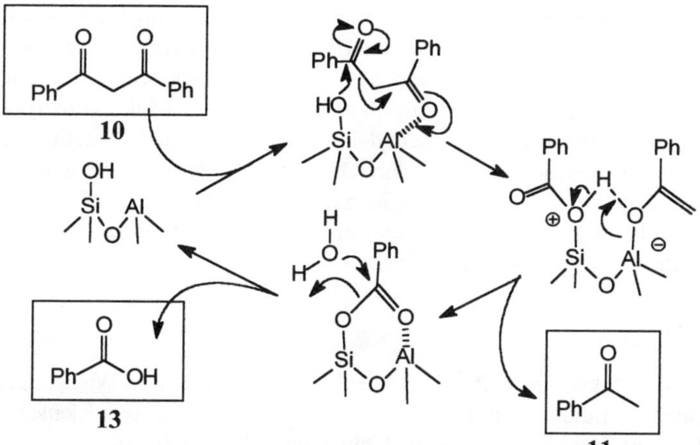

Scheme 6

Dehydration of benzoylacetone **14** under catalytic – FVT conditions was also investigated. Again, the inertness of this type of compound under thermal conditions was confirmed since only minor amounts of acetophenone (**11**) were formed at 600°C.

As for **10**, the HA-SHPV catalyst showed good activity towards **14** leading to complete conversion from 450°C on. More importantly, two dehydration products were obtained in about 50% total yield at 500°C They were identified as 3-methyl-1-indenone

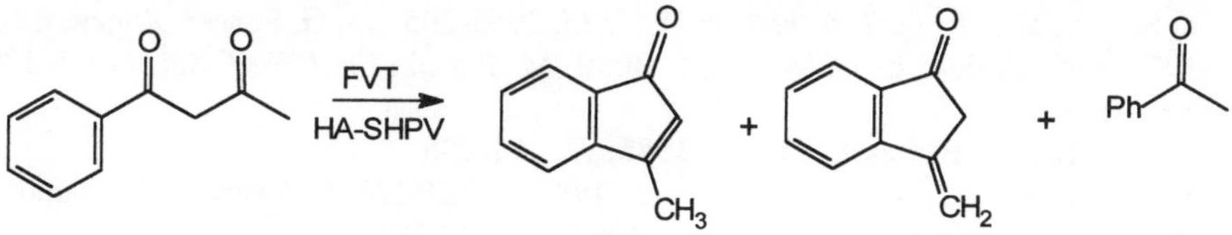

Scheme 7

(**15**) and isomeric 3-methylene-1-indanone (**16**) after purification (Scheme 7). A slight preference for **15** over its *exo*-methylene isomer **16** was observed in the entire temperature range. A similar mechanism as depicted in Scheme 5 can be envisaged for this formation of these indenones. As with dibenzoylmethane (**10**), acetophenone (**11**) was found to be the major fragmentation product at lower temperatures. Again no alkynones were produced at all.

4 CONCLUSION

In conclusion, the acid-catalyzed rearrangement of *cis*- and *trans*-stilbene oxide (**1**) was successfully achieved under catalytic flash vacuum thermolysis conditions using amorphous silica-alumina. With a small amount of catalyst isomerization of either stilbene oxide gave 88% of diphenylacetaldehyde (**2**) in the 100-250°C range, whereas at higher temperature (250-500°C) deoxybenzoin (**3**) is the main product (75%). With a larger amount of catalyst per unit substrate a yield of 65% of deoxybenzoin (**3**) was obtained at 350°C. Further temperature increase furnished diphenylacetylene (**6**) almost quantitatively (98%) at 550°C *via* a novel dehydration reaction. A mechanism is proposed that involves dehydration of neighboring surface silanol groups to form reactive siloxane sites.

As the proposed mechanism predicts that enolizable carbonyl compounds in general could undergo this novel reaction, two 1,3-diketones were thermolyzed under catalytic-FVT conditions. However, the expected linear alkynones were not obtained. Dibenzoylmethane (**10**) gave 3-phenyl-1-indenone (**12**), while benzoylacetone (**14**) produced a mixture of isomeric 3-methyl-1-indenone (**15**) and 3-methylene-1-indanone (**16**). Although the intermediate formation of alkynones can not yet be unequivocally excluded, the formation of indenones is most likely the result of an acid-catalyzed intramolecular electrophilic aromatic substitution. Apparently, this reaction is preferred over the dehydration to the isomeric alkynones.

5 REFERENCES

1. This work is taken from the PhD thesis of A.C.L.M. v.d. Waals, 'Flash Vacuum Thermolysis in Industrial Perspective. Development of Pourous Element Heating and Application of Solid Acid Catalysts, University of Nijmegen, 1997.
2. R.F.C Brown, 'Pyrolytic Methods in Organic Chemistry', Organic Chemistry Monographs, Vol. 41, Academic Press, New York, 1980.
3. See for a recent application of FVT in natural product synthesis: J. Zhu, J-Y. Yang, A.J.H. Klunder, Z-Y. Liu and B. Zwanenburg, *Tetrahedron* 1995, **51**, 5847-5870.
4. a) G. Paparatto, G. Gregorio, *Tetrahedron Lett.,* 1988, **29**, 1471-1472; b) T. Bharati Rao, J. Madhusudana Rao, *Synth. Commun.* 1993, **23**, 1527-1533; c) C. Lemini, M. Ordoñez, J. Péraz-Flores, R. Cruz-Almanza, *Synth. Commun.,* 1995, **25**, 2695-2702.
5. a) S. Dev, V.S. Joshi, *Tetrahedron* **1977**, *33*, 2955-2957; b) G. Posner, *Angew. Chem.* **1978**, *90*, 527-536; c) Á. Molnár, I. Bucsi, M. Bartók, *Stud. Surf. Sci. Catal.* **1991**, *59*, 549-556.
6. E. Ruiz-Hitzky, B. Casal, *J. Catal.* **1985**, *92*, 291-295.
7. M. Nomura, Y.Fujihara, *Chem. Express* **1992**, *7*, 121-124; D. Brunel, M. Chamoumi, P. Geneste, P. Moreau, *J. Mol. Catal.* **1993**, *79*, 297-304.
8. HA-SHPV was obtained from AKZO-Nobel Chemicals, The Netherlands
9. A.O. Oyewale, R.A. Aitken, *Russ. Chem. Bull.* **1995**, *44*, 919-922.
10. a) T. Tsuji, Y. Watanabe, T. Mukaiyama, *Chem. Lett.* **1979**, 481-482; b) J.J. Harrison, *J. Org. Chem.* **1979**, *44*, 3578-3580; c) T. Ando, J. Yamawaki, *Chem. Lett.* **1979**, *45*, 755-758; d) T. Kitazume, N. Ishikawa, *Chem. Lett.* **1980**, 1327-1328.
11. G. Posner, *Angew. Chem.* **1978**, *90*, 527-536; Á. Molnár, I. Bucsi, M. Bartók, *Stud. Surf. Sci. Catal.* **1991**, *59*, 549-556. 11.   E. Ruiz-Hitzky, B. Casal, *J. Catal.* **1985**, *92*, 291-295.

CATALYTIC REACTIONS OF MESOPOROUS MOLECULAR SIEVES

A.Corma

Universidad Politecnica de Valencia, Spain

ABSTRACT

Mesoporous materials with well defined pore sizes can offer a great opportunity for people looking for catalysts to process large molecules. Their pore sizes can vary between 15 and 100 angstroms and therefore allow easy diffusion of reactants of interest in organic synthesis. The synthesis of the materials will be described and their possibilities as supports for acids (heteropolyacids), bases (alkali cations and anchored amines), metals (Pt) and transition metal complexes active for hydrogenation and oxidation will be described together with their associated reactions.

The possibility of introducing acid sites in the walls, their nature and strength will be presented, and the possibility for catalysing reactions going from less demanding to more demanding (acetalization, Friedel-Crafts alkylations, isomerization, condensation, cracking) will be outlined.

Finally the catalytic properties of mesoporous materials containing transition metal catalysts will also be presented.

ETHYLENE INTERACTION WITH PARAMAGNETIC RHODIUM SPECIES IN SILICOALUMINOPHOSPHATE TYPE 11 MOLECULAR SIEVES

Tomasz Wasowicz and Larry Kevan

University of Houston
Department of Chemistry
Houston, Texas   77204-5641   USA

1. INTRODUCTION

Rhodium species in molecular sieves and other microporous materials have been reported to be effective catalyst sites for processes such as hydrogenation and dimerization of ethylene,[1] methanol carbonylation[2] and propylene hydroformylation[3]. Most of the studies were concerned with the oxidation   state and chemical nature of the active rhodium species. Depending on the specific catalytic process, paramagnetic Rh(0) and nonparamagnetic Rh(I) were proposed as active species.

Silicoaluminophosphate (SAPO-n) molecular sieves belong to a new class of microporous materials.[4,5] The modification of these materials by isomorphous replacement of framework cations by transition metal ions or by incorporation of such ions into extraframework positions can play a significant role in catalytic reactions. Since the location and oxidation state of transition metal ions can affect the catalytic activity, it is important to study the chemical environment and location of such transition metal ions. The SAPO-11 molecular sieve is composed of 10-ring straight channels surrounded by four 4-ring and six 6-ring channels. This structure has the AlPO-11 structure[6,7] except that some of the framework tetrahedral phosphorus (P) and aluminum (Al) sites are partially substituted by silicon (Si). This substitution produces a net negative framework charge which is balanced by H^+ ions which can be ion-exchanged to some extent by transition-metal cations.

Recently paramagnetic rhodium species were stabilized in SAPO-11 by solid state ion exchange to give Rh-SAPO-11 and by direct synthesis to give RhAPSO-11.[8] After activation by heating under vacuum at 500 °C the observed paramagnetic rhodium species is assigned to Rh(0) in both materials but it is more stable in RhSAPO-11. Here we report ethylene interactions with these two materials which lead to new paramagnetic species which are potential catalytic intermediates. The results also support different sites for Rh in these two materials.

2. EXPERIMENTAL

Rh-SAPO-11 and RhAPSO-11 were synthesized as described previously and calcined in flowing oxygen at 550 °C for 18 h.[8] Analysis showed 0.3 wt % Rh in both sample types. Metal ion concentrations were determined by electron microprobe analysis with a Jeol JXA 8600 spectormeter. Electron spin resonance (ESR) spectra were obtained with a Bruker ESP 300 X-band spectrometer at 77 K.

3. RESULTS AND DISCUSSION

For all Rh-SAPO-11 samples a prominent narrow ESR singlet with $\Delta H_{pp} \approx 5$ G at g = 2.002 was observed which is due to matrix defects formed in all samples during the calcination process. This signal appears also in SAPO-11 samples after calcination. Most probably the defects are formed as a result of dehydration during calcination of the SAPO-11 material. In the figures the top and the bottom part of this signal was cut off to allow a better presentation of the spectra. The other ESR signals are assigned to rhodium species formed by thermal reduction during the solid state ion-exchange heating. The rhodium signals are very weak which indicates that only a very small part of the Rh(III) can be reduced and stabilized in paramagnetic form.

Figure 1a shows the weak ESR spectrum of rhodium species B in Rh-SAPO-11 after activation as indicated in the figure caption. This is assigned to Rh(0).[8]

After exposure to 50 torr ethylene for 3 minutes at room temperature there appears a weak signal Et1 at g=2.368 (Figure 1b). A small amount of rhodium species A also appears. After four days, species Et1 and A decrease and form signal Et2 at g = 2.25 which is visible as a shoulder (Figure 1c). After evacuation of the ethylene from Rh-SAPO-11 and heating in vacuum at 100 °C the intensity of signal Et2 (which overlaps with the parallel component of signal B) decreases but remains present in the spectrum (Figure 1d). Similar signals were reported by Bass and Kevan at g=2.25 after exposure of ion-exchanged RhCa-X zeolite to ethylene[9], at g = 2.26 after exposure of RhCa-L zeolite to ethylene[10] and at g=2.26 after exposure of Rh/SiO$_2$ to ethylene.[11] Oxidation of Rh-SAPO-11 at 500 °C restores the initial spectrum observed before ethylene adsorption (Figure 1e).

The change in the Rh(0) spectra caused by ethylene adsorption and the transformations which occur in the spectrum during long ethylene exposure clearly indicates a Rh(0) interaction with ethylene. The initial interaction complex Et1 seems unstable to longer ethylene exposure. After a longer ethylene adsorption time, the spectrum changes and the Rh(0) signal B starts to decrease, while an additional ethylene complex forms at g = 2.25 which overlaps with the parallel component of signal B. This signal partially decreases after evacuation of ethylene which suggests that the signal is connected to a species formed with ethylene but which has a barrier to ethylene desorption. The reoxidation at 500 °C of the sample after ethylene desorption restores the spectrum which was observed before ethylene adsorption.

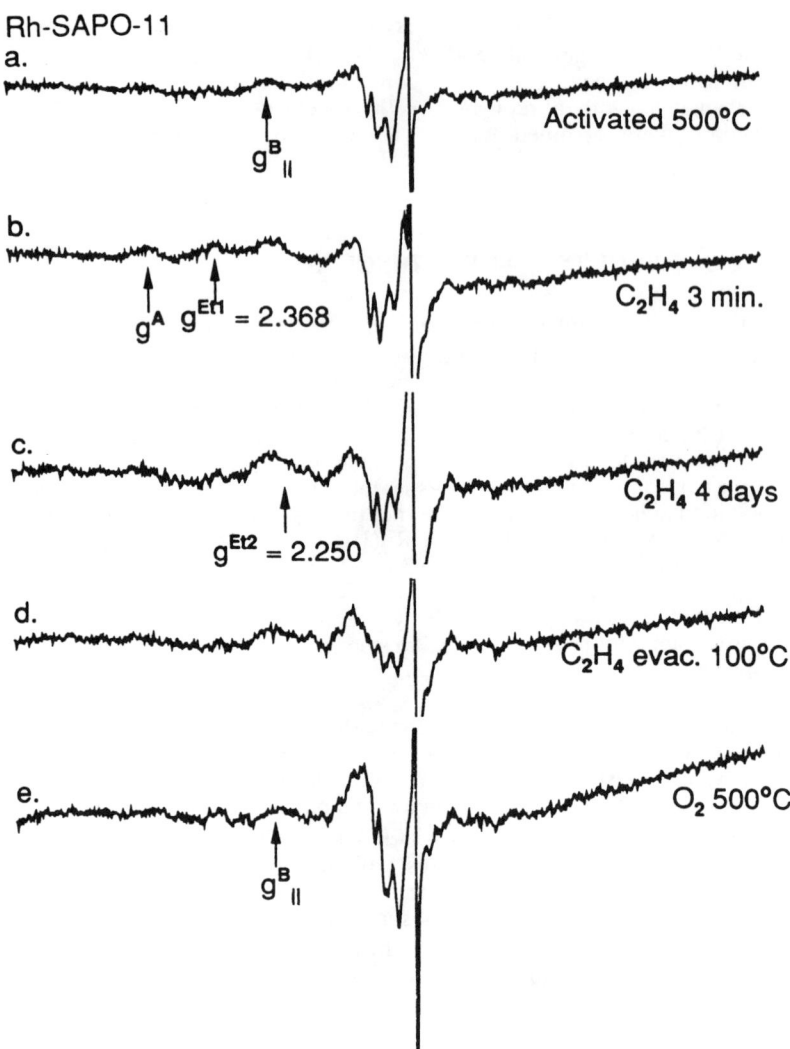

Rh-SAPO-11

a. Activated 500°C

$g^B_{||}$

b. C_2H_4 3 min.

g^A $g^{Et1} = 2.368$

c. C_2H_4 4 days

$g^{Et2} = 2.250$

d. C_2H_4 evac. 100°C

e. O_2 500°C

$g^B_{||}$

Figure 1.   ESR spectra at 77 K of solid state ion-exchanged Rh-SAPO-11 (a) after heating under vacuum, oxidation with O_2 and evacuation at 500 °C, (b) after 3 minutes of ethylene adsorption (50 torr) at room temperature, (c) after 4 days of ethylene adsorption, (d) after subsequent evacuation at 100 °C and (e) after subsequent oxidation at 500 °C.

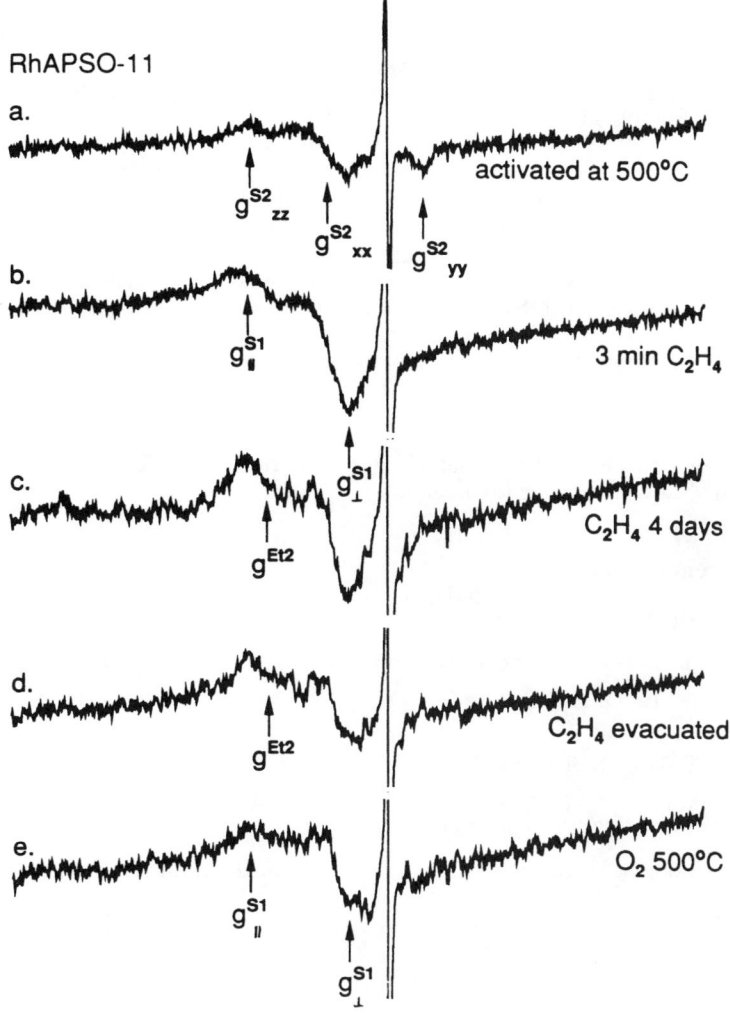

Figure 2. *ESR spectra at 77 K of RhAPSO-11 (a) after evacuation under vacuum, oxidation with O_2 and evacuation at 500 °C (b) after 3 minutes of ethylene adsorption (40 torr) at room temperature, (c) after 4 days of ethylene adsorption, (d) after subsequent evacuation at 100 °C, and (e) after subsequent oxidation at 500 °C. Compare these spectra with those in Figure 1.*

In RhAPSO-11, evacuation at 100 °C generates paramagnetic Rh(0) species S1 with $g_{\parallel}$ = 2.27 and $g_{\perp}$ = 2.07; this signal is stable to at least 500 °C.[4] After heating in 500 torr O_2 species S1 is converted to species S2 as shown in Figure 2a. If RhAPSO-11 showing signal S2 is exposed to 40 torr ethylene for 3 minutes at room temperature the S2 signal transforms to an axial signal (Figure 2b) with $g_{\parallel}$ = 2.281 and $g_{\perp}$ = 2.068 which are close to the parameters of signal S1. In the presence of ethylene for four days the overall spin concentration in the sample slightly decreases and a new weak Et2 signal appears (Figure 2c). Evacuation and heating to 100 °C of this sample does not cause any decrease of the Et2 signal (Figure 2d). Subsequent oxidation at 500 °C after ethylene evacuation only partially restores the initial S2 signal intensity (Figure 2e) in contrast to the behavior in Rh-SAPO-11. It can be concluded that an ethylene complex also forms with the rhodium species in RhAPSO-11. The different behaviors of Rh-SAPO-11 and RhAPSO-11 with respect to ethylene interaction supports the conclusion that Rh is in different sites in these two materials.

Acknowledgment. This research was supported by the Robert A. Welch Foundation and the National Science Foundation.

References
1.   Y. Okamoto, N. Ishida, T. Imanaka and S. Teranishi, *J. Catal.* 1979, **58**, 82.

2.   S. Lars, T. Andersson and M. S. Scurrell, *J. Catal.* 1981, **71**, 233.

3.   M. E. Davis, E. Rode, D. Taylor and B. E. Hanson, *J. Catal.* 1984, **86**, 67.

4.   Wilson, S. T.; Lok, B. M.; Flanigen, E. M. U.S. Patent 4 310 440, 1982.

5.   Wilson, S. T.; Lok, B. M.; Messina, C. A.; Cannan, T. R.; Flanigen, E. M. *J. Am. Chem. Soc.* **1982**, *104*, 1146.

6.   Bennett, J. M.; Richardson, J. W.; Pluth, J. J.; Smith, J. V. *Zeolites* **1987**, *7*, 160.

7.   Richardson, J. W.; Pluth, J. J.; Smith, J. V. *Acta Cryst.* **1988**, *B44*, 367.

8.   T. Wasowicz and L. Kevan in *Modern Applications of EPR/ESR: From Biophysics to Materials Science*, C. Rudowicz and H. Hiraoka, eds., Springer Verlag, Berlin, 1997, in press.

9.   S. J. Bass and L. Kevan, *J. Phys. Chem.* **1990**, *94*, 1483.

10.   S. J. Bass and L. Kevan, *J. Phys. Chem.* **1990**, *94*, 4640.

11.   S. J. Bass and L. Kevan, *J. Chem. Soc. Faraday Trans.* **1990**, *86*, 3015.

PREPARATION, CHARACTERIZATION, AND APPLICATION OF NEW, SUPPORTED, SUPERACIDIC HETEROPOLY ACID CAESIUM SALTS

Á. Molnár[a]*, T. Beregszászi[a], Á. Fudala[c], B. Török[a,b], M. Rózsa-Tarjáni[a] and I. Kiricsi[c]

[a]Department of Organic Chemistry and [b]Organic Catalysis Research Group, [c]Applied Chemistry Department, József Attila University
Dóm tér 8, Szeged, H-6720, Hungary

1 INTRODUCTION

Caesium salts of heteropoly acids have been thoroughly studied in recent years. Of these $Cs_{2.5}H_{0.5}[PW_{12}O_{40}]$, an acidic, non-stoichiometric salt of 12-tungstophosphoric acid has been found to be the best catalyst in electrophilic transformations of organic compounds in liquid-phase heterogeneous systems. It exhibits excellent catalytic activity in various organic transformations. The reactions most thoroughly studied are Friedel–Crafts alkylation[1-6] and acylation.[5,6] In addition, alkane isomerization,[3,6,7] alcohol dehydration,[2,9] formation of ethers[8] and esters,[4] hydrolysis[8] and decomposition[2,8] of esters, transformation of dimethyl ether to lower hydrocarbons,[2,9,10] and alkylation of isobutane with butenes[11] were also investigated.

This outstanding performance is attributed to the surprisingly high surface area and the micro- and mesoporous structure of this material.[2,3,5-8,12] It was shown[2,3,5-8] that the BET surface area of caesium salts of 12-tungstophosphoric acid – and, in fact, other salts of heteropoly acids – decreases with increasing caesium content until the composition of $Cs_2H[PW_{12}O_{40}]$. Further increase in caesium content brings about a marked increase in surface area reaching a value of about 150 m^2 g^{-1} for $Cs_3[PW_{12}O_{40}]$. The highest specific surface acidity and, consequently, the highest activities are achieved at a stoichiometry of $Cs_{2.5}$.

A problem associated with the use of this salt as catalyst material is its easy solubilization in water and in many organic solvents. The colloidal solution thus formed is difficult to separate which complicates the work-up procedure. This difficulty can be eliminated by immobilizing $Cs_{2.5}H_{0.5}[PW_{12}O_{40}]$ into a silica matrix.[5,13] The resulting material exhibits strong acidity and effectively catalyzes the hydrolysis of ethyl acetate. We reasoned that a viable alternative could be the use of high-surface-area support materials which may accommodate caesium salts. This method may even allow to increase the effective surface of other caesium salts of low surface area thereby transforming them into more effective catalysts. This can be of great significance in developing environmentally friendly practical applications.

A comparative study, therefore, has been carried out using the well-studied $Cs_{2.5}H_{0.5}[PW_{12}O_{40}]$ and two sets of newly prepared supported samples containing low and high amounts of caesium. Herein, we report our results with respect to the preparation, characterization, and catalytic application of these new supported caesium salts of 12-tungstophosphoric acid.

2 EXPERIMENTAL

2.1 Materials and their preparation

Reactants originating from Aldrich or Fluka were of at least 99% purity and used without further purification. A literature method was applied to synthesize phenyl acetate, while *meso*-2,5-hexanediol was available from our previous studies.[14]

$Cs_{2.5}H_{0.5}[PW_{12}O_{40}]$ (denoted as $Cs_{2.5}PW$) was prepared by adding dropwise an appropriate amount of aqueous $CsNO_3$ (Aldrich) to aqueous $H_3[PW_{12}O_{40}]$ (Serva) under continuous stirring. The resulting precipitate was aged for 24 h at room temperature, separated by centrifugation and washed with distilled water.

Davisil silica gel (Aldrich, grade 363, 35-60 mesh), and the mesoporous molecular sieve MCM-41 (synthesized according to the method by Kim *et al.*[15]) were used to prepare the supported caesium salts applying a literature method.[16] The supports were pretreated at 573 K for 24 h before use. First $CsNO_3$ was deposited on 5 g of the support by impregnation of an aqueous solution. After drying (393 K, 12 h) and calcination (773 K, 2 h) $H_3[PW_{12}O_{40}]$ (1.6 g) was impregnated followed by drying (373, K, 2 h and 393 K, 12 h). The quantity of $CsNO_3$ was chosen to have less than monolayer loadings, and the nominal compositions $Cs_1H_2[PW_{12}O_{40}]$ and $Cs_{2.5}H_{0.5}[PW_{12}O_{40}]$.

2.2 Catalyst characterization

Instrumental methods of characterization included X-ray diffraction (XRD; DRON 2, operating under computer control), differential scanning calorimetry (DSC; Perkin Elmer DSC 2 equipment), and FTIR spectroscopy (Mattson Genesis 1) in the framework vibration range using the KBr pellet technique. ^{31}P MAS-NMR spectra were recorded on a Bruker AM400 spectrometer. Prior to measurements the samples were pretreated at 573 K for 2 h in vacuo (better than 10^{-5} Torr) or in flowing helium.

BET measurements were carried out in a conventional volumetric adsorption apparatus at liquid-N_2 temperature (77 K).

IR spectroscopy was applied to determine the acidity of the samples by monitoring the adsorption of pyridine. Self-supporting wafers of the samples were pretreated in an IR cell then cooled to room temperature and treated with 10 Torr of pyridine for 1 h. After evacuation (1 h) spectra were recorded at 293 K, 373 K, 473 K and 573 K.

Chemical methods for characterization included the isomerization of 1-butene in a recirculatory batch reactor. After pretreatment the samples (50 mg) were cooled to reaction temperature (423 K) then 1-butene (520 Torr) was loaded. Sampling was carried out at regular time intervals and product distributions were determined by GC (HP-5710, 4.5 m dimethylsulfolane column, 293 K).

2.3 Catalytic studies

The Friedel–Crafts reactions were carried out in a two-necked flask (about 10 ml) with vigorous stirring. Water formed in benzylation was continuously removed by a Dean-Stark trap. Exact details of experimental conditions are given in the corresponding tables (Tables 2-4). The product cyclic ethers formed in the cyclodehydration of 2,5-hexanediol were continuously distilled off from the reaction flask and collected at 250 K.

Product compositions were determined by means of gas chromatography (Carlo Erba Fractovap G, TCD, 1.2 m SE 52 or Reoplex 400 column, and HP-5890, FID, 30 m DB-5 column). Hexadecane was applied as internal standard in Friedel–Crafts acetylation. Compounds formed were identified via the GC retentions of authentic samples and GC-MS spectroscopy (HP 5890 GC coupled with a HP 5970 MSD). None of the supports was found to exhibit any activity under the experimental conditions applied.

3 RESULTS AND DISCUSSION

3.1 Preparation and characterization of supported caesium salts

Characteristic ^{31}P MAS-NMR spectra of $Cs_{2.5}PW$ and the supported samples indicate that the actual compositions of the supported samples are different than expected (spectra not shown). The main peak at -14.9 ppm is assigned to the Keggin anion having no proton attached (Cs_3 species).[1,3,8] The contribution of the peak at -13.4 ppm (the Cs_2 species) in the samples of high caesium content is only very minor. It can be estimated, therefore, that the stoichiometry is about $Cs_{2.9}$ (the samples will be denoted as $Cs_{2.9}/SiO_2$ and $Cs_{2.9}/MCM41$). The two samples of low caesium content, in contrast, have a stoichiometry of about Cs_2. This can be estimated from the significant additional contribution of the peak at -10.8 ppm which is assigned to $H_3[PW_{12}O_{40}]$ (samples Cs_2/SiO_2 and $Cs_2/MCM41$).

These observations may be explained by the specific method of preparation as opposed to the usual synthesis found in the literature. Salts of heteropoly acids are prepared, in general, by titrating the aqueous solution of the acid with $CsCO_3$, followed by evaporation of the resulting slurry. In our case, due to the specific requirements of the method applied, the solution of $H_3[PW_{12}O_{40}]$ was added to $CsNO_3$ adsorbed on various supports. It appears that such modifications result in the formation of caesium salts with stoichiometries higher than expected. The supported samples thus prepared are stable up to about 800 K (DSC) and show full crystallinity similar that of neat $Cs_{2.5}PW$ (XRD).

The characteristic data of the supports and supported caesium salts are collected in Table 1. It is seen that upon adsorption of the salts a certain decrease in the BET surface areas occurred. The resulting supported samples, nevertheless, still have high enough surface area to ensure the formation of well-dispersed catalytic materials.

Table 1 *Characteristic data of the supports and supported caesium salts*

Sample	BET/ $m^2 g^{-1}$	Acidity/ m^{-2}		Activity in isomerization of 1-butene	
		Broensted	Lewis	w_0/ mmol m^{-2} s^{-1}	$t_{1/2}$/ s
SiO_2	555	0.02	1.27	-	-
$Cs_{2.9}/SiO_2$	326	0.04	2.58	0.019	930
Cs_2/SiO_2	312	0.27	1.95	0.128	150
MCM41	773	0.07	0.83	0.016	450
$Cs_{2.9}/MCM41$	609	0.07	1.71	0.024	300
$Cs_2/MCM41$	544	0.22	2.34	0.032	180

All samples display both Broensted and Lewis acidity. The acidity of the samples was determined by calculating the integral adsorption of bands at 1560 cm^{-1} characteristic of Broensted acidity and that at 1450 cm^{-1} assigned as pyridine bonded to Lewis acid sites. These values were divided by the mass of the samples and the corresponding BET areas. In this way comparable acidity data for the kinetic experiments were obtained. It is clearly seen that the deposition of $Cs_2H[PW_{12}O_{40}]$ on the supports results in a significant increase in Broensted acidity which is in harmony with the NMR results. Significant increases in Lewis acidity, in contrast, are measured for all supported samples.

Table 1 also contains activity data for the isomerization of 1-butene. The transformation was found to take place over all four supported samples and the MCM-41 support, whereas SiO_2 showed no activity at all. Characteristic plots of consumption of 1-butene observed over the silica-supported samples are given in Figure 1.

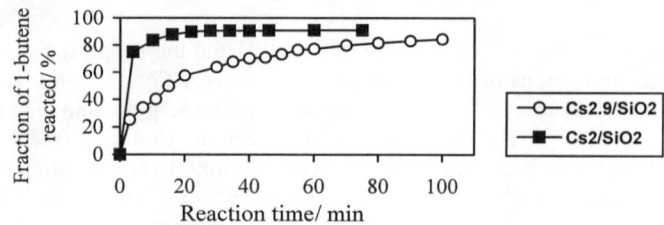

Figure 1 Consumption of 1-butene during isomerization as a function of time

3.2 Catalytic studies

Four well-studied reactions, Friedel–Crafts acylation and alkylation (acetylation, benzylation and adamantylation), and cyclodehydration were chosen to test the performance of the supported caesium salts in comparison with that of neat $Cs_{2.5}PW$. Results of these experiments (conversions, activities and selectivities) are summarized in Tables 2-4. It is important to point out, that solubilization during these applications was not observed except in the case of $Cs_{2.9}/MCM41$.

Friedel–Crafts reactions catalyzed by $Cs_{2.5}PW$ have been thoroughly studied.[1-6] In our case the supported caesium salts exhibit comparable activity, but the activity of $Cs_{2.5}PW$ is about one order of magnitude higher. Noteworthy is the behaviour of the supported samples in repeated experiments with recovered catalysts since they display only slight deactivation.

Table 2 *Catalytic activity of caesium salts in Friedel–Crafts reactions*

Catalyst	Acetylation of anisole[a,b] Conversion/ %	Activity/ [d]	Benzylation of benzene[c] Conversion/ %	Activity/ [d]
$Cs_{2.5}PW$	85	0.196	100	0.170
$Cs_{2.9}/SiO_2$	24	0.012	6	0.004
$Cs_{2.9}/MCM41$	63	0.017	94	0.030
Cs_2/SiO_2	54	0.014	100 (98, 95)[e]	0.062
$Cs_2/MCM41$	67	0.010	95 (90, 63)[e]	0.034

[a] 50 mmol anisole, 2.5 mmol acetic anhydride, catalyst quantity: 32-134 mg, 363 K, 3 h.
[b] Only the *para* isomer was detected.
[c] 50 mmol benzene, 2.5 mmol benzyl alcohol, catalyst quantity: 65 mg, 353 K, 2 h.
[d] mmol m^{-2} h^{-1}
[e] Repeated experiments with recovered catalyst.

A characteristic set of data for adamantylation of anisole with 1-bromoadamantane is given in Table 3. This reaction is brought about by electrophilic catalysts with regioselectivities strongly depending on the acid strength: increasing *meta* selectivity is observed with increasing acidity. The transformation requires the use of a larger amount of the caesium salts than in other reactions studied. In this reaction smaller activity values are observed. The relative activities of the various catalysts, however, are similar to those in other Friedel–Crafts reactions. Since the *meta/para* ratio was shown to correlate with the acid strength of catalysts in adamantylation,[17] $Cs_{2.5}PW$ is the strongest acid, but all supported samples exhibit comparable acidity.

Table 3 *Activity and selectivity in the adamantylation of anisole*[a]

	Conversion/ %	Activity/ 10^3 mmol m^{-2} h^{-1}	Selectivity/ m:p
$Cs_{2.5}PW$	100	20.6	56:43
$Cs_{2.9}/SiO_2$	5	0.086	22:78
$Cs_{2.9}/MCM41$	2	0.015	11:89
Cs_2/SiO_2	15	0.160	24:76
$Cs_2/MCM41$	72	0.439	21:79

[a] 1 g anisole, 50 mg 1-bromoadamantane, catalyst quantity: 50-100 mg, 425 K, reaction time: 2-7 h.

Cyclodehydration of 2,5-hexanediol (isomeric mixture with a racem.:*meso* ratio 47:53) yields exclusively a mixture of *cis*- and *trans*-2,5-dimethyltetrahydrofuran with the same composition (no traces of dienes were detected) (Table 4). The process is also stereospecific: the *meso* isomer (98% isomer purity) gives *trans*-2,5-dimethyltetrahydrofuran with the same purity. A comparison of activity data, again, shows similar tendencies observed in the other reactions studied.

Table 4 *Activity and selectivity in the cyclodehydration of 2,5-hexanediol*[a]

	Isomeric mixture		*meso*-Isomer
Catalyst (quantity used)	Conversion/ %	Activity/ mmol m^{-2} h^{-1}	Selectivity/cis:trans
$Cs_{2.5}PW$ (13 mg)	51	8.680	2:98
$Cs_{2.9}/SiO_2$ (26 mg)	6	0.177	
$Cs_{2.9}/MCM41$ (26 mg)	42	0.663	
Cs_2/SiO_2 (67 mg)	100(100)[b]	1.196	
$Cs_2/MCM41$ (67 mg)	100(100)[b]	0.687	2:98

[a] 25 mmol diol, 473 K, 1 h.
[b] Conversion in fifth repeated experiment with recovered catalyst.

4 CONCLUSION

Caesium salts of 12-tungstophosphoric acid supported on silica or MCM-41 have been found to be active, selective and recyclable catalysts in organic transformations requiring the use of electrophilic catalysts. The activity of the various catalyst preparations is, in most cases, comparable with that of neat $Cs_{2.5}H_{0.5}[PW_{12}O_{40}]$. This indicates that the active species are in a well-dispersed form. The excellent performance of these catalysts in the test reactions promises their further successful utilization in organic synthesis.

These new observations clearly call for further studies. Additional instrumental characterization will certainly provide sufficient information to get a full understanding of the phenomena observed. Our findings may open new possibilities for the practical application of these and other supported heteropoly acid salts. With such catalysts in hand the development of more economic and environmentally friendly processes can be expected.

Acknowledgements. This work was supported by the Hungarian National Science Foundation (OTKA Grants T016941 and F023674). Thanks are due to Prof. J.B. Nagy for the ^{31}P NMR measurements.

References

1. T. Okuhara, T. Nishimura, H. Watanabe and M. Misono, *J. Mol. Catal.*, 1992, **74**, 247.
2. T. Okuhara, T. Nishimura and M. Misono, *Studies in Surface Science and Catalysis*, 1996, **101**, 581.
3. T. Okuhara, T. Nishimura, H. Watanabe, K. Na and M. Misono, *Studies in Surface Science and Catalysis*, 1994, **90**, 419.
4. Y. Izumi, M. Ono, M. Ogawa and K. Urabe, *Chem. Lett.*, 1993, 825.
5. Y. Izumi and K. Urabe, *Studies in Surface Science and Catalysis*, 1994, **90**, 1.
6. Y. Izumi, M. Ogawa and K. Urabe, *Appl. Catal., A*, 1995, **132**, 127.
7. K. Na, T. Okuhara and M. Misono, *J. Chem. Soc., Faraday Trans.*, 1995, **91**, 367.
8. N. Essayem, G. Coudurier, M. Fournier and J.C. Védrine, *Catal. Lett.*, 1995, **34**, 223.
9. S. Tatematsu, T. Hibi, T. Okuhara and M. Misono, *Chem. Lett.*, 1984, 865.
10. T. Hibi, K. Takahashi, T. Okuhara, M. Misono and Y. Yoneda, *Appl. Catal.*, 1986, **24**, 69.
11. T. Okuhara, M. Yamashita, K. Na and M. Misono, *Chem. Lett.*, 1994, 1451.
12. J.L. Bonardet, J. Fraissard, G.B. McGarvey and J.B. Moffat, *J. Catal.*, 1995, **151**, 147.
13. Y. Izumi, M. Ono, M. Kitagawa, M. Yoshida and K. Urabe, *Microporous Mater.*, 1995, **5**, 255.
14. Á. Molnár, K. Felföldi and M. Bartók, *Tetrahedron*, 1981, **37**, 2149.
15. J.M. Kim, J.H. Kwak, S. Jun and R. Ryoo, *J. Phys. Chem.*, 1995, **99**, 16742.
16. J.-M. Tatibouët, C. Montalescot and K. Brückman, *Appl. Catal., A*, 1996, **138**, L1.
17. G.A. Olah, B. Török, T. Shamma, M. Török and G.K.S. Prakash, *Catal. Lett.*, 1996, **42**, 5.

NEW DEVELOPMENTS IN SELECTIVE AROMATIC SUBSTITUTION WITH THE HELP OF SOLID CATALYSTS

Keith Smith

Department of Chemistry
University of Wales Swansea
Swansea SA2 8PP, UK

1 INTRODUCTION

Electrophilic aromatic substitution reactions[1] are mainstays of the chemical industry but suffer several severe disadvantages that render them increasingly vulnerable in modern times. For example, such reactions often require catalysis by mineral or Lewis acids and in many cases the catalyst is required in large amounts, sometimes even more than a stoichiometric amount, in order for the reaction to proceed efficiently. The catalysts are frequently corrosive and require aqueous treatment during the work-up. This can cause additional problems with waste water streams that have to be treated and may mean that the catalysts cannot be recovered. Finally, many electrophilic aromatic substitution reactions give mixtures of regioisomers, which results in separation problems as well as a large amount of unwanted by-product.

The chemicals industry is coming under increasing pressure to reduce its environmental impact and it is important that processes resulting in large amounts of waste be improved.[2] In this context electrophilic aromatic substitution reactions are prime targets. Use of solids for control of such reactions is attractive.[3-6] Many solids can act as Brønsted or Lewis acids; they are usually easy to recover and if necessary to regenerate by heating; no aqueous work-up is necessary; and in some cases, particularly with zeolites, the selectivity of the reaction may be substantially improved. We have sought to develop a number of such reactions and present three of them here, namely nitration, bromination and methanesulfonylation reactions.

2 NITRATION

Nitration of aromatic substrates is one of the most important and widely studied of chemical reactions.[7] Despite this, industry still largely relies upon early technology involving mixtures of nitric and sulfuric acids. Mixed acid nitration systems, however, are not very selective, particularly if the *para*-isomer is the commercially more desirable isomer. The acids are corrosive and used in excess; over-nitration or oxidised by-products often result; and the need for an aqueous washing stage results in a waste inorganic acid stream that is environmentally unfriendly or costly to treat.

In recent years there has been a spate of activity aimed at the development of new nitration methods using solid acid catalysts in an attempt to overcome these disadvantages. The use of Nafion-H and other polysulfonic acid resins reduces the corrosive nature of the reaction mixtures but does not substantially improve regioselection for the *para*-isomer.[8] Toluene has been nitrated *para*-selectively with benzoyl nitrate over a zeolite catalyst to give mononitrotoluenes in almost quantitative yield, of which 67% is the *para*-isomer,[9] but the method is not commercially attractive because the reagent is expensive and a chlorinated hydrocarbon solvent is used. Similar or equally potent disadvantages apply to a number of other methods developed recently.[10] Unfortunately, therefore, none of the methods that exhibit *para*-regioselectivity are attractive for large scale use.

As part of our continuing research into the development of useful synthetic methods that make use of solids as catalysts or controlling agents[11] we have undertaken further investigations of nitration in an attempt to develop a method that would allow high yield and selectivity with inexpensive reagents and without the need for large quantities of solvent. We now report success in this endeavour.[12]

Following the success of our previous work with benzoyl nitrate as the reagent[9] we decided to try nitration of toluene using acetyl nitrate generated *in situ* from nitric acid and acetic anhydride according to Equation 2.[7] This would provide a cheaper reagent and avoid the need for its isolation or extraction.

$$Ac_2O \quad + \quad HNO_3 \quad \longrightarrow \quad AcONO_2 \quad + \quad AcOH \qquad (2)$$

In the initial experiments we used a large excess of acetic anhydride to ensure complete conversion of all of the nitric acid into acetyl nitrate. In order to discover whether the order of addition of the materials had a significant effect on the *para*-selectivity of the process the addition sequence was systematically varied. The best order involved pre-mixing the zeolite with nitric acid followed by addition of the anhydride and lastly the toluene. This is consistent with the formation of acetyl nitrate predominantly within the pores of the zeolite, where reaction subsequently takes place with the toluene. However, the amount of *para* product was only 7% less when the anhydride was mixed with the zeolite followed by the addition of the nitric acid and then the toluene. This could still provide a very useful procedure if necessary.

Using the optimum order of addition, a range of different zeolites was tested for efficacy at catalysing the reaction. In view of our earlier findings with benzoyl nitrate we looked mainly at large pore zeolites, but ZSM-5, a medium pore zeolite, was included for comparison. The results are shown in Table 1.

Table 1 *The effect of zeolite type on the nitration of toluene* [a]

Zeolite	Si/Al	Time/min	Yield (%)	ortho	meta	para
HMord	35	30	17	53	3	44
HZSM-5	80	30	26	64	0	36
HZSM-5	300	30	19	57	0	43
HY	5	30	4	61	0	39
HY	40	5	>99	52	3	45
HY	80	5	>99	53	3	44
NaBeta	13	30	3	50	0	50
FeBeta	13	5	>99	25	3	72
AlBeta	13	5	>99	25	3	72
HBeta	13	5	>99	25	3	72

[a] Nitric acid (70%, 2.5 mmol), zeolite (0.1 g), acetic anhydride (20 ml), toluene (0.23 g, 2.5 mmol), 20 °C.

The rate of reaction and *para*-selectivity with the medium pore zeolite, ZSM-5, were both low, suggesting that little of the reaction was occurring inside the pores. The situation with the three larger pore zeolites, however, was very interesting. Mordenite gave a low yield but with somewhat better *para*-selectivity than the homogeneous reaction. Zeolite Y with a high Si/Al ratio allowed a rapid reaction but with limited improvement in *para*-selectivity. The result with the proton form of zeolite beta remained the best - a rapid reaction and an outstanding degree of *para*-selectivity. Apparently, the geometrical constraints are such as to allow relatively easy diffusion but to impose some order on the transition state. Several cation exchanged forms of zeolite ß gave results that were almost identical to those of the proton form, while the sodium form gave a poor yield. Therefore, Hß was selected for more detailed study.

A series of reactions in which the ratio of the anhydride to the nitric acid was varied revealed that no deleterious effect was observed until the amount of acetic anhydride was reduced to below the stoichiometric amount required to convert all the nitric acid into acetyl nitrate and all the water into acetic acid. At the stoichiometric point the nitration of toluene (35 mmol) was effected by nitric acid (2.5 g, 90%, 35 mmol) and acetic anhydride (5.0 ml, 53 mmol) over Hß (1.0 g) and the yield and *para*-selectivity were excellent. Furthermore, following removal of the acetic acid and product from the reaction vessel by reduced pressure distillation the zeolite was left in a state where it could be used again with only a small lowering of performance. This represents a very useful, environmentally friendly way of nitrating toluene with excellent *para*-selectivity. Therefore, the reaction was applied to a range of aromatic substrates (equation 3, Table 2).

$$PhR \xrightarrow[H^+\beta]{HNO_3,\ Ac_2O} \text{(ortho-)} \mathbf{2} + \text{(meta-)} \mathbf{3} + \text{(para-)} \mathbf{4} \qquad (3)$$

1 **2** **3** **4**

Table 2 *The nitration of PhR according to Equation 3* [a]

R	Time (min)	Yield (%)	2	3	4
F	30	>99	6	0	94
Cl	30	>99	7	0	93
Br	5	>99	13	0	87
H	30	>99	-------	-------	-------
Me	30	>99	18	3	79
Et	10	>99	15	3	82
iPr	30	>99	9	3	88
tBu	30	92	8	trace	92
Ph [d]	30	70	trace	0	>99

[a] Hß (1 g), HNO$_3$ (2.5 g of 90%, 35 mmol), Ac$_2$O (5.0 ml, 53 mmol), PhR (35 mmol), ambient temperature for the indicated time followed by distillation under reduced pressure.

As the results in Table 2 show, the new method is applicable to a range of substrates of moderate activity, which are the ones that often give selectivity problems in traditional nitrations. Indeed, the present results represent the highest *para*-selectivities yet achieved in high yielding nitration reactions for the entire range of substrates. 1,2-Disubstituted benzenes often present even more problems of selectivity in nitrations than monosubstituted benzenes and it was therefore of interest that the new system also proved to be more successful than traditional mixed acid nitration for a number of such substrates.

3 BROMINATION

Production of *para*-bromotoluene by bromination of toluene is particularly troublesome. Separation of the regioisomers by distillation is extremely difficult, so that the commercial approach has usually involved nitration of toluene and separation of the isomers at that stage, followed by reduction to 4-nitroaniline and then diazotization and treatment with copper(I) bromide. Even with the improvements in nitration reported above this would still be a very expensive and environmentally undesirable approach, and this accounts for the very high cost of such a simple chemical. Clearly it would be better to gain control over the direct bromination of toluene to produce the 4-bromo isomer cleanly.

Work by Sasson and co-workers showed that zeolite NaY brought about a substantial rate enhancement in the reaction of bromine with toluene to give a product that was almost entirely *para*, but the yield was low, about 13 %.[13] Higher yields at lower selectivity could be obtained over a prolonged reaction period. When we investigated Sasson's method further we discovered that the extent of the rapid, highly selective reaction increased with the amount of zeolite added, which suggested that a stoichiometric reaction (eq 4, R = Me) was occurring. The simple device of using a large quantity of the zeolite allowed the synthesis in high yield and with high *para*-selectivity of bromination products from a range of simple substrates (Table 3).[14]

$$\text{(4)}$$

The main problem with this method is the large quantity of zeolite needed because of the stoichiometric nature of the reaction. However, the solid recovered from the reaction comprises the proton form of the zeolite intimately mixed with sodium bromide. When this is heated to a high temperature hydrogen bromide is driven off and NaY having the same activity as the original material is recovered (eq 5). Therefore, the same sample of zeolite can be used many times to render the process more economical.

Table 3 *Yields of products formed according to equation (4)*

R	Yield (%)	Product proportions (%)		
		ortho	meta	para
Me	99	1	trace	99
Et	98	1	trace	99
i-Pr	98	trace	trace	100
t-Bu	97	trace	trace	100
F	93	trace	trace	100
Cl	82	trace	trace	100
Br	66	trace	trace	100

$$NaBr + HY \xrightarrow{400\ ^\circ C} NaY + HBr \qquad (5)$$

4 METHANESULFONYLATION

The best known method for methanesulfonylation of toluene involves aluminium chloride as the catalyst and methanesulfonyl chloride as reagent but affords a yield of only 52% with an *o:m:p* isomer distribution of 53:14:33.[15] The best *para*-selectivity prior to our work, reported in the patent literature, was obtained by the use of the protonated form of zeolite beta as catalyst with methanesulfonyl chloride as reagent, which in the most favourable cases gave methyl tolyl sulfone in a yield of 14%, of which 81% was the *para*-isomer.[16] We therefore undertook a broader study of the latter reaction in the hope of finding a better system for the selective *para*-methanesulfonylation of toluene.

Initially, different methanesulfonylating agents were tested with a small number of selected solid catalysts in refluxing toluene. Methanesulfonic anhydride gave much better yields than methanesulfonyl chloride in these tests. With this reagent the yield of methyl tolyl sulfones was greatest when the catalyst was an amorphous silica-alumina. Under standard conditions (0.6 g of catalyst, 1.2 g of anhydride in 15 ml of refluxing toluene, 18h) the yield was 85%, but it could be raised to 91% by the use of a dry bag during transfer of the anhydride and by addition of a little phosphorus pentoxide to the reaction mixture. Unfortunately, although the yield (based on anhydride) was excellent, the regioselectivity was unexceptional (*o:m:p* = 56:10:34). Attention was therefore turned to the use of zeolite catalysts in the hope of gaining shape-selectivity.

Preliminary screening of a range of proton-form zeolites confirmed that zeolite ß offered the best *para*-selectivity (*o:m:p* = 37:11:53) and a good yield (79%). Therefore, a range of cation-exchanged forms of zeolite ß was tested as catalysts in the reaction (equation 6). The results are reported graphically in Figure 1.

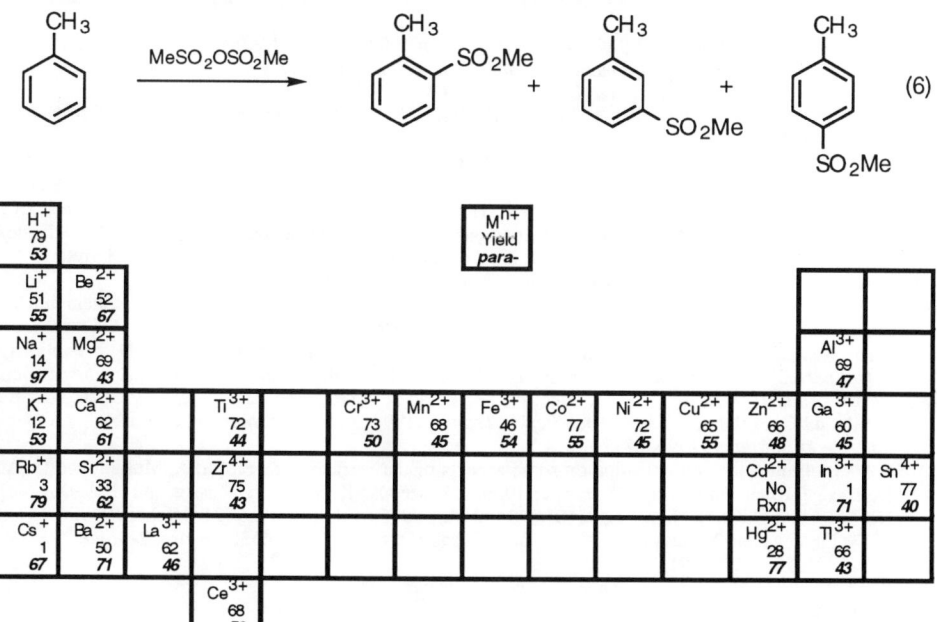

Figure 1   *Total yield and proportions of products formed according to equation 6*

It is clear from Figure 1 that overall there is a broad correlation between cation size and both yield and selectivity. Unfortunately, there is also a trade-off between the two, with those cases providing highest *para*-selectivity also giving rather poor yields. Nevertheless, the sodium-exchanged case offers the possibility of very high *para*-selectivity if the yield based on the anhydride is not critical (methanesulfonic acid can be recovered after work-up). Alternatively, reasonable *para*-selectivity can be achieved in good yield by use of the proton exchanged form of zeolite ß. In this case the *para*-selectivity and the yield are much better than for the published reaction with aluminium chloride and methanesulfonyl chloride. With the additional advantages of easy catalyst recovery and the lack of a metal-containing waste stream, this approach should prove attractive for sulfonylation reactions.[17]

5 ACKNOWLEDGEMENTS

The original experimental work was carried out by Adam Musson, Dawoud Bahzad and Gordon Ewart, research students in Swansea. We thank Zeneca, the Kuwait Institute for Scientific Research and the EPSRC for financial support to the students, the EPSRC and the University of Wales for grants for the purchase of NMR equipment used in this study, and PQ Zeolites for gifts of zeolites.

6 REFERENCES

1.   R. Taylor, *Electrophilic Aromatic Substitution*, John Wiley and Sons, Chichester, 1990.
2.   *Chemistry of Waste Minimisation*, ed. J. H. Clark, Chapman and Hall, London, 1995.
3.   *Solid Supports and Catalysts in Organic Synthesis*, ed. K.Smith, Ellis Horwood, Chichester, 1992.
4.   *Preparative Chemistry Using Supported Reagents*, ed. P. Laszlo, Academic Press, London, 1987.
5.   H. Van Bekkum, E. M. Flanigan and J. C. Jansen, *Stud. Surf. Sci. Catal.*, 1991, **58.**
6.   K. Smith in *Catalysis of Organic Reactions*, ed. M. G. Scaros and M. L. Prunier, Marcel Dekker, New York, 1994, 91.
7.   G. A. Olah, R. Malhotra, and S. C. Narang, *Nitration: Methods and Mechanisms*, VCH, New York, 1989 ; pp 5-7; K. Schofield, *Aromatic Nitration*, Cambridge University Press, Cambridge, 1980.
8.   G. A. Olah, R. Malhotra, and S. C. Narang, *J. Org. Chem.*,1978, 43, 4628.
9.   K. Smith, *Bull. Soc. Chim. Fr.*, 1989, 272; K. Smith, K. Fry, M. Butters, and B. Nay, *Tetrahedron Lett.*, 1989, 30, 5333.
10. S. M. Nagy, K. A. Yarovoy, M. M. Shakirov, V. G. Shubin, L. A. Vostrikova, and K. G. Ione, *J. Mol. Catal.*,1991, 64, L31; T. J. Kwok, K. Jayasuriya, R. Damavarapu, and B. W. Brodman, *J. Org. Chem.*, 1994, 59, 4939; P. Laszlo, *Acc. Chem. Res.*, 1986, 19, 121; P. Laszlo, J. Vandormael, *Chem. Lett.*, 1988, 1843; P. Laszlo, A. Cornelis, A. Gerstmans, *Chem. Lett.*, 1988, 1839; A. Cornelis, L. Delaude, A. Gerstmans and P. Laszlo, *Tetrahedron Lett.*, 1988, 29, 5657; F. J. Waller, A. G. M. Barrett, D. C. Braddock and D. Ramprasad, *J. Chem. Soc., Chem. Commun.*, 1997, 613.
11. K. Smith, M. Butters, W. E. Paget and B. Nay, *Synthesis*, 1985, 1155; K. Smith, M. Butters and B. Nay, *ibid.*, 1157; K. Smith, *Stud. Surf. Sci. Catal.*, 1991, **59**, 55; K. Smith and K. B. Fry, *J. Chem. Soc., Chem. Commun.*, 1992, 187; K. Smith, D. M. James, A. G. Mistry, M. R. Bye and D. J. Faulkner, *Tetrahedron*, 1992, **48**, 7479; K. Smith, D. M. James, I. Matthews and M. R. Bye, *J. Chem. Soc., Perkin Trans 1*, 1992, 1877; L. Delaude, P. Laszlo and K. Smith, *Acc. Chem. Res.*, 1993, **26**, 607; K. Smith and G. Pollaud, *J. Chem. Soc., Perkin Trans 1*, 1994, 3519; K. Smith and D. Bahzad, *J. Chem. Soc., Perkin Trans. 1*, 1996, 2793.
12. For a preliminary communication on some aspects of the work, see K. Smith, A. Musson and G. A. DeBoos, *J. Chem. Soc., Chem. Commun.*, 1996, 469; see also K. Smith, A. Musson and G. A. DeBoos, UK Patent Application Numbers, 9510166.3 (19 May 1995) and 9521705.5 (24 October 1995).
13. F. De La Vega and Y. Sasson, *J. Chem. Soc., Chem. Commun.*, 1989, 653; see also, *Zeolites*, 1989, 9, 418; 1991, **11**, 617; 1991, **13**, 341.
14. K. Smith and D. Bahzad, *J. Chem. Soc., Chem. Commun.*, 1996, 467.
15. G. A. Olah, S. Kobayashi and J. Nishimura, *J. Am. Chem. Soc.*, 1973, **95**, 564.
16. S. Daley, K. A. Trevor, K. R. Randles and B. D. Gott, PTC Int. Appl. WO93 18.000. (16.09.93); *Chem. Abstr.*, **120**, P54320u.
17. K. Smith, G. M. Ewart and K. R. Randles, *J. Chem. Soc., Perkin Trans 1*, 1997, 1085.

HETEROGENEOUS CATALYSTS FOR LIQUID PHASE OXIDATIONS

R.A. Sheldon, I.W.C.E. Arends and H.E.B. Lempers

Laboratory for Organic Chemistry and Catalysis
Delft University of Technology
Julianalaan 136, 2628 BL Delft,
The Netherlands

1 INTRODUCTION

The catalytic epoxidation of olefins by alkyl hydroperoxides (reaction 1) was independently discovered by Halcon[1] and Atlantic Richfield[2] workers in the early sixties. Soluble compounds of *inter alia* molybdenum, vanadium, tungsten and titanium were shown to catalyse reaction 1, molybdenum being the most effective.

$$R^1 \diagdown\!\!\!\diagup\!\!\!\diagdown \quad + \quad R^2O_2H \quad \xrightarrow{\text{catalyst}} \quad R^1 \diagdown\!\!\!\triangle^O \quad + \quad R^2OH \tag{1}$$

Shell workers[3,4] on the other hand, developed a heterogeneous, Ti(IV)/SiO$_2$, catalyst for reaction 1 which exhibits selectivities comparable to homogeneous molybdenum and (for a heterogeneous catalyst) high activities.[4-6] This contrasts with soluble titanium compounds which were shown to be mediocre catalysts for the epoxidation of 1-octene with tert-butylhydroperoxide (TBHP).[5,6] The superior catalytic activity of Ti(IV)/SiO$_2$ was attributed to site isolation of discrete Ti(IV) centers on the silica surface preventing oligomerization to less reactive μ-oxo species, which readily occurs with soluble Ti(IV) compounds, and to an increased Lewis acidity of Ti(IV) resulting from electron withdrawal by silanoxy ligands.[4-6] Furthermore, it was shown that the combination of titanium(IV) with silica was unique: all other combinations, e.g. Mo(VI), W(VI), V(V) etc. on silica, led to rapid leaching of the metal ion, producing a homogeneous catalyst. These discoveries were subsequently developed into commercial processes for the production of propylene oxide using homogeneous molybdenum (ARCO) or heterogeneous Ti(IV)/SiO$_2$ (Shell) in combination with TBHP or ethylbenzene hydroperoxide (EBHP) as the oxidant.[7]

1.1 Mechanism of oxygen transfer

The high yields of epoxide obtained with nucleophilic olefins and the retention of (cis or trans) stereochemistry of the olefin in reaction 1 are consistent only with a heterolytic mechanism involving electrophilic attack on the olefinic double bond. We proposed [4,8] a mechanism involving oxygen transfer from an electrophilic alkylperoxo-metal complex to the olefin (Figure 1, mechanism a). Chong and Sharpless [9] subsequently proposed a variant (Figure 1, mechanism b), which is now generally accepted, involving coordination of the distal rather than the proximal oxygen to the metal center.

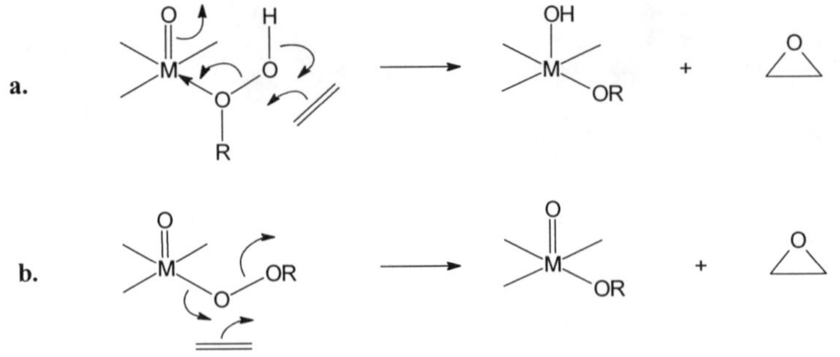

a.

b.

Figure 1 *Mechanism of Oxygen Transfer*

The primary function of the metal ion is to increase the electrophilic character of the peroxidic oxygens by withdrawing electrons from the O–O bond. Hence, superior catalysts are strong Lewis acids in their highest oxidation state and relative activities follow the order of Lewis acidities, i.e. $MoO_3 \gg WO_3 > TiO_2$, V_2O_5. The metal ion should also be a weak oxidant to minimize competing hydroperoxide decomposition via one-electron redox processes.[4-6] This explains why chromium(VI), although a strong Lewis acid, is generally a poor catalyst for reaction 1.

Reaction 1 is one example of a variety of metal-catalysed oxidations involving a peroxometal pathway.[10] Alternatively, metal-catalysed oxidations with H_2O_2 or RO_2H can involve an oxometal species (M=O) as the key oxidant (Figure 2). Early transition elements, e.g. Mo, W, Ti, favor peroxometal pathways while later and/or first row transition elements, e.g. Cr, Mn, Fe, Ru, Os, tend to favor oxometal pathways. Vanadium is intermediate in character and can involve either pathway, depending on the substrate.[10] Typical reactions involving peroxometal pathways are epoxidation, oxidation of N and S compounds and alcohol oxidations. Oxometal oxidants, in addition to the above oxidations, are responsible for other types of transformations, e.g. benzylic and allylic oxidations and alkane oxidations. When the latter reactions are observed with typical peroxometal-type catalysts they are usually the result of competing homolytic processes.

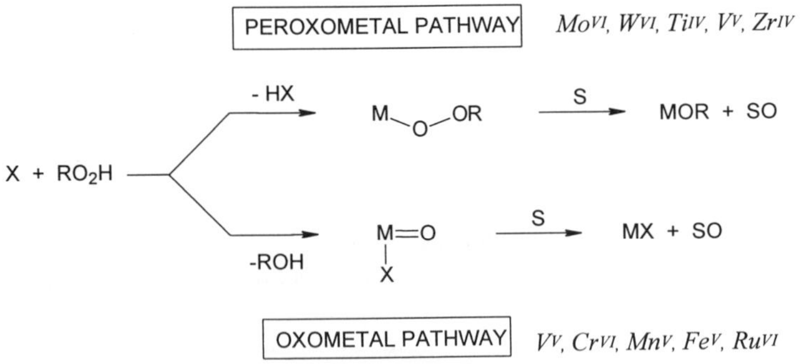

Figure 2 *Oxometal vs. Peroxo Metal Pathways*

2 EPOXIDATION WITH H_2O_2: TITANIUM SILICALITE (TS-1)

One feature which homogeneous titanium catalysts and the Ti(IV)/SiO_2 catalyst have in common is deactivation by protic molecules, the extent of which increases in the order t-BuOH < EtOH < MeOH < H_2O. Water has a particularly deleterious effect which means that both homogeneous catalysts and the heterogeneous Ti(IV)/SiO_2 are unsuitable catalysts for epoxidations with H_2O_2. Hence, the appearance, in the mid-eighties, of publications from Enichem workers [11] describing the remarkable activity of titanium(IV)silicalite (TS-1) as a catalyst for a variety of oxidations, including olefin epoxidation, with 30% aqueous hydrogen peroxide, under very mild conditions, was initially greeted with some scepticism. Thus, two materials (Ti(IV)/SiO_2 and TS-1) having the same elemental composition, i.e. 2% TiO_2/98% SiO_2, exhibited totally different catalytic properties. Indeed, attempts by various groups to reproduce the Enichem results were largely unsuccessful until it became clear that certain parameters in the synthesis of TS-1, e.g. the necessity for an alkali metal-free template, are critical. It was subsequently shown, using various characterization techniques, [12] that TS-1 contains titanium isomorphously substituted for silicon in the framework of silicalite-1, a hydrophobic, medium pore molecular sieve possessing a three dimensional pore system with pore diameters of 5.3x5.5 Å and 5.1x5.5 Å.

The remarkable activity of TS-1 is attributed to site isolation of Ti(IV) centres in the hydrophobic micropores of silicalite. The hydrophobic environment of the active site favors the adsorption of hydrophobic substrates, thus circumventing the deactivation by adsorbed water molecules observed with the hydrophilic Ti(IV)/SiO_2 catalyst. Moreover, confinement of the active site in a cavity of molecular dimensions creates essentially solvent-free conditions that are conducive to rapid reactions, analogous to the situation in the active site of enzymes.

The unprecedented activity of TS-1 as a catalyst for olefin epoxidation with aqueous H_2O_2 is illustrated in Table 1. Relatively unreactive olefins, such as propylene or even allyl chloride, are readily epoxidized, at temperatures close to ambient in methanol as solvent. [13]

3 REDOX MOLECULAR SIEVES

The success of TS-1 stimulated a flourish of activity in the synthesis of metal-substituted molecular sieves based on the expectation that TS-1 was the progenitor of a broad family of novel catalytic materials, so-called redox molecular sieves, for use in selective liquid phase oxidations. One could envisage the incorporation of titanium or other redox active metals, e.g. vanadium, chromium etc., in a variety of molecular sieves e.g. silicalites, zeolites, aluminophosphates (AlPOs) and silicoaluminophosphates (SAPOs).

Table 1 *TS-1 catalysed epoxidations with 60% aq. H_2O_2* [a,b]

Olefin	T (°C)	Time (min)	H_2O_2 conv. (%)	Epoxide sel. (%)
Propylene	40	72	90	94
1-Hexene	25	70	88	90
1-Octene	45	90	81	91
Cyclohexene	25	90	10	n.d.
Allyl chloride	45	30	98	92

a) Adapted from ref. 13. b) Olefin/H_2O_2 molar ratio = 5; MeOH solvent

Indeed, in the last ten years a vast array of metal-substituted molecular sieves has been synthesized and the catalytic properties investigated.[14] We emphasize, however, that, in contrast to TS-1, many of these materials have not been adequately characterized, and hence, isomorphous substitution has not been confirmed. Moreover, the stability of many of these materials towards leaching of the metal ion has not been demonstrated (see later).

4 TITANIUM-SUBSTITUTED MOLECULAR SIEVES

A shortcoming of TS-1 as an oxidation catalyst is that reactions are limited to substrates with kinetic diameters < 5.5 Å. For example, 1-hexene is readily epoxidized with aq. H_2O_2 at 50°C while cyclohexene is essentially unconverted (Table 1). Molecular graphics indicate that although cyclohexene could just fit into the pores of TS-1 it would be difficult to accommodate the transition state for oxygen transfer from a peroxotitanium(IV)species to the double bond. By the same token TS-1 does not catalyse epoxidations with the more bulky oxidant TBHP, which means that TS-1 exhibits complementary properties to the $Ti(IV)/SiO_2$ catalyst.

To circumvent the size restriction larger titanium-substituted molecular sieves have been synthesized. For example, Corma and coworkers [15] synthesized titanium-substituted zeolite-beta, which has pore diameters of 7.6 Å, and showed that it catalysed the H_2O_2 oxidation of cyclohexene and 1-hexene at roughly the same rate (Table 2). However, in contrast to reactions with TS-1, the main product was the glycol monomethyl ether, formed by acid-catalysed ring opening of the epoxide by the methanol solvent.

We subsequently showed [16] that the epoxide could be obtained in high selectivity, in the oxidation of 1-octene, simply by neutralizing the Brønsted acid aluminium sites by reaction with an alkali metal acetate and recalcination (Table 3). Similarly, 1-octene afforded the epoxide in high selectivity when treated with TBHP in the presence of alkali metal exchanged titanium-beta.[17]

Table 2 *Ti-Beta Catalysed Epoxidations with H_2O_2* [a]

Olefin	Catalyst	H_2O_2 conv. (%)	Product sel. (%)	
			epoxide	glycol ether
1-Hexene	TS-1	98	96	4
	Ti-beta[c]	80	12	80
Cyclohexene	TS-1	<5	100	
	Ti-beta[c]	80	0	100

a) Adapted from ref. 15 b) Olefin/H_2O_2 molar ratio = 12; MeOH solvent; 25°C c) 4% Ti; Si/Al = ca. 200.

Table 3 *Ti-Beta Catalysed Epoxidation of 1-Octene: Effect of Alkali Metal Exchange* [a,b]

Catalyst	H_2O_2 conv. (%)	Product Sel. (%)	
		epoxide	glycol ether
TS-1	95	76	24
Li-TS-1	85	98	0
Ti-Beta	48	0	97
Li-Ti-Beta	31	87	5
Na-Ti-Beta	22	84	6

a) Adapted from ref. 16. b) 35% aq. H_2O_2; MeOH solvent; 100 min at 40°C.

The development of mesoporous molecular sieves,[18] such as MCM-41, containing a regular array of hexagonal, ordered mesopores with diameters in the range 15-100 Å, further extended the scope of molecular sieve catalysis to much larger molecules. Corma et al.,[19] for example, showed that Ti-MCM-41 catalyses epoxidations with H_2O_2 and TBHP, including the epoxidation of the bulky olefin, norbornene with TBHP.

Another approach to synthesizing redox molecular sieves involves grafting of metal complexes to the internal surface by reaction with pendant silanol groups. For example, Thomas and coworkers[20] prepared a titanium-grafted MCM-41 by reaction with bis-(cyclopentadienyl)titanium(IV)dichloride followed by calcination. The resulting material was shown to catalyse the epoxidation of cyclohexene and pinene with TBHP at 40°C. A possible advantage of grafting or tethering (via a spacer group) of metal complexes, compared to framework substitution, could be improved accessibility of the metal centre, leading to higher activities.[21]

5 TITANIUM AEROGELS AND XEROGELS

Baiker and coworkers [22] have reported the synthesis of amorphous titania-silica aerogels containing up to 20% titanium. These materials resemble the Shell catalysts in being amorphous but the method of synthesis allows for the incorporation of ten times as much titanium. Hence, this material was significantly more active (per gram of catalyst) than Ti(IV)/SiO_2 in epoxidations with cumene hydroperoxide (CHP) (Table 4).[23] Drying of the gel by semicontinuous extraction of the solvent (aqueous ethanol) with supercritical CO_2 at low temperature (to produce a so-called aerogel) was found to be crucial for obtaining a high-surface area, mesoporous structure with a high degree of Ti-O-Si connectivity. A strong correlation between the concentration of Ti-O-Si linkages and reaction rate was established.

Maier and coworkers,[24] on the other hand, synthesized titania-silica xerogels using a conventional drying procedure. The resulting materials contained up to 17% Ti and had high surface areas (ca. 400-500 m^2g^{-1}) and a narrow pore size distribution (effectively 6.5-7.7 Å) comparable with large-pore zeolites. They were shown to be effective catalysts for the epoxidation of linear and cyclic alkenes with TBHP.

Both the titanium aerogels and xerogels described above are hydrophilic in nature and, analogous to the Ti(IV)/SiO_2 catalyst, are not effective catalysts for epoxidations with aqueous H_2O_2. More recently, Maier and coworkers [25] have described the synthesis of hydrophobic xerogels by using mixtures of $MeSi(OEt)_3$ and $Si(OEt)_4$, instead of $Si(OEt)_4$, in combination with $Ti(OPr-i)_4$ for the gel preparation. A xerogel with the composition

Table 4 *Comparison of Titanium Catalysts in the Epoxidation of Cyclohexene* [a]

Catalyst	Oxidant	Temp. (°C)	Activity (g/g/h)	Conv. (%)	Sel. (%)
Aerogel	CHP	90	90	50	100
Aerogel	CHP	60	32	50	100
Ti(IV)/SiO_2	CHP	60	5.9	80	100
TiO_2/MCM-41	TBHP	40	10	50	95
Ti-Beta	TBHP	50	0.1	34	99
Ti-MCM-41	TBHP	60	0.01	14	93

a) Data taken from ref. 23.

1 TiO_2 : 50 $MeSiO_{1.5}$: 49 SiO_2 was shown to have a hydrophobic index approaching that of TS-1. Interestingly, inhibition by the tert-butyl alcohol product was shown to decrease with increasing SiMe content in the gel, i.e. with increasing hydrophobicity.[25] This led the authors to conclude that these hydrophobic titanium xerogels should be effective catalysts with aqueous H_2O_2. Unfortunately, to establish this they performed comparisons with TS-1 in the epoxidation of cyclohexene with H_2O_2.[25] As noted earlier, TS-1 is not an effective catalyst for the epoxidation of cyclohexene. Hence, confirmation of the catalytic properties of these materials in epoxidations with H_2O_2 awaits the choice of a suitable substrate, e.g. 1-hexene.

6 CHROMIUM-SUBSTITUTED MOLECULAR SIEVES AND THE QUESTION OF LEACHING

We have previously shown [26,27] that the incorporation of chromium(VI) into silicalite or aluminophosphate-5 and -11 produces materials that catalyse a variety of oxidations with TBHP or O_2 as the terminal oxidant. The as-synthesized materials contain chromium(III) which is probably isomorphously substituted for Si or Al. On calcination the chromium is oxidized to the hexavalent chromyl (CrO_2^{2+}) state which is attached to the framework by only two metal-oxygen bonds and, hence, cannot be isomorphously substituted. As noted earlier, chromium(VI) typically catalyses oxidations via an oxometal mechanism, in which $Cr^{VI}=O$ is the active oxidant.

Hence, Cr-APO-5 was shown to catalyse reactions typical of oxometal oxidants, i.e. benzylic and allylic oxidations and (cyclo)alkane oxidations.[26,27] The catalytic cycle is assumed to involve reoxidation of Cr^{IV} to $Cr^{VI}=O$ by TBHP or, in the case of reactions with O_2, by the hydroperoxide derived from (chromium-catalysed) autoxidation of the substrate.

One of the first reactions which we studied with chromium-containing molecular sieves was the decomposition of cyclohexylhydroperoxide to a mixture of cyclohexanone and water (reaction 2). Both CrAPO-5 and CrS-1 were shown[28] to be effective catalysts for this reaction, which is of importance in the manufacture of cyclohexanone by the autoxidation of cyclohexane. CrAPO-5 was particularly effective in giving high rates and high selectivity to cyclohexanone.

$$(2)$$

catalyst	conv(%)	sel. (%)	
		ketone	alcohol
CrAPO-5	87	86	13
CrS-1	98	64	36

In order to confirm that the reaction was taking place in the micropores of CrAPO-5 we performed experiments with the bulky tertiary alkyl hydroperoxide, triphenylmethyl-hydroperoxide (TPMHP) which is too large to enter the micropores. Both the homogeneous $Cr(acac)_3$ and supported $CrO_2Cl_2/SiO_2\text{-}Al_2O_3$ were effective catalysts (75% and 72% decomposition in 2 hrs at 70°C in dichloroethane). CrAPO-5 gave only 1%

decomposition.[28] This led us to conclude that reaction was exclusively taking place within the micropores and that no leaching of the chromium occurred during the reaction.

More recently, we returned to this question of leaching in a study of the allylic oxidation of alpha-pinene with TBHP in the presence of chromium-containing molecular sieves (reaction 3).[29]

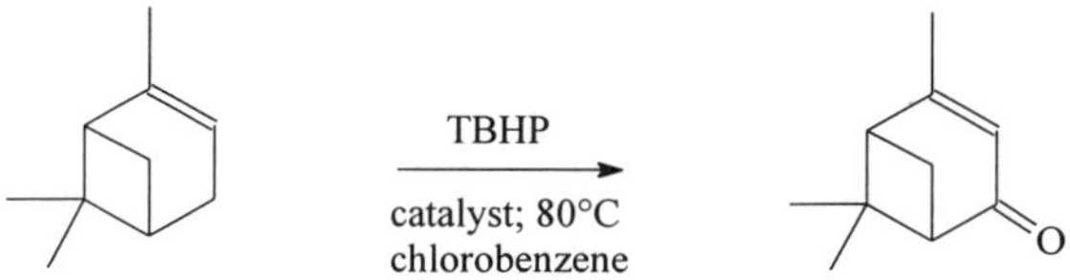

(3)

catalyst	conv (%)	sel. (%)
CrAPO-5	85	77
CrAPO-11	100	75
CrS-1	100	69

To test for leaching we filtered the catalyst (CrAPO-5) after 30 minutes, which corresponded to ca. 20% pinene conversion, and allowed the filtrate to react further. The catalyst filtration was performed at the reaction temperature (80°C) in order to avoid possible readsorption of solubilized chromium on cooling. Indeed, we found that after hot filtration the filtrate (mother liquor) reacted further at roughly the same rate as that observed when the catalyst was not filtered. In contrast, when the mixture was allowed to cool to room temperature before catalyst filtration, the procedure generally followed when testing for leaching, little further reaction was observed. On the basis of these results we concluded that chromium is leached from the catalyst and that the observed catalysis is due to homogeneous chromium.

In a second test for leaching we studied the allylic oxidation of alpha-pinene with a bulky hydroperoxide (TMPHP) and of a bulky olefin (valencene) with TBHP (reactions 4 and 5, respectively).[29]

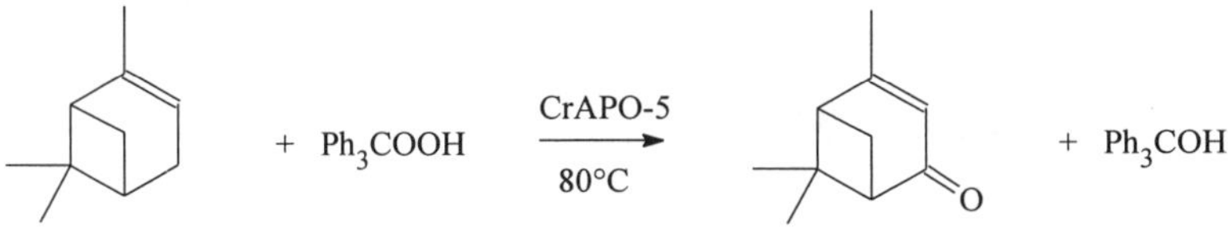

(4)

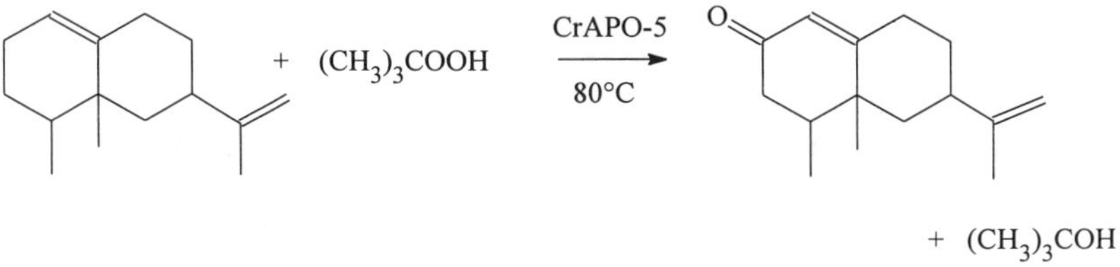

(5)

If the reaction was occurring within the micropores one would expect to observe no reaction in both cases, since both TPMHP and valencene are too bulky to access the pores.

However, we observed facile conversion of valencene with TBHP but virtually no conversion of pinene with TPMHP. The only explanation which is consistent with these observations is that soluble chromium is responsible for the observed catalysis and that <u>it is leached by reaction with the hydroperoxide</u>. That this is the case was confirmed in experiments where CrAPO-5 was pretreated with TBHP, filtered, and pinene added to the mother liquor resulting in facile oxidation. Based on these observations the results obtained[28] in the CrAPO-5-catalysed decomposition of TPMHP (see earlier) were re-interpreted to mean exactly the opposite of our initial interpretation.

The next question which we addressed was: how much chromium is being leached from the catalyst? Since conventional ICP analysis of the filtrate obtained from CrAPO-5-catalysed oxidation showed that the chromium content was below the detection limit we employed diphenylcarbazide in a sensitive colorimetric analysis of chromium(VI). We found that 0.3% of the 0.88% chromium (i.e. 0.0026% of total catalyst weight) present in CrAPO-5 was leached under the above mentioned conditions.[30] This corresponds to a substrate/catalyst (S/C) ratio of 17000 or ca. 1 ppm Cr in the filtrate.

Analogous experiments with CrAPO-11 and samples of CrS-1 prepared by different procedures showed that the extent of chromium leaching varied between 0.3 and 34% and increased markedly with decreasing crystallite size. Furthermore, we showed that if the filtrate was allowed to cool to room temperature the chromium was reduced to the trivalent state. Subsequent experiments in which pyridinium dichromate, $(py)_2Cr_2O_7$, was used as a homogeneous catalyst (at a S/C ratio of 17000) in the TBHP oxidation of pinene afforded the same rate as that observed in the CrAPO-5-catalysed reaction. In contrast, when $Cr(acac)_3$, i.e. trivalent chromium, was used as the catalyst long induction periods (e.g. 5 hrs at S/C = 10000) were observed. In short, these results unequivocally demonstrate that the observed catalysis with CrAPO-5, CrAPO-11 and CrS-1 is homogeneous in nature and due to chromium being leached from the catalyst. They also show that conventional cooling of the reaction mixture prior to catalyst filtration could lead to erroneous conclusions. Moreover, we showed that in homogeneous catalysis by chromium (VI) the relationship between reaction rate and catalyst concentration is nonlinear: turnover frequencies increased with decreasing catalyst concentration suggesting possible dissociation of dimeric or oligomeric chromium species in solution. Here again, this could lead to erroneous conclusions in comparison of homogeneous with "heterogeneous" catalysts.

7 CONCLUSIONS & FUTURE PROSPECTS

We hope that the above discussion has shown that there are many exciting developments in the design of heterogeneous catalysts for liquid phase oxidations. Redox molecular sieves, particularly those containing titanium, have considerable potential as selective recyclable catalysts for oxidations with RO_2H and H_2O_2. Pore size, hydrophobic/hydrophilic character and Brønsted acidity are important parameters that influence the catalytic properties of these materials. Microporous mixed oxides with tunable surface hydrophobicity, prepared by the sol-gel method, also show promise as recyclable heterogeneous catalysts, perhaps even with aqueous H_2O_2.

However, it has become increasingly clear that rigorous proof of heterogeneity is needed for many of these materials. Since homogeneous titanium is a poor oxidation catalyst there would appear to be no doubt that catalysis by titanium-substituted molecular sieves, and related materials, is heterogeneous in nature. However, we note that this does not exclude the leaching of titanium <u>to give inactive soluble titanium</u>, particularly with the less-studied materials. As we have shown, in the case of chromium-substituted molecular sieves the observed catalysis can be attributed to soluble chromium leached from the surface by reaction with RO_2H. Moreover, the conventional test for heterogeneity - filtering the catalyst and recycling it without any apparent loss of activity - is totally inadequate. Thus, the CrAPO-5 catalysts could have been recycled 10 or even 100 times (losing 3 or 30% of the available chromium, respectively) and would still, presumably, have exhibited the same activity. Rigorous proof of heterogeneity involves, as we have shown, filtration of the catalyst <u>at the reaction temperature</u> and examination of the filtrate for catalytic activity. Indeed, application of this test to the V-APO-5-catalysed oxidations with TBHP similarly showed that the observed catalysis is due to leached vanadium.[31]

Hence, we conclude that other metal-substituted molecular sieves, e.g. Zr, Sn, V, Co, etc., that are claimed to be heterogeneous catalysts for liquid phase oxidations should be subjected to the same rigorous proof to establish if they are truly heterogeneous and stable.

References

1. R. Landau, Hydrocarbon Process., 1967, <u>46</u>, 141; J. Kollar, US Pat. 3350422 (1967) to Halcon.
2. M.N. Sheng and J.G. Zajacek, *Advan. Chem. Ser.*, 1968, <u>76</u>, 418; M.N. Sheng and J.G. Zajacek, Brit. Pat., 1136923 (1968) to Atlantic Richfield.
3. Brit. Pat. 1248185 and 1249079 (1971) to Shell Oil; H.P. Wulff, US Pat., 3923843 (1975) to Shell Oil.
4. R.A. Sheldon, *J. Mol. Catal.*, 1980, <u>7</u>, 107.
5. R.A. Sheldon and J.A. van Doorn, *J. Catal.*, 1973, <u>31</u>, 427.
6. R.A. Sheldon, in 'Aspects of Homogeneous Catalysis', R. Ugo, Ed., D. Reidel, Dordrecht, 1981, Vol. 4, p. 3.
7. R.A. Sheldon, in 'Applied Homogeneous Catalysis by Organometallic Compounds', B. Cornils and W.A. Herrmann, Eds., VCH. Weinheim, 1996, Vol. 1, p. 411; J.R. Valbert, J.G. Zajacek and D.I. Orenbuch, in 'Encyclopedia of Chemical Processing and Design', J. McKeth, Ed., Marcel Dekker, New York, 1993, p. 88.
8. R.A. Sheldon, *Rec. Trav. Chim. Pays Bas*, 1973, <u>92</u>, 253 and 367.
9. A.O. Chong and K.B. Sharpless, *J. Org. Chem.*, 1977, <u>42</u>, 1587.
10. R.A. Sheldon, *Bull. Soc. Chim. Belg.*, 1985, <u>94</u>, 651; R.A. Sheldon, Top. Curr. Chem., 1993, <u>164</u>, 21.
11. B. Notari, *Stud. Surf. Sci. Catal.*, 1988, <u>37</u>, 413; M. Taramasso, G. Perego and B. Notari, US Pat., 4410501 (1983) to Snamprogetti.
12. G. Bellussi and M.S. Rigutto, *Stud. Surf. Sci. Catal.*, 1994, <u>85</u>, 177 and references cited therein.
13. M.G. Clerici and P. Ingallina, *J. Catal.*, 1993, <u>140</u>, 71.
14. I.W.C.E. Arends, R.A. Sheldon, M. Wallau and U. Schuchardt, *Angew. Chem. Int. Ed. Engl.*, 1997, <u>36</u>, 1144 and references cited therein.
15. A. Corma, M.A. Camblor, P. Esteve, A. Martinez and J. Perez-Pariente, *J. Catal.*, 1994, <u>145</u>, 151.
16. T. Sato, J. Dakka and R.A. Sheldon, *Stud. Surf. Sci. Catal.*, 1994, <u>84</u>, 1853.
17. T. Sato, J. Dakka and R.A. Sheldon, *J. Chem. Soc. Chem. Commun.*, 1994, 1887.

18. J.S. Beck, J.C. Vartuli, W.J. Roth, M.E. Leonowicz, C.T. Kresge, K.P. Schmitt, C.J.W. Chu, D.H. Olsen, E.W. Sheppard, S.B. McCullen, J.B. Higgins and J.L. Schenker, *J. Am. Chem. Soc.*, 1992, 114, 10834.
19. A. Corma, M.T. Navarro and J. Perez Pariente, *J. Chem. Soc. Chem. Commun.*, 1994, 147.
20. T. Maschmeyer, F. Rey, G. Sankar and J.M. Thomas, *Nature*, 1995, 378, 159.
21. J.M. Thomas, *J. Mol. Catal. A: Chemical*, 1997, 115, 371.
22. R. Hutter, T. Mallat and A. Baiker, *J. Catal.*, 1995, 135, 177.
23. Anonymous, CATTECH, 1997, 1, 50.
24. W.F. Maier, J.A. Martens, S. Klein, J. Heilman, R. Parton, K. Vercruysse and P.A. Jacobs, *Angew. Chem. Int. Ed. Engl.*, 1996, 35, 180; S. Klein, J.A. Martens, R. Parton, K. Vercruysse, P.A. Jacobs and W.F. Maier, *Catal. Lett.*, 1996, 38, 209.
25. S. Klein and W.F. Maier, *Angew. Chem. Int. Ed. Engl.*, 1996, 35, 2230.
26. R.A. Sheldon, *J. Mol. Catal. A: Chemical*, 1996, 107, 75 and references cited therein.
27. J.D. Chen and R.A. Sheldon, *J. Catal.*, 1995, 153, 1; R.A. Sheldon, J.D. Chen, J. Dakka and E. Neeleman, *Stud. Surf. Sci. Catal.*, 1994, 82, 515 and 1994, 83, 407.
28. J.D. Chen, J. Dakka and R.A. Sheldon, *Appl. Catal., A: General*, 1994, 108, L1.
29. H.E.B. Lempers and R.A. Sheldon, *Stud. Surf. Sci. Catal.*, 1997, 105, 1061; see also H.E.B. Lempers and R.A. Sheldon, *Appl. Catal. A: General*, 1996, 143, 137.
30. H.E.B. Lempers and R.A. Sheldon, submitted for publication.
31. M.J. Haanepen, A.M. Elemans-Mehring and J.H.C. van Hooff, *Appl. Catal. A: General*, 1997, 152, 203.

THE DESIGN OF NEW SOLID INORGANIC CATALYSTS

Sir John M Thomas

The Royal Institution, London and Peterhouse, Cambridge

ABSTRACT

Success in the design of solid oxide as well as other (e.g. heterogenized organometallic) catalysts predicates precise knowledge of the structure of the active site that is to be targeted. The greatest possible precision is therefore required in determining, under operating conditions, the structure of the catalyst in general and of the active site in particular. Combined x-ray absorption spectroscopy and x-ray diffraction are premier tools for such *in situ* investigations.

Armed with these techniques, as well as with recent advances in preparative and computational procedures, it is possible to assemble a number of high-performance new catalysts for the selective oxidation of hydrocarbons and the shape selective dehydration of methanol to light olefins. It is also possible to prepare well-defined, active, bimetallic hydrogenation catalysts anchored within the mesopores of siliceous supports.

LIQUID-PHASE OXIDATION OF CYCLOHEXANE TO ADIPIC ACID CATALYSED BY MANGANESE-CONTAINING ZEOLITES

I. Belkhir[*], A. Germain, F. Fajula

Laboratoire de Matériaux Catalytiques et Catalyse en Chimie Organique, UMR-CNRS 5618, ENSCM, 8, Rue de l'Ecole Normale, 34296 Montpellier Cedex 5, France.

and E. Fache

Rhône-Poulenc Industrialisation, CRIT-Carrières, 85, Avenue des Frères Perret, BP 62, 69192 Saint-Fons Cedex, France.

1 INTRODUCTION

Adipic acid, an important intermediate for the manufacture of nylon 66, can be obtained by direct aerial oxidation of cyclohexane in acetic acid using cobalt acetate as a homogeneous catalyst[1,2]. CoAPO has been proposed as a solid catalyst[3-5], and we have recently studied the catalytic oxidation with cobalt-containing zeolites[6]. However, acetic acid, in which the diacid products are obtained, is responsible for the leaching of part of the cobalt[7] and the observed activity is the result of a homogeneous catalysis[6].

Unlike cobalt acetate, manganese acetate, used in the same conditions, demonstrates no catalytic activity[1,8]. Nevertheless, suspension and alumina-supported MnO_2 oxide are active towards the formation of adipic acid in pure cyclohexane[9]. In order to explain this contradiction, we have studied the activity of manganese-containing zeolites.

2 EXPERIMENTAL

2.1 Catalysts

BEA and MOR stand for β-zeolites and mordenites respectively. The numbers after the structure type code of zeolites denote Si/Al ratio (determined by analysis).

Manganous acetate tetrahydrate, acetic acid purex and cyclohexane for analysis were used as received. Synthetic powder MOR 15 in the proton form was supplied by PQ Corporation. MOR 100, obtained by dealumination, was a gift from Zeocat (reference: ZM-980). BEA 15 zeolite was synthesized according to the procedure described by Wadlinger et al.[10] Dealuminated BEA 2000 zeolite was obtained by treating a BEA 15 zeolite with concentrated nitric acid[11].

The zeolites (MOR and BEA) were loaded with 1 to 2% Mn using a manganous acetate-water solution[12,13]. After evaporation until dryness at 70°C, the solids were calcined at 823 K for 6 hours. Depending on the composition of the zeolite, such a procedure led to Mn cations at exchange sites and to oxide particles on the surface and inside cavities.

X-ray diffraction patterns were recorded on a CGR Theta 60 instrument using Cu $K\alpha_1$ filtered radiation. Temperature Programmed Reduction (TPR) of the catalysts was performed with an in house manufactured apparatus in a quartz flow cell equipped with a thermal conductivity detector of a Shimadzu GC8 chromatograph. The reduction temperature was programmed at a rate of 10 K.min^{-1} and the experiments were conducted in a flow of 20 cm^3.min^{-1} of an H_2-Ar (3% H_2) gas mixture.

2.2 Oxidation Procedure

Cyclohexane oxidations were carried out in glacial acetic acid, in a semi-batch reactor at 110°C and at 21 bar of total pressure under a constant flow of oxygen and nitrogen (10/90). Small amounts of acetaldehyde were used as promoter. The reaction products, analyzed by gas chromatography after esterification by methanol, consisted of adipic, succinic, glutaric and 6-hydroxycaproic acids, cyclohexanol, cyclohexanone and butyrolactone. A typical procedure used for the oxidation was described in detail in a previous work[6].

3 RESULTS AND DISCUSSION

3.1 Characterization of Manganese – containing Zeolites

Manganese loading was performed on zeolites featuring a wide range of exchange capacity (Si/Al between 15 and 2000). In all cases, the aluminosilicate network after treatment was not modified as confirmed by X-ray diffraction.

To ascertain the manganese species formed on zeolites, X-ray diffraction and temperature programmed reduction (TPR) were used and permitted the distinction of two types of manganese-containing zeolites:

- High silica solids (Si/Al>30) (samples b, c and e): The TPR patterns of these manganese-containing zeolites comprised a single broad reduction peak at T_{max} between 661 and 693 K. Assuming that MnO is the final state after the reduction[14], the total H_2 consumption was converted into an average H_2/Mn molar ratio on the solid and the results are reported in Table 1.

TABLE 1 *Characteristics of the Catalysts*

Catalyst	Si/Al atomic ratio	Mn content (weight %)	Exchange degree *	Reduction temperature (K)	H_2/Mn atomic ratio **
a Mn/BEA 15	15	1.1	0.57	>1370	0
b Mn/BEA 2000	2000	1.6	78	661	0.32
c Mn/BEA 2000S	2000	12	590	727	0.33
d Mn/MOR 15	15	2.1	1.10	663	0.033
e Mn/MOR 100	100	1.4	3.50	671	0.31
Mn_3O_4		72		763	0.33
Mn_2O_3		70		783	0.50

* Exchange degree = 2.Mn/Al. ** Total molar H_2 consumption per mol of Mn

The average stoichiometry of these Mn-zeolites was equal to 0.33 and corresponded to that of Mn_3O_4. The broad convolution of the reduction peak and its lower temperature than bulk oxide[14] suggested a highly dispersed system with small particles. Furthermore, the increase in the exchange degree of the solids resulted in an increase of the reduction temperature T_{max} which became close to that of the unsupported Mn_3O_4 oxide. These results are consistent with those reported for alumina-supported Mn_3O_4 oxide catalysts[15-17] obtained by incipient wetness impregnation. XRD was used to confirm the presence of crystalline phase of Mn_3O_4 on zeolites. Detectable Mn_3O_4 species were observed in the case of Mn/BEA 2000S in which the manganese loading was higher than 10% (Figure 1).

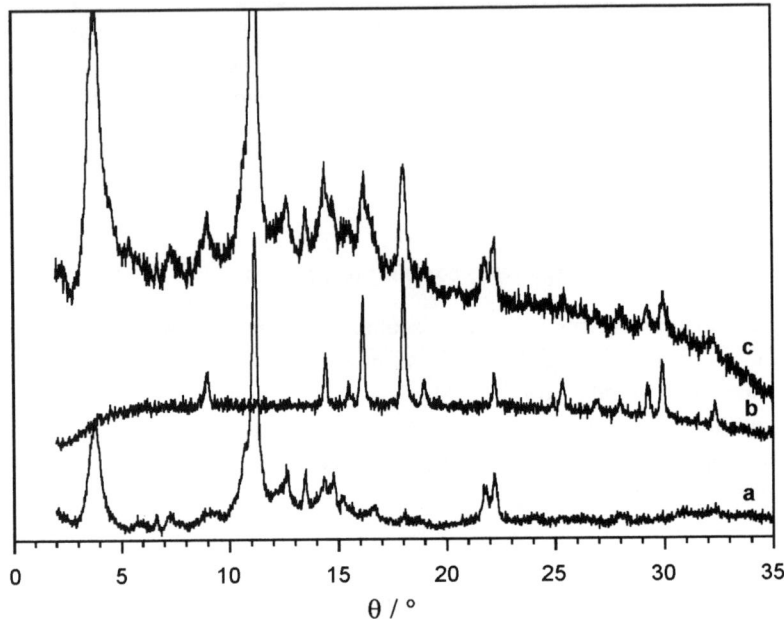

Figure 1 *X-ray diffraction patterns of a H/BEA, b Mn₃O₄, c Mn/BEA 2000S*

- Zeolites with intermediate alumina content (Si/Al=15) (samples a and d in Table 1): In the TPR experiments, no reduction was observed for sample a. As the exchange degree of this zeolite was lower than 1 and as manganese species in the oxidation state II were difficult to reduce, we deduced that the manganese was incorporated as Mn^{2+} exchanged cations. Part of the manganese species present in the overexchanged mordenite d and corresponding to the excess of metallic cations introduced was reduced at 663 K. Both Mn^{2+} cations and Mn_3O_4 oxides coexisted.

When the amount of metallic cations exceeded the exchange capacity of the support, the excess appeared as Mn_3O_4 oxides on the surface and inside the pores of the solid, after calcination at 823 K.

3.2 Activity of Mn-Zeolites in the Cyclohexane Oxidation

The manganese-containing zeolites exhibited two different catalytic behaviours according to the aluminium content of the support (Table 2). When the aluminic acid sites of the zeolite were not completely compensated by Mn^{2+} cations, the reaction rate of the cyclohexane oxidation was thus lowered as compared to that of the reaction without catalyst. This result had already been described in a previous work[6] and was attributed to an inhibition of the oxidation due to the free aluminic acid sites.

TABLE 2 *Activity of Mn-Zeolites in the Oxidation of Cyclohexane*

Catalysts	Mn content (mmol)	Acid sites * (mmol)	Reaction rate** (mmol/min)	Cyclohexane conversion *** (%)	Adipic acid selectivity *** (%)
None	0	0	0.36	6.6	16
Mn/BEA 15	0.24	0.32	~0	0.2	0
Mn/BEA 2000	0.29	0	1.3	22	52
Mn/MOR 15	0.38	0	1.1	27	56
Mn/MOR 100	0.25	0	1.3	23	53

Cyclohexane: 690 mmol; acetaldehyde: 5 mmol; acetic acid: 68 ml; N_2/O_2: 90/10; 21 bars; flow: 20 $l.h^{-1}$; 110°C.
* Overall free aluminium content in the reaction medium
** Rate of oxygen consumption measured after 2 hours of reaction.
*** The reaction lasted for 6 hours.

In contrast, the zeolite-supported manganese oxide solids were effective catalysts as they increased the reaction rate of the oxidation. In these cases, the adipic acid selectivity was 3 to 4 times higher than that observed in the absence of catalysis. The activity was independent on the structure of the solid.

To explain the type of catalysis occurring, the Mn-zeolites were treated in acetic acid, refluxed overnight and after separation of the solid, the filtrates were fed into the reaction. Activities of these filtrates towards the oxidation of cyclohexane are reported in Table 3.

TABLE 3 *Activity of Mn-Zeolite Filtrates in the Oxidation of Cyclohexane*

Filtrates	Mn content (mmol)	Reaction rate (mmol/min)	Cyclohexane conversion (%)	Adipic acid selectivity (%)	Mn dissolved (mmol/l)
None	0	0.36	6.6	16	0
Mn/BEA 15	0.18	1.3	19	55	0.55
Mn/BEA 2000	0.29	1.2	26	58	2.0
Mn/MOR 15	0.40	1.4	34	60	0.89
Mn/MOR 100	0.25	1.2	24	48	1.4

Same conditions as Table 2.

The Mn-zeolite filtrates were active in the catalysis of the cyclohexane oxidation and high adipic acid selectivities were obtained. The activities of the filtrates were similar to those of the mother zeolites. Although the amount of dissolved manganese was low, these results showed that the catalysis in the presence of supported Mn_3O_4 oxide solids was homogeneous and resulted from the manganese leached in the reaction medium.

3.3 Activity of Manganous Acetate

As Mn-zeolites are effective catalysts in the cyclohexane oxidation and manganous acetate is said to be unactive, we decided to explore the activity of Mn(II) acetate over a wide range of concentrations, to understand this contradiction. The results are represented in Figure 2.

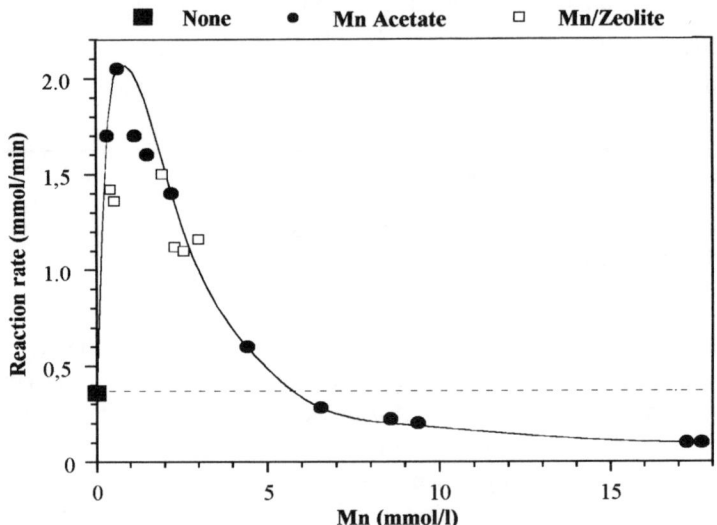

Figure 2 *Activity of Mn(II) acetate as a function of its concentration*

The study of the activity of Mn(II) acetate towards the oxidation of cyclohexane emphasized two distinct phenomena related to the manganese concentration. Although Mn(II) catalysed the oxidation of cyclohexane at very low concentration, it was an inhibitor at higher concentration. As the concentration was about 1 mmol/l, a sharp maximum of activity was obtained. This phenomenon, called "catalyst-inhibitor conversion"[18], had already been observed in low polarity media[19], but neither had in acetic acid. Such a behaviour is in agreement with the absence of catalytic activity showed by manganese acetate at concentration higher than 17 mmol/l[1,8].

Catalytic activities of manganese-containing zeolites and their filtrates as a function of the determined concentration of Mn in the reaction medium are also reported in Figure 2. Activities similar to those in the presence of manganous acetate were obtained. These results confirmed that the catalysis using Mn-zeolites is homogeneous and suggested the formation of manganese acetate during the leaching of Mn from these solids.

4 CONCLUSION

When the aluminic acid sites were completely compensated, manganese-containing zeolites were effective catalysts in the oxidation of cyclohexane to adipic acid in acetic acid. After calcination, the excess of manganese was converted into supported Mn_3O_4. In actual fact, the catalytic activity resulted from the leaching of part of the manganese. However, manganese acetate exhibited a "catalyst-inhibitor conversion" phenomenon and it was demonstrated that the manganese-containing zeolites behaved as a distributor of suitable amounts of manganese for observable catalytic activities.

References

1.   K. Tanaka, *Hydrocarbon Proc.*, 1974, **53**, 114.
2.   J. Kollar, W.O. Pat. 94.07833 (1993).
3.   S. S. Lin and H. S. Weng, *Appl. Catal. A*, 1993, **105**, 289; *Appl. Catal. A*, 1994, **118**, 21; *J. Chem. Eng. Jpn*, 1994, **27**, 211.
4.   B. Kraushaar-Czarnetzki and W. G. M. Hoogervorst, Eur. Pat., 519.569 (1992).
5.   D. L. Vanoppen, D. E. de Vos, M. J. Genet, P. G. Rouxhet and P. A. Jacobs, *Angew. Chem.*, Int. Ed. Engl., 1995, **34**, 560.
6.   I. Belkhir, A. Germain, F. Fajula and E. Fache, Proceedings of the 3rd World Congress on Oxidation Catalysis, San Diego, USA, 1997.
7.   B. Kraushaar-Czarnetzki, W. G. M. Hoogervorst and W. H. J. Stork, *Stud. Surf. Sci. Catal.*, 1994, **84**, 1869.
8.   A. Onopchenko and J. G. D. Schulz, *J. Org. Chem.*, 1975, **40**, 3338.
9.   J. Rouchaud and B. Chantraine, *Bull. Soc. Chim. Fr.*, 1968, 1329.
10. R.L. Wadlinger, G.T. Kerr, E.J. Rosinski U.S. Patent 3,308,069 (1967).
11. E. Bourgeat Lami, F. Fajula, D. Anglerot and T. Des Courieres, *Microporous Mater.*, 1993, **1**, 237.
12. R. P. Townsend, *Stud. Surf. Sci. Catal.*, 1991, **58** , 359.
13. J. M. Stencel, V. U. S. Rao, J. R. Diehl, K. H. Rhee, A. G. Dhere and R. J. De Angelis, *J. Catal.*, 1983, **84**, 109.
14. F. Kapteijn, A. D. van Langeveld, J. A. Moulijn, A. Andreïni, M. A. Vuurman, A. M. Turek, J. Jehng and I. E. Wachs, *J. Catal,* **150**, 94, 1994; F. Kapteijn, L. Singoredjo, A. Andreïni and J. A. Moulijn, *Appl. Catal. B*, **3**, 173, 1994.
15. W. Wang, Y. Yongnian and J. Zhang, *Appl. Catal. A*, 1995
16. L. Singoredjo, R. B. Korver, F. Kapteijn and J. A. Moulijn, *Appl. Catal. B*, **1**, 297, 1992.
17. Y. Yongnian, H. Ruili, C. Lin and Z. Jiayu, *Appl. Catal. A*, **101**, 233, 1993.
18. R. A. Sheldon and J. K. Kochi in "Metal-catalysed oxidations of organic compounds", Academic Press, New York, 1981.
19. J. F. Black, *J. Am. Chem. Soc.*, **100**, 527, 1978.

MICELLE-TEMPLATED SILICA-BOUND MANGANESE (III) SCHIFF BASE COMPLEXES: SOLID CATALYSTS FOR EPOXIDATION OF OLEFINS

P. Sutra and D. Brunel

Laboratoire de Matériaux Catalytiques et Catalyse en Chimie Organique
UMR 5618 ENSCM/CNRS
8, Rue Ecole Normale 34296 Montpellier Cedex 5, France.

1 INTRODUCTION

Oxidation reactions promoted by transition-metal complexes are currently of considerable interest because of their relevance to organic chemistry. A great number of transition metal complexes has been designed in order to selectively catalyze olefin epoxidations. Among them, systems based on manganese (III) Schiff base complexes combined with NaOCl as oxygen donor, first developed by Jacobsen,[1] showed high selectivities and activities. Immobilization of such complexes onto solid supports can provide heterogeneous catalysts which are easy to handle and which may exhibit improved activities owing to the support environment. Thus, Schiff base transition-metal complexes have already been immobilized through covalent linkages on polymers[2] and they showed promising activities in heterogeneous oxido-reduction catalysis. Anchoring such complexes on inorganic supports through covalent bond is of great interest for two main reasons. On the one hand, the support is not soluble in organic solvents. On the other hand, it can provide a high dispersion of catalytic sites due to high surface area. Moreover, the catalytic sites are resistant toward solvent leaching. Synergic effect induced by adsorption properties of the inorganic support and catalytic activities of the metal complexes can indeed be expected for these organic-inorganic hybrid materials. We first reported[3] the covalent grafting of Mn^{III} Schiff base complexes onto the surface of mesoporous MCM-41 type silica. In addition to their high specific surface area, these materials offer regular hexagonal arrays of channels. The monodisperse pore diameter of the tubes can be precisely adjusted in the range of 30 to 100 Å thanks to a well controlled synthesis process.[4] Until now, only a few examples[5] of covalent linkage of such complexes on the surface of mesoporous silica have been related. Herein, the covalent anchoring of a Mn^{III}-salen type complex family onto the internal surface of a mesoporous micelle-templated silica (MTS) and the behaviour of these heterogeneous catalysts under different reaction conditions as a function of the oxygen donor in the styrene epoxidation are described.

2 COVALENT GRAFTING OF THE COMPLEXES

2.1 Support Synthesis

The solid support MTS was synthesized by a micelle-templated pathway.[4] The hexagonal arrays of uniformly sized channels with pore diameter of 32 Å was obtained after silicate condensation around cylindrical micelles of organic surfactants. After calcination of the organic template, the MTS material exhibited high surface area (900 m^2/g). According to Brunauer, Emmet and Teller classification,[6] nitrogen sorption experiment showed a type IV isotherm typical of a mesoporous solid. The X-ray diffraction pattern (lines 100, 110, 200 and 210), using Cu-Kα radiation, was consistent with an hexagonal array typical of MCM-41 type silica.

2.2 Ligand Grafting

(3- Chloropropyl)triethoxysilane was previously linked to the activated MTS (1) surface using alkoxysilane reaction with surface siloxane groups (Scheme 1). Ligand bounding was then carried out through partial nucleophilic displacement of chlorine atom of Cl-MTS (2) by the basic amino group of the 3-[N,N'-bis-3-salicylidenamino propylamine] or 3-[N,N'-bis-3-(3,5-di-*tert*-butyl salicylidenamino) propylamine] ligand (Salpr or tSalpr).

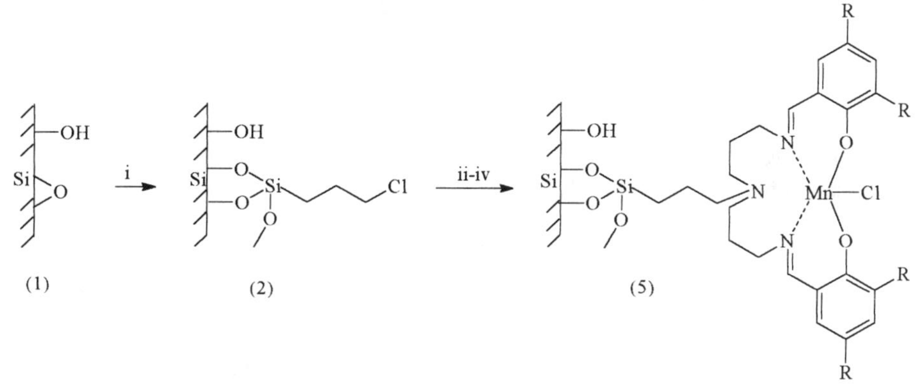

Scheme 1: *Mn^{III} complexes grafting on MTS materials. Reagents : i, $(EtO)_3Si(CH2)_3Cl$; ii, Salpr or tSalpr; iii, Mn(acac)2; iv, Air oxidation in saturated brine.*

The infrared spectrum (Figure 1) of Cl-MTS (2) showed the C-H stretching vibration bands around 2900-3080 cm^{-1} due to 3-chloropropylsilane moities. In addition to these bands, Salpr-MTS and tSalpr-MTS (3) exhibited signals characteristic of Salpr and tSalpr, respectively. Near the C=N stretching band (1635 cm^{-1}), the spectrum shows a strong absorption signal (1688 cm^{-1}) assigned to a carbonyl group. This group had already been observed by Smith et al[7] on UV spectra of salen-type Schiff base in polar solvents, and it was attributed to the conjugated form of the ligand (Scheme 2).

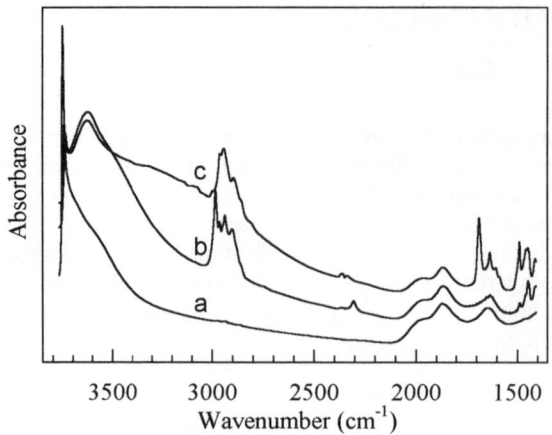

Figure 1 *Infrared spectra of MTS (a), Cl-MTS (b), Salpr-MTS (c)*

In the case of modified solids, the polar environment could result from the presence of neighbouring silanol groups. Salpr-MTS or tSalpr-MTS spectra presented also the charateristic pair of C=C stretching bands (1496 and 1453 cm^{-1}) of ligand aromatic nuclei.

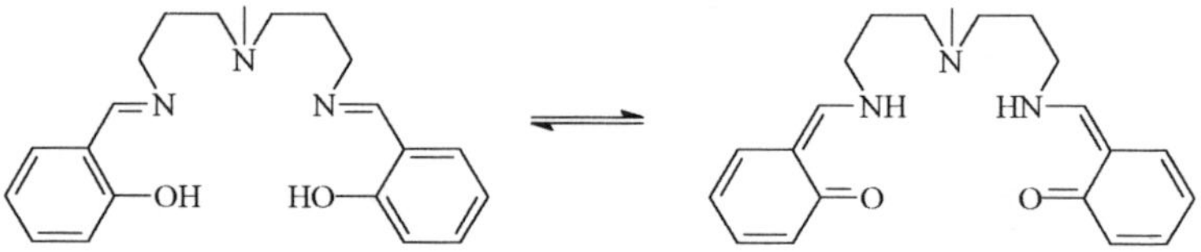

Scheme 2: *Conjugated forms of the ligand*

13C NMR spectra of the three grafted MTS were also in good accordance with covalent linkages of the organic moities on the support surface, when compared to the corresponding compounds spectra in CDCl$_3$ solution.

X-ray diffraction and nitrogen sorption analysis showed that the mesoporous texture did not collapse during the grafting experiments. Nitrogen sorption isotherms indicated a decrease of the mesoporous volume with respect to the surface modifications of MTS materials.

2.3 Metal Complexation

The metal chelation process was first developed on the homogeneous free ligand 3-[N,N'-bis-3-salicylidenamino methyl propylamine] (Smdpt). The methyl group of Smdpt simulated the propyl chain of the ligand when immobilized on MTS. According to Jacobsen's synthesis, Mn^{II}(OAc)$_2$ was the manganese atom source. Nevertheless, manganese complexation on Smdpt required the use of a strong base in order to form the phenolate salt of the Schiff base. We observed a dramatic   change in the regular porosity of the MTS system that showed the poor stability of the support in basic conditions.

Table 1 *Organic Moieties and Manganese Content of Modified MTS*

	3-chlorosilane $\times 10^4$ (mol.g^{-1})	Ligand content $\times 10^4$ (mol.g^{-1})	Mn $\times 10^4$ (mol.g^{-1})
MTS	***	***	***
Cl-MTS	10.0	***	***
Salpr-MTS	7.0[a]	3.0	***
tSalpr-MTS	9.0[a]	1.0	***
Mn^{III}-Salpr-MTS	7.0[a]	1.3[b]	1.7
Mn^{III}-Salpr-MTS	9.0[a]	0.0[b]	1.0

a. *3-Chlorosilane content after ligand grafting or metal complexation.*
b. *Non-complexed ligand content after metal chelation.*

Metal atoms were then introduced thanks to a mild exchange between Mn^{II}(acac)$_2$ and the Schiff base ligands in refluxing methanol, under inert atmosphere. After successive washing with MeOH, DMF and EtOH under argon, the EPR spectrum of grafted Mn^{II}-Salpr or Mn^{II}-tSalpr (4) showed a strongly distorted sextuplet signal at g = 2, typical of an unsymmetrical Mn^{II} complex. Determination by gas chromatographic analysis of the amount of 2,4-pentanedione formed during the reaction gave the yield of the ligand exchange. Suspension of the grafted Mn^{II} complexes in saturated brine were then air-oxidized to give the desired Mn^{III}-Salpr or Mn^{III}-tSalpr complexes, covalently anchored on the surface of MTS (5). The amount of Mn^{III} which was determined by elemental analysis corresponded with the one estimated by GC. The Mn^{III} complexes had been evidenced by their UV-Vis diffuse reflectance spectra, as illustrated in figure 2. Comparison of the spectra recorded before and after complexation revealed the same absorption bands as the corresponding complexes Mn^{III}-Smdpt in CH$_2$Cl$_2$ solution.

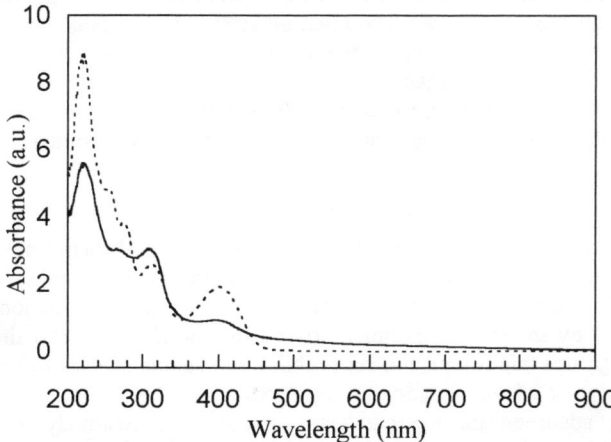

Figure 2 *UV-Vis diffuse reflectance spectra of Salpr-MTS (---) and Mn^{III}-Salpr-MTS (——)*

Organic compositions of the modified solids summed up in table 1 were determined by elemental analysis combined with thermogravimetric analysis. It was noticeable that ligand grafting through the coupling reaction with 3-chlorosilane was partial. The steric hindrance of the tSalpr may induce a more homogeneous dispersion of the ligands characterized by a nitrogen adsorption step in a restricted range of P/P_0. This behaviour towards nitrogen sorption demonstrated a monodisperse pore diameter *i.e.* an homogeneous dispersion of the ligand on the surface. This homogeneous dispersion could also explain the quantitative yield obtained for the complexation of manganese in the case of Mn^{III}-tSalpr MTS whereas Salpr-MTS led to a moderate complexation yield.

3 CATALYTIC BEHAVIOUR

Both activity and stability of the immobilized complexes were investigated in the epoxidation of styrene as a function of the oxygen donor. In a typical oxidation process, 2 mmol (80 eq) of styrene were added under inert atmosphere to a suspension of 200 mg (1 eq) of activated catalyst and 4 mmol (160 eq) of final oxydant in 20 mL of an adequate solvent. The reaction progress at room temperature was monitored by GC analysis. Results are detailed in table 2.

3.1 Sodium Hypochlorite

The reaction was carried out using NaOCl in a buffered solution (pH=11.3) according to the homogenous conditions developed by Jacobsen[1] with salen-type complexes. After 46 hours of reaction, no epoxide was observed. The nitrogen sorption analysis of the solid after reaction indeed showed a type III isotherm featuring a macroporous silica-type solid. This was attributed to a collapse of the inorganic mesoporous system due to hydrolysis of Si-O-Si bridges followed by silica recondensation in basic conditions.

3.2 Hydrogen Peroxide

Hydrogen peroxide is a clean oxygen donor due to the formation of water as a by-product during the epoxidation reaction. Acetonitrile was the solvent used in this process. Unfortunately, H_2O_2 was consumed at once in the course of its addition and no epoxide appeared. This phenomenon suggested a Fenton's type reaction around the metallic center generating hydroxyl radicals catalyzing the peroxide decomposition. Imidazole, which is known to limit this by-side reaction with manganese (III) complexes[8], was added to the reaction mixture in order to avoid the homolytic cleavage. However, different imidazole concentrations (10 to 50 eq) failed to inhibit the autodecomposition of the final oxidant.

3.3 Iodosylbenzene

Iodosylarenes constitute another class of oxygen donors widely used with manganese (III) salen-type complexes.[9] In spite of its very low solubility in most solvents, a suspension of iodosylbenzene in acetonitrile was used to perform the styrene epoxidation. The epoxidation was influenced by the steric hindrance around the metal center and the MTS environment. Tert-butyl groups might limit the accessibility of organic species to the catalytic sites inside the mesoporous system. In addition, by-products of the reaction were mainly polymers which were adsorbed on the catalysts surface. The relatively low selectivities could be explained by a copolymerization reaction between styrene and epoxystyrene due to specie diffusion limitations in the porous MTS. Larger mesoporous systems could be an interesting alternative to avoid the diffusion limitations of the species.

Table 2 *Heterogeneous Epoxidation of Styrene using various Oxygen Donors*

Catalyst	Solvent	Oxygen donor	Conversion (%)	Selectivity (%)
Mn^{III}-Salpr-MTS	Buffered solution pH=11.3	NaOCl	0	***
Mn^{III}-Salpr-MTS	Acetonitrile	H$_2$O$_2$	0	***
Mn^{III}-Salpr-MTS Imidazole 10eq	"	"	"	"
Mn^{III}-Salpr-MTS Imidazole 50eq	"	"	"	"
None	Acetonitrile	PhIO	0	***
Mn^{III}-Salpr-MTS	"	"	53	58
Mn^{III}-Salpr-MTS	"	"	44	24

Developments on the design of the catalytic systems and their re-activation after reaction are still in progress.

4 CONCLUSION

Manganese (III) salen-type complexes providing various steric hindrances had been covalently anchored on the surface of MTS with respect to the regularity of their mesoporous structure. The homogeneous dispersion of the complexes obtained with the tSalpr ligand lost its benefits owing to diffusion restrictions of organic species in the mesopore system during the catalyzed reaction. Nevertheless, conversions and selectivities obtained in the styrene epoxidation when complexes were associated with PhIO were promising. Synthesis processes now lead to well controlled mesopores, in the range of 80 Å diameter, with a great stability of textural properties towards surface covering reaction conditions. Further studies on grafting manganese complexes on the surface of large mesopore materials are under investigation together with a clean hydrogen peroxide compatible catalytic system.

References

1. E. N. Jacobsen, 'Catalytic Asymmetric Synthesis', VCH Publishers, New York, 1993.
2. B. B. De, B. B. Lohray, S. Sivaram and P. K. Dhal, *Tetrahedron Asymmetry*, 1995, **6**, 2105.
3. P. Sutra and D. Brunel, *Chem. Commun.*, 1997, 2485.
4. C. T. Kresge, M. E. Leonowicz, W. J. Roth, J. C. Vartuli and J. S. Beck, *Nature*, 1992, **359**, 710.
5. S. Brunauer, P.H. Emmett, E. Teller, *J. Am. Chem. Soc.*, 1938, **60**, 309.
6. Y. V. S. Rao, D. E. De Vos, T. Bein and P. A. Jacobs, *Chem. Commun.*, 1997, 355.
7. H. E. Smith, *Chem. Rev.*, 1983, **83**, 359.
8. P. Battioni, J.P. Renaud, J.F. Bartoli, M. Reina-Artiles, M. Fort and D. Mansuy, *J. Am. Chem. Soc.*, 1988, **110**, 8462.
9. T. Hamada, R. Irie, T. Katsuki, *Synlett Letters*, 1994, 479.

OXIDE-SUPPORTED ORGANIC LAYERS AS POTENTIAL FRIEDEL-CRAFTS CATALYSTS

Christopher H. Barclay and John M. Winfield

Department of Chemistry
University of Glasgow
Glasgow G12 8QQ

1 INTRODUCTION

Although aluminium(III) chloride is probably the most widely used catalyst in Friedel-Crafts reactions,[1] its lifetime is limited in many applications, and large scale operations result in the requirement to destroy unacceptably large quantities of toxic material.[2] For these reasons alternative acids are being evaluated, those that can be used under heterogeneous conditions being of particular interest.[3] The work described below was prompted by the recent development of heterogeneous catalysts that are active for room temperature F-for-Cl halogen exchange involving conversion of hydrochlorocarbons to hydrochlorofluorocarbons (HCFCs),[4] a process which, like Friedel-Crafts alkylation or acylation, depends on the ability of the catalyst to exhibit strong Lewis acid behaviour.

2 CATALYSTS

Catalysts derived from the defect spinel γ-alumina were prepared by the methods developed previously in this laboratory. Calcined γ-alumina was either chlorinated with CCl_4 or $OCCl_2$ at 325°C or fluorinated with SF_4, nominally at ambient temperature. In all cases replacement of surface hydroxyl groups and bound H_2O occurs and HX, X = Cl or F, is retained by the solids, resulting in a pool of labile surface halogen.[5] Enhancement of the surface Lewis acidity is apparent, shown by the facile dehydrochlorination of CH_3CCl_3 at room temperature to give volatile $CH_2=CCl_2$ and (using fluorinated γ-alumina) CH_3CCl_2F.[4a] Highly coloured organic layers derived from $CH_2=CCl_2$ and comprising mixtures of heavily chlorinated C_4, C_6 and C_8 species[4b] are deposited on the solids. These materials are termed oxide-supported organic layer catalysts. Other materials that have been examined as potential catalysts are calcined γ-alumina, CCl_4-chlorinated and SF_4-fluorinated γ-aluminas, CCl_4-chlorinated γ-alumina subsequently exposed to various chloro-olefins and β-aluminium(III) fluoride; the latter material has also been shown to be an active catalyst for halogen exchange reactions that lead to HCFCs.[6]

3 FRIEDEL-CRAFTS ALKYLATION

Alkylation of toluene and benzene by *t*-butyl chloride (mole ratio 10:1) at room temperature has been studied in the presence of the potential catalysts under heterogeneous, liquid-solid conditions, the reactions being followed by gas chromatography. Product-data, expressed as % yield based on the quantities of *t*-BuCl, are summarised for reactions involving toluene and benzene in Tables 1 and 2 respectively. Calcined γ-alumina showed no activity in either case; the remainder appear to show greater activity towards toluene as expected and in these cases, because of the stoichiometry employed, reaction is limited essentially to mono-alkylation, Table 1. Apart from CCl_4-chlorinated γ-alumina, the materials show reduced or no activity towards benzene alkylation but mixtures of mono- and di-alkylated products are observed using SF_4-fluorinated γ-alumina or β-AlF_3, Table 2.

On the basis of simple electronegativity considerations it would have been predicted that fluorinated γ-alumina would be more active than its chlorinated analogue which is inconsistent with the results but the observation of an orange colouration on the SF_4-fluorinated γ-alumina surface during the reactions may indicate the formation of an organic layer *in situ*. Reaction profiles of chlorinated γ-alumina reactions are characterised by rapid loss of *t*-BuCl from the liquid phase and its complete consumption within 1-2 h. In less active situations, with fluorinated substrates, although the loss of *t*-BuCl from the liquid phase after 1 h is substantial, it is not complete even after 24 h. The behaviour of the supported organic layers is similar.

4 [^{36}Cl]-RADIOTRACER EXPERIMENTS

Interactions between the Friedel-Crafts reactants and catalyst surfaces have been probed using [^{36}Cl]-labelled *t*-BuCl and a Geiger Müller direct monitoring technique,[7] an approach that has been used successfully to clarify the catalytic role of an oxide-supported organic layer catalyst in halogen exchange.[4b] All experiments were carried out at room temperature *in vacuo* under vapour-solid conditions.

Exposure of CCl_4-chlorinated γ-alumina to *t*-Bu^{36}Cl vapour results in the immediate detection of [^{36}Cl] radioactivity from the surface due to adsorption and [^{36}Cl] exchange between *t*-Bu^{36}Cl and the labile surface Cl (*cf.* ref. 5a). The two effects cannot be separated from this experiment. The surface [^{36}Cl] count rate decreases marginally on removal of volatile material by pumping but is unaffected by subsequent addition of benzene vapour, Fig. 1. Exposure of chlorinated γ-alumina to benzene vapour prior to addition of *t*-Bu^{36}Cl leads to similar behaviour, Fig. 2. The effects of these operations on the oxide-supported organic layer catalyst derived from the dehydrochlorination-oligomerization reactions of CH_3CCl_3 on chlorinated γ-alumina are shown in Figs. 3 and 4. The surface count rates in these cases will reflect those events occurring at the organic surface, since self-absorption of β$^-$ radiation from [^{36}Cl] will

Supported Reagents and Catalysts in Chemistry

Table 1 *Reaction of t-butyl chloride with toluene*

CATALYTIC SUBSTRATE	YIELD /%[a]	PRODUCT Monoalkylated /%	DISTRIBUTION Dialkylated /%
Calcined γ-alumina	0	0	0
CCl$_4$ chlorinated γ-alumina	100	63 : 35	2
SF$_4$ fluorinated γ-alumina	49	97 : 3	0
O.S.O.L. catalyst derived from 1,1,1-trichloroethane	33	78 : 22	0
O.S.O.L. catalyst derived from trichloroethene	35	53 : 47	0
β-AlF$_3$	36	84 : 16	0

[a] Based on GC analysis of product mixtures

Table 2 *Reaction of t-butyl chloride with benzene*

CATALYTIC SUBSTRATE	YIELD /%[a]	PRODUCT Monoalkylated /%	DISTRIBUTION Dialkylated /%
Calcined γ-alumina	0	0	0
CCl$_4$ chlorinated γ-alumina	100	92	5 : 3
SF$_4$ fluorinated γ-alumina	8	89	11
O.S.O.L. catalyst derived from 1,1,1-trichloroethane	0	0	0
O.S.O.L. catalyst derived from trichloroethene	0	0	0
β-AlF$_3$	2	79	18 : 3

[a] Based on GC analysis of product mixtures

Figure 1 *[³⁶Cl]-Surface count rate from chlorinated alumina after exposure to t-Bu³⁶Cl then benzene*

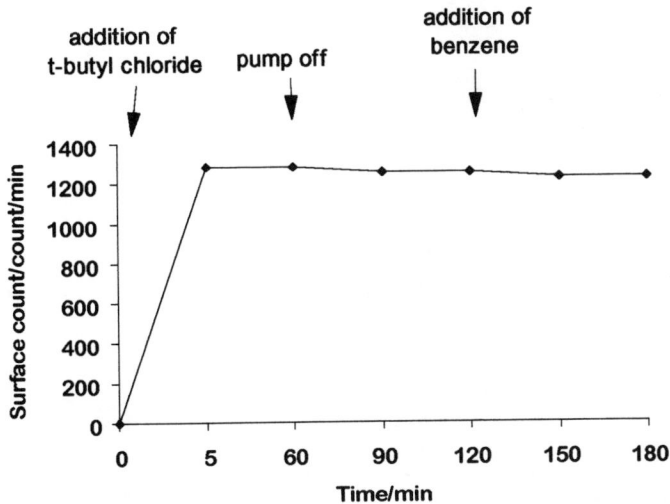

Figure 2 *[³⁶Cl]-Surface count rate from chlorinated alumina exposed to benzene prior to t-Bu³⁶Cl*

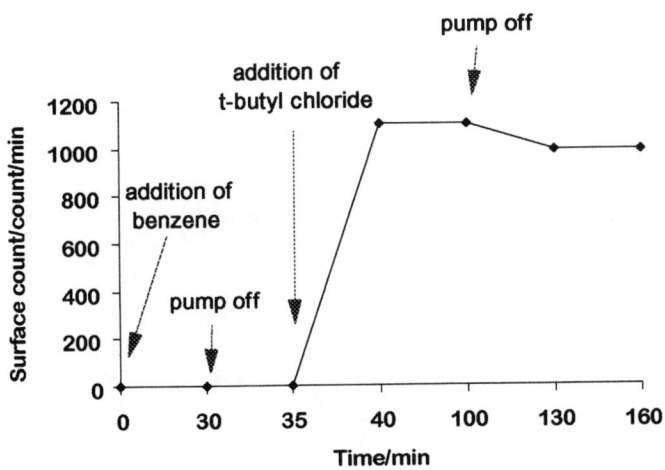

Figure 3 *[³⁶Cl]-Surface count rate from an oxide supported organic layer catalyst after exposure to t-Bu³⁶Cl then benzene*

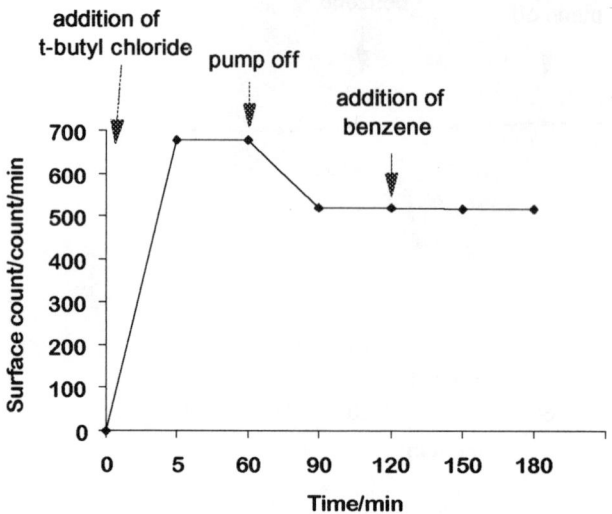

Figure 4 *[³⁶Cl]-Surface count rate from an oxide supported organic layer catalyst exposed to benzene prior to t-Bu³⁶Cl*

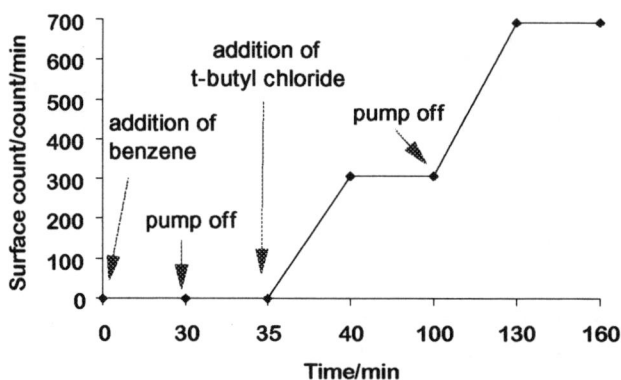

screen the effects of events occurring *within* the layer or at the inorganic-organic interface. The behaviour of the surface count rate data, Figs. 3 and 4, indicates a significant difference which depends upon the order of exposure of the reagents to the supported organic layer. In particular, Fig. 4 implies strongly that both benzene and *t*-BuCl are held *within* the organic layer, the latter diffusing to the surface on pumping.

5 CONCLUSIONS

The enhanced Lewis acidity of γ-alumina that has been chlorinated by CCl_4 or $OCCl_2$ results in this material being an active catalyst for the room temperature alkylation of benzene and toluene. Deposition of a chlorocarbon organic layer results in reduced activity under laboratory conditions but because of the possibility of reagent turn-over within the layer, the potential exists for use under flow conditions.

References

1. G.A. Olah, 'Friedel-Crafts Chemistry' Wiley and Son, New York, 1973.
2. M.J. Braithwaite 'Chemistry of Waste Minimisation' J.H. Clark (ed), Blackie, London 1995, ch 2.
3. *eg*. A. Cornélis, A. Gerstmans, P. Laszlo, A. Mathy and I. Zieba, *Catal. Lett.*, 1990, **6**, 103; S.J. Barlow, J.H. Clark, M. Darby, A.P. Kybett, P. Landon and K. Martin, *J. Chem. Res. (S)*, 1991, 74; R.S. Drago, S.C. Petrosius and P.B. Kaufman, *J. Mol. Catal.*, 1994, **89**, 317; J.H. Clark, K. Martin, A.J. Teasdale and S.J. Barlow, *J. Chem. Soc., Chem. Commun.*, 1995, 2037.
4. (*a*) J. Thomson, G. Webb, J. Winfield, D. Bonniface, C. Shortman and N. Winterton, *Appl. Catal. A*, 1993, **97**, 67; (*b*) A. Bendada, D.W. Bonniface, F. McMonagle, R. Marshall, C. Shortman, R.R. Spence, J. Thomson, G. Webb, J.M. Winfield and N. Winterton, *Chem. Comm.*, 1996, 1947.
5. (*a*) J. Thomson, G. Webb and J.M. Winfield, *J. Mol. Catal.*, 1991, **67**, 117; *ibid.*, 1991, **68**, 347; (*b*) A. Bendada, G. Webb and J.M. Winfield, *Eur. J. Solid State Inorg. Chem.*, 1996, **33**, 907.
6. A. Hess, E. Kemnitz, A. Lippitz, W.E.S. Unger and D.-H. Menz, *J. Catal.*, 1994, **148**, 270.
7. A.S. Al-Ammar and G. Webb, *J. Chem. Soc., Farad. Trans. 1*, 1978, **74**, 195.

AEROBIC OXIDATION OF α-SUBSTITUTED ALCOHOLS OVER PROMOTED PLATINUM METAL CATALYSTS

T. Mallat, L. Seyler, M. Mir Alai and A. Baiker

Department of Chemical Engineering and Industrial Chemistry,
Swiss Federal Institute of Technology,
CH-8092 Zürich, Switzerland

1 INTRODUCTION

The Pt metal catalysed oxidation of alcohols and polyols to the corresponding carbonyl compounds or carboxylic acids is an attractive, environmentally friendly process. Selectivities above 90 % can be obtained under very mild conditions in aqueous medium and with molecular oxygen as oxidant.[1,2] The discovery of the promoting effect of heavy metals (e.g. Bi, Pb) represents a breakthrough in the development of the method.[3] Promoters can increase both the reaction rate and selectivity (i) by a geometric blocking of a fraction of Pt^0 or Pd^0 active sites and suppressing side reactions, or (ii) by forming new adsorption sites for the alcohol or oxygen.[4-7]

Considerably lower rates and selectivities were reported for the oxidation of alcohols where the product carbonyl compound possessed an electron-withdrawing group in α position. Examples are the oxidation of lactic acid to pyruvic acid and L-sorbose to 2-keto-L-gulonic acid, a key intermediate in vitamin C synthesis.[8,9] We have shown recently that in the latter reaction the performance of supported Pt can be substantially improved by modification with some strongly adsorbed N- or P-compounds.[10]

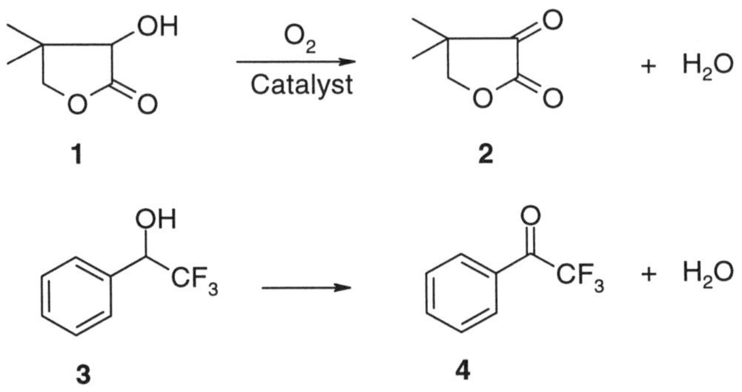

Scheme 1 Oxidation of pantolactone (**1**) and 1-phenyl-2,2,2-trifluoroethanol (**3**)

Here we present another possibility for enhancing the reaction rate and yield in the oxidation of functionalised secondary alcohols to ketones. Working under water-free conditions and above 100 °C, in order to promote the rapid removal of water formed during the oxidative dehydogenation reaction, this provides an interesting alternative to the conventional aqueous phase process. The influence of reaction medium, temperature and metal promotion will be illustrated using the selective oxidation of pantolactone (**1**) and 1-phenyl-2,2,2-trifluoroethanol (**3**) as model reactions (Scheme 1).

2 EXPERIMENTAL

Purum or puriss grade reagents and distilled water (after ion-exchange) were used. The 5 wt% Pd/C (Engelhard 5008), 4 wt% Pd - 1 wt% Pt - 5 wt% Bi/C (Degussa 7304) and 2.7 wt% Pb - 5 wt% Pd/CaCO$_3$ (Engelhard 5088) catalysts were used as delivered. Other Bi- or Pb-promoted catalysts have been prepared by modification of a 5 wt% Pt/alumina (Engelhard Escat 24), 5 wt% Pd/C (Doduco) or 5 wt% Pt/C (Engelhard Escat 21). Details of the deposition of Pb or Bi by hydrogen reduction of Pb(OAc)$_2$ and Bi(NO$_3$)$_3$, respectively, in weakly acidic aqueous medium have been discussed elsewhere.[4,9]

Surface composition of the Pb-promoted Pt/C catalysts was determined by XPS. The spectra were recorded with a Leybold-Heraeus LHS 11 instrument using MgK$_\alpha$ (1253.6 eV) radiation. The base pressure of the apparatus was lower than 5×10^{-10} mbar.

The oxidation reactions were performed in a 100 ml flat bottomed, thermostated glass reactor equipped with gas inlet and outlet, reflux condenser, thermometer and magnetic mixing. In the general procedure for *pantolactone oxidation*, 0.2 g catalyst was pre-reduced *in situ* under nitrogen atmosphere (10 min) with the reactant (1 g) in 10 cm^3 solvent. The prereduction of the promoted catalysts was carried out with hydrogen for 15 min at room temperature before addition of the reactant and heating up to the reaction temperature. During reaction the air flow rate was 10 cm^3 min^{-1} and the mixing rate 750 min^{-1}.

For the study of the stability of **2** under reaction conditions, 0.5 g of **2** and 0.5 cm^3 pentadecane (internal standard) were mixed in ethylene glycol diacetate in the presence of 5 wt% Pd/C (Engelhard 5008) at 137 °C.

Before the oxidation of *1-phenyl-2,2,2-trifluoroethanol* (**3**), 0.1 g catalyst was prereduced by 1.0 g reactant in 20 cm^3 solvent under nitrogen, and heated up to the reaction temperature. In aqueous solution 50 mg dodecylbenzenesulfonic acid Na salt was used as detergent and a weakly basic pH was set with 40 mg Li$_2$CO$_3$. The reaction was started by substituting nitrogen to air or oxygen (5 cm^3 min^{-1}). The stability of **4** was investigated in BuOAc at 128 °C over 0.8 wt% Bi-5 wt% Pt/Al$_2$O$_3$.

The products (**2** and **4**) were analysed by gas chromatography. During oxidation only the alcohol/ketone ratio was determined. At the end of the reaction an internal standard was added for determining conversion and selectivity.

3 OXIDATION OF PANTOLACTONE (1)

The influence of various solvents and the reaction temperature on the rate of oxidation of pantolactone (D,L-2-hydroxy-3,3-dimethyl-γ-butyrolactone, **1**) to ketopantolactone (dihydro-4,4-dimethyl-2,3-furanedione, **2**) is shown in Table 1. The reactions were catalysed by a Pb-promoted Pt/C catalyst. In the preliminary kinetic analysis the rate was followed by determining the molar ratio of **2/1** (Fig. 1). Due to the extensive formation of byproducts, this ratio is only a qualitative measure of the progress of the reaction.

At around 60 °C the initial reaction rate was the highest in water, but the catalyst rapidly deactivated and after 30 min the ketone/alcohol ratio decreased. This product loss is seemingly due to the decomposition of ketopantolactone. The formation of strongly acidic byproducts was indicated by a drop in pH of the aqueous solution to around 2. The byproduct acids adsorbed strongly on the Pt surface and hindered the oxidation reaction.

Catalyst deactivation could not be avoided in other polar or apolar solvents, when working below 100 °C. Solvent polarity is characterised by the empirical solvent parameter E_T^N in Table 1. Note that the choice of solvents is limited by the strong oxidative conditions. The highest rate was obtained when working above 100 °C in weakly polar solvents, such as ethylene glycol diacetate. Under these conditions water, which is formed in the reaction according to Scheme 1, is rapidly removed from the catalyst surface.

It is interesting that working at reflux temperature (e.g. in MeOAc, EtOAc or BuOAc, Fig. 1) the equilibrium oxygen partial pressure is zero. In practice, 10 cm^3min^{-1} air flow rate was enough to reach an (unknown) steady-state oxygen concentration in the liquid phase; above this value the air flow rate had only minor influence on the reaction rate. As a result of low oxygen concentration in the reaction mixture, there was no sign of oxygen poisoning (catalyst over-oxidation).

Promotion by Pb, which metal alone is inactive in this aerobic oxidation reaction, had a strong influence on the rate and selectivity of the reaction. As illustrated in Table 2, 66 % ketopantolactone selectivity at 95 % pantolactone conversion could be achieved in ethylene glycol diacetate. Some other bi- and trimetallic catalysts were more selective but considerably less active. The rapid decomposition of ketopantolactone (**2**) in the best solvent, and in the presence of oxygen and catalyst is illustrated in Fig. 2, where the relative concentration of **2** (i.e. the actual concentration related to the initial concentration) is plotted as a function of time. Almost half of the initial amount of **2** decomposed in 210 min (reaction time in Table 2), rationalizing the moderate selectivities achieved in the partial oxidation of **1**.

Table 1   *Effect of solvent and temperature on the oxidation of pantolactone (**1**) to ketopantolactone (**2**) (Pb-Pt/C catalyst with a Pb/Pt = 0.15 atomic ratio, 30 min reaction time)*

Solvent	E_T^N	Temperature °C	**2/1** mol/mol
H_2O	1.00	60	0.081
2-butanone	0.33	60	0.011
2-butanone	0.33	78	0.024
ethylene glycol diacetate	≈0.3	137	1.10
MeOAc	0.29	57	0.054
BuOAc	0.24	60	0.013
BuOAc	0.24	125	0.32
EtOAc	0.23	76	0.057
heptane	0.01	96	0.10

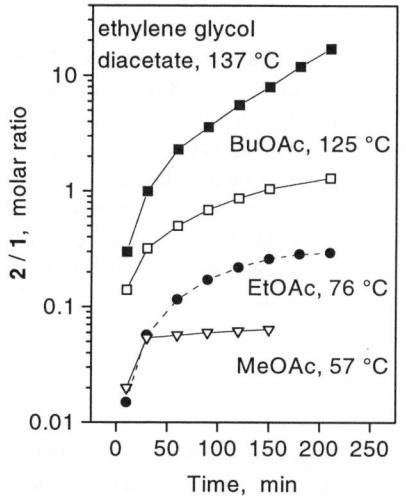

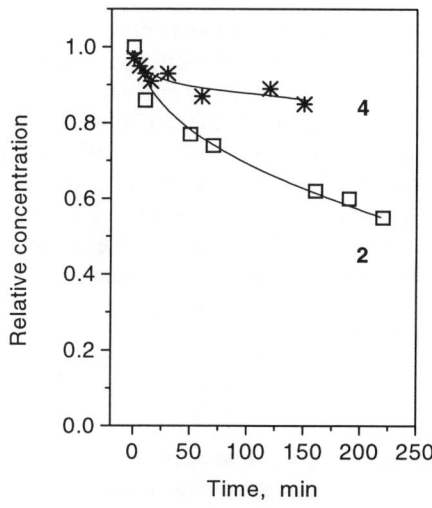

Figure 1 *Influence of solvent and temperature on the oxidation of pantolactone (1) to ketopantolactone (2) (Pb-promoted Pt/C, Pb:Pt = 0.15)*

Figure 2 *Oxidative decomposition of 2 in ethylene glycol diacetate at 137 °C over 5 wt% Pd/C, and 4 in BuOAc at 128 °C over a 0.8 wt% Bi - 5 wt% Pt/Al$_2$O$_3$ catalyst*

Table 2 *Oxidation of pantolactone (1) to ketopantolactone (2) over various promoted and unpromoted noble metal catalysts (137 °C, 210 min reaction time)*

Catalyst	Promoter/Pt (Pd) Atomic Ratio		Conversion %	Selectivity %
	Bulk	Surface		
5 wt% Pt/C	0	0	50.0	40.5
0.8 wt% Pb-5 wt% Pt/C	0.15	0.23	96.5	58.0
1.05 wt% Pb-5 wt% Pt/C	0.20	0.28	95.0	66.0
1.3 wt% Pb-5 wt% Pt/C	0.25	0.33	91.5	54.0
0.8 wt% Bi-5 wt% Pt/Al$_2$O$_3$	0.15	-	77.0	58.5
5 wt% Bi-4 wt% Pd-1 wt% Pt/C	0.56	-	15.0	71.0
2.7 wt% Pb-5 wt% Pd/CaCO$_3$	0.28	-	39.5	72.0
2.9 wt% Pb-5 wt% Pd/C	0.30	-	27.5	49.0

4 OXIDATION OF PHENYL-TRIFLUOROETHANOL (3)

In preliminary experiments the oxidation of **3** was attempted in a broad range of polar and apolar solvents over various supported Pt and Pd catalysts, at temperatures between 50 and 120 °C, but without any success. Only traces of **4** were produced, seemingly due to rapid catalyst deactivation. Bi- and Pb-promoted Pt catalysts were more active on condition that the reaction temperature was above 100 °C. Some representative examples are collected in Table 3. It is clearly seen that there is an optimum in the coverage of surface active sites by inactive Bi species; a too high coverage results in a drop of conversion. This is an indication that the positive effect of promotion is due to site blocking by the inactive promoter.[2,4] The decreasing size of active site ensembles suppresses the side reactions to a larger extent than the alcohol oxidation reaction. Note that the selectivity to **4** was around 90 %, independent of the Bi/Pt (or Pb/Pt) ratio.

Table 3 *Oxidation of phenyl-trifluoroethanol (3) to trifluoroacetophenone (4) over Pb-and Bi-promoted 5 wt% Pt catalysts (5 cm^{3}min^{-1} O$_2$, 120 min reaction time)*

Catalyst	Promoter	Prom./Pt Atomic Ratio	Solvent	Temperature °C	Conversion %
Pt/C	-	0	BuOAc	120	0
Pt/C	Pb	0.15	BuOAc	120	5
Pt/C	Pb	0.25	BuOAc	120	7
Pt/Al$_2$O$_3$	Bi	0	BuOAc	120	0
Pt/Al$_2$O$_3$	Bi	0.09	BuOAc	120	10
Pt/Al$_2$O$_3$	Bi	0.12	BuOAc	120	13.5
Pt/Al$_2$O$_3$	Bi	0.15	BuOAc	120	23
Pt/Al$_2$O$_3$	Bi	0.18	BuOAc	120	14
Pt/Al$_2$O$_3$	Bi	0.15	chlorobenzene	120	20
Pt/Al$_2$O$_3$	Bi	0.15	heptane	95	0

A kinetic analysis of the reactions presented in Table 3 revealed that the rate of formation of **4** was almost constant up to 1-2 h (due to mass transport limitation), and that the reaction suddenly stopped. Catalyst reactivation in a nitrogen atmosphere (i.e. reduction of the oxidized metal surface by the reactant) resulted in only minor further conversion. This is an indication that deactivation was caused by poisoning of the Pt surface (strong adsorption of byproducts), rather than "over-oxidation" of the Pt^0 sites to inactive Pt^{n+} sites.[2] The stability of the product **4** was found to be significantly higher than that of **2**, as shown in Fig. 2. Note that traces of strongly adsorbed byproducts are sufficient for deactivating the catalyst.

A study of the influence of reaction conditions revealed that the crucial parameters were the temperature, reactant concentration and the rate of oxygen transport. Optimization of these parameters provided 98 % selectivity to trifluoroacetophenone at 51 % conversion of phenyl-trifluoroethanol. The fast reaction (30 min) was performed at 128 °C in refluxing BuOAc. The rate of oxygen supply was tuned to the actual rate of surface chemical reaction by decreasing the oxygen partial pressure during reaction, from 1 to 0.21 bar.

5 DISCUSSION

The Pt- and Pd-catalysed oxidation of pantolactone (1) and phenyl-trifluoroethanol (3) to the corresponding ketones revealed two crucial requirements for achieving reasonable conversions and selectivities:
(i) The aerobic oxidations have to be performed under "water-free" conditions; the water formed in the oxidative dehydrogenation reaction has to be rapidly removed by working at elevated temperature in weakly polar water-immiscible organic solvents;
(ii) The active metal (Pt or Pd) has to be partially covered by an inactive "promoter" metal (Pb or Bi) in order to minimise the formation of bulky and strongly adsorbed poisoning species.

In both reactions the product ketones 2 and 4 possess an electron-withdrawing (activating) functional group in α position. This type of carbonyl compounds is rapidly hydrolysed in aqueous medium; the equilibrium reaction is catalysed by acids and bases.[11] The equilibrium constant (K_d = [ketone]/[hydrate]) decreases in the presence of electron-withdrawing substituents (de-stabilisation of the carbonyl compound). The hydrate is more reactive for further oxidation or degradation than the carbonyl compound,[4] which explains the poor performance of Pt and Pd in these reactions, as compared to the oxidation of other types of alcohols.[1,2]

Acknowledgement

Thanks are due to U. W. Göbel for the XPS measurements. Financial support by F. Hoffman - La Roche AG and the Komission zur Förderung der wissenschaftlicher Forschung is gratefully acknowledged.

References

1. H. van Bekkum, in F. W. Lichtenthaler, *Carbohydrates as Organic Raw Materials*, VCH, Weinheim, 1990, p. 289.
2. T. Mallat, and A. Baiker, *Catal. Today*, 1994, **19**, 247.
3. H. Fiege and K. Wedemeyer, *Angew. Chem.*, 1981, **93**, 812.
4. T. Mallat, Z. Bodnar, P. Hug and A. Baiker, *J. Catal.*, 1995, **153**, 131.
5. M. Besson, F. Lahmer, P. Gallezot, P. Fuertes and G. Flèche, *J. Catal.*, 1995, **152**, 116.
6. A. Abbadi and H. Van Bekkum, *Appl. Catal.*, 1996, **148**, 113.
7. H. Kimura, K. Tsuto, T. Wakisaka, Y. Kazumi and Y. Inaya, Appl. Catal. A, 1993, **96**, 217.
8. T. Tsujino, S. Ohigashi, S. Sugiyama, K. Kawashiro and H. Hayashi, *J. Mol. Catal.*, 1992, **71**, 25.
9. C. Brönnimann, Z. Bodnar, P. Hug, T. Mallat and A. Baiker, *J. Catal.*, 1994, **150**, 199.
10. T. Mallat, C. Brönnimann and A. Baiker, *J. Mol. Catal. A: Chem.*, 1997, **117**, 425.
11. J. Zabicky (Ed.), The Chemistry of the Carbonyl Group, Intersci., London, 1970, Vol. 2, p. 3.

A NOVEL ALL PALLADIUM VEHICLE EMISSIONS CONTROL LEAN-BURN CATALYST

J.A.Cairns[1] and J.Thomson[2*]

[1] Department of Applied Physics and Electronic and Mechanical Engineering,

[2] Department of Chemistry, University of Dundee, Dundee, DD1 4HN, Scotland, UK

([*]To whom communications should be addressed)

A novel ceria stabilised zirconia supported palladium-zinc catalyst performs efficiently as a three-way-catalyst (TWC) between stoichiometric to lean burn conditions, under simulated vehicle exhaust gas and engine test conditions. Flowing nitric oxide in the presence of dioxygen shows that the removal of NO gas is by way of formation of NO_2 through a Langmuir-Hinshelwood mechanism. The formation of nitrogen dioxide is observed to have site specificity with competitive adsorption between NO and O_2. These sites are shown to be distinct from the adsorption sites required for hydrocarbon and carbon monoxide respectively. Engine studies indicate a removal of NO_x at air/fuel ratios up to 40:1.

Introduction:

The catalytic removal of vehicle exhaust gases is an important process in the control of pollutant gases in the urban environment. The conventional vehicle exhaust catalyst contains i) a platinum or platinum/palladium function which converts uncombusted hydrocarbons and carbon monoxide, ii) a rhodium function which reacts NOx gases with reducing adsorbates such as carbon monoxide through a disproportionation mechanism, to generate carbon dioxide and nitrogen, iii) a zirconia stabilised ceria function which stabilises the metal particles against sintering and mobilizes oxygen during fluctuating oxygen partial pressures. All of these catalyst functions are supported usually on γ-alumina and coated on a cordierite monolith [1-3].

The industry has been moving to an all palladium catalyst to lower costs and enhance the hydrocarbon combustion process [4]. However the formation of bulk palladium oxide during the high combustion temperatures generated during catalysis may result in loss of palladium from the washcoat, and has therefore been a severe limitation to the development of an all palladium catalyst. The increasing use of lean burn engines poses further problems for the catalyst designer because there tends to be insufficient CO or HC available to cause NO_x removal by a disproportionation mechanism The use of γ-alumina supported barium oxide in conjunction with either platinum or rhodium acting as a NOx scrubber for lean burn systems, or a zeolite impregnated with copper metal to decompose oxides of nitrogen are alternative NO_x removal technologies being presently evaluated [5].

We report here the development of a novel all palladium catalyst which can operate as a TWC either under stoichiometric or under lean burn conditions [6,7]. A combination of catalyst rig and engine studies show that NOx is removed from stoichiometric to lean burn conditions.

Experimental:

Samples of a 16 mol% ceria stabilised zirconia supported Pd-Zn catalyst was prepared by impregnation of the stabilised oxide support (MEL Chemicals Ltd.) with palladium and zinc nitrates (Pd/Zn= 1:2, Aldrich Chemical Co), dried and fired to 823K in air. Standard blank cordierite 400 segment monolith sections were coated from the nitrate precursor materials prior to firing. The catalyst samples were transferred to a catalyst test rig (Grade 316 stainless steel) or mounted "close coupled" to a Golf gti 1.6 litre engine as required. Laboratory catalyst testing was performed under flow conditions and gas samples were analysed using on-line gas chromatographic techniques. X-ray analyses of prepared powder samples were performed using a Philips PW1010 X-ray generator fitted with a PW 2213/20 1.5kW copper tube running at 40kv. Differential Scanning Calorimetric Analysis was obtained using a Mettler Toledo TA 8000 thermal analysis system fitted with a Mettler DSC 25 Module. Prepared samples were analysed by DSC in a flow of dry air at a heating rate of 10^0C min^{-1} over the temperature range 40-625^{0}C. Samples were heated to 150^{0}C in dinitrogen and held for 15 min. prior to cooling to 30°C at a rate of 10^0C min^{-1}. Temperature programme reduction was performed using 10% dihydrogen in oxygen free nitrogen (BOC gases, flowrate 50 ml min^{-1}). The samples were heated over the temperature range 30-650^{0}C at a rate of 10^0C min^{-1} using a Mettler Toledo TA8000 thermal analysis system.

Reactor studies were performed using a coated Dow Corning cordierite monolith cut into 0.5in. x 0.5in. x 6in. sections. Simulated vehicle exhaust gas mixture comprised of a stoichiometric mixture of 10% propylene in dinitrogen, dioxygen, 10% nitric oxide in dinitrogen, carbon monoxide (BOC gases, volume hourly space velocity of 44,000 h^{-1}). Engine testing was performed using a 4 in. diameter x 6in. cordierite monolith coated with the prepared catalyst washcoat (Pd loading 11g per cubic ft).

Results and Discussion:

X-Ray analysis of the 16 mol% ceria stabilised zirconia supported palladium-zinc catalyst system confirms that the addition of zinc to palladium reduces the mean particle size and stabilises the metallic character of the palladium and zinc functions. The Pd/Ce-ZrO$_x$ system shows a prominent peak at the 2θ value of 40.2^0 attributed to the low index Pd[110] face of FCC crystal structure. The addition of zinc to the formulation reduces the peak area of this peak by 62% with a corresponding increase in the peak width at half peak height and shift to a 2θ value of 40.6^0. Estimation of the crystallite dimensions using the Scherrer equation indicates that the mean particle volume diameter is reduced from *ca* 350 Å to 220 Å. Analysis of peak shifts attributed to palladium reflections indicates restructuring of the palladium crystallites upon the addition of zinc. Analysis of the XRD data is consistent with restructuring of the FCC structure of the palladium crystallites to a HCP orientation. The reduction of the mean crystallite volume diameter of palladium crystallites owing to the addition of zinc results in an increase in the relative density of $\geq$ 2-fold unsaturated palladium environments with a corresponding increase in the palladium/zinc junctions per unit area. We have recently reported the reversible oxygen switching mechanism from an oxide support to zinc and ultimately to palladium with the formation of labile surface palladium(II) oxide [7]. An increase in the concentration of palladium-zinc junctions will correspond to an increase in the efficiency of the catalyst to mobilise oxygen from the oxide support relative to the palladium supported ceria-zirconia system.

DSC analysis in air of the Pd/Ce-ZrOx material exhibits four discrete exotherms at 410, 430, 448 and 469 °C, the last two peaks being generated by the stabilised zirconia support. Hence the two peaks at 410 and 430°C are consistent with two thermodynamically discrete palladium environments, able to complex with oxygen at the catalyst surface. Thermochemical analysis of the Pd-Zn/Ce-ZrOx material shows that the addition of zinc to palladium produces a single ΔT_{max} at 445°C, confirming homogenisation of the palladium environments in which the modified palladium sites are thermodynamically distinct from those observed for the Pd/Ce-ZrOx samples. A second exotherm is again observed at 678°C, attributed to the formation of bulk palladium(II) oxide. Hence the oxygen hysteresis for the palladium/zinc system covers 233°C, compared with a hysteresis of 210°C and the oxidation onset at 600°C for Pd/γ-alumina [8]. Thermodynamic analysis shows that the addition of zinc results in stabilisation of the Pd^0 crystallites towards oxidation relative to the Pd/Ce-ZrO$_x$ samples by 9.1 kJ (g catalyst)$^{-1}$ through the temperature range 310-362°C. The relative ratio of the $k_{formation}$ of PdO for the Pd/Ce-ZrO$_x$ and Pd-Zn/Ce-ZrO$_x$ respectively is 0.997, and hence the presence of zinc retards the formation of surface oxide on bulk metallic palladium crystallites, confirming the findings of our previous studies [7]. Zinc(II) oxide is thermodynamically unstable, decomposing to Zn^0 and $\frac{1}{2}O_2$ at temperatures above 250°C [9]. Hence at temperatures above this threshold, thermodynamics prevail to stabilise the formation of metallic zinc, even in the presence of a relatively high partial pressure of oxygen. Where Pd^0 is coupled to Zn^0 , electron density is available from the reducing Zn^0 component to induce electron density to the Pd^0 function, resulting in an enhanced n-type character from the palladium-zinc couples relative to supported palladium. We have reported previously a lowering of the binding energy of the Pd$_{3d\ 5/2}$ electron of 0.4eV owing to the addition of zinc to palladium, confirming the electron rich environments associated with the palladium function [7,9].

Monolith sections coated with the Pd-Zn/Ce-ZrOx catalyst, and exposed to simulated vehicle exhaust gas under stoichiometric conditions, exhibited 95% conversion efficiency for hydrocarbons and carbon monoxide, and 65% efficiency for NO$_x$. Subjecting the catalyst to a flow of nitric oxide and dioxygen results in the formation of nitrogen dioxide at 359°C. Analysis of the eluent as a function of partial pressure of oxygen yields Langmuir-Hinshelwood kinetics. At $p_{o2} > p_{no}$ reduction in the p_{NO2} is observed, confirming that competitive adsorption takes place between dioxygen and nitric oxide. Hence the rate expression for the formation of NO$_2$ within the system is as follows:-

$$\frac{\delta[NO_2]}{\delta t} = \frac{k'\, p_{NO}\, p_{o2}^{\frac{1}{2}}}{(1 + b_{no}p_{no} + b_{o2}^{\frac{1}{2}}p_{o2}^{\frac{1}{2}})} \qquad [Equ\ 1]$$

[where $k' = k.b_{no}b_{o2}^{\frac{1}{2}}$]

and the percentage conversion ($\xi\%$) of nitric oxide is

$$\xi\% = \frac{Kt\theta_{NO}\theta_{O2}^{\frac{1}{2}}}{1+Kt\theta_{NO}\theta_{O2}^{\frac{1}{2}}} \times 100$$

[where K = rate constant $l^2\,mole^2\,s^{-1}$, t = residence time of adsorbed species, θ = surface coverage]

As surface coverage of oxygen tends to unity, kinetics show $\frac{1}{2}$ order in oxygen molecularity. First order kinetics are observed for the consumption of NO, consistent with associative adsorption of nitric oxide. Studies of the combustion of propene in dioxygen at steady state conditions, show zero order kinetics for hydrocarbon and carbon monoxide respectively, and a $\frac{1}{2}$ order in dioxygen. Pulsing propene in a flow of O_2 in the presence of excess NO has no effect on the molecularity of NO_2 formation, and hence the results are consistent with site specificity between the HC/CO and NO/O_2 respectively.

The overall rate can be expressed as follows:-

$$r = \frac{k''p_{no}p_{o2}^{\frac{1}{2}}}{\{(1+ b_{HC}p_{HC} + b_{O2}^{\frac{1}{2}}p_{o2}^{\frac{1}{2}})(1+ b_{NO}p_{NO} + b_{O2}^{\frac{1}{2}}p_{o2}^{\frac{1}{2}})\}} \qquad [Equ\ 2]$$

[where $k'' = kb_{NO}b^{\frac{1}{2}}_{O2}$]

The apparent activation energy (E_{app}) for propene combustion is 69.8 kJ mol^{-1} under stoichiometric conditions. (This value includes ΔH_{ads} of oxygen.) Combined GC and Fourier Transform Infra-red Spectroscopy (FTIR) of the gas phase has identified the presence of nitrogen dioxide, dinitrogen pentoxide and nitric acid . Nitric oxide is observed in trace quantities. Removal of NO_x species is primarily through the reaction of N_2O_5 with hydrocarbon. The addition of carbon monoxide to the feed does not affect the surface equilibria leading to the removal of N_2O_5.

Engine testing of the Pd-Zn/Ce-ZrO$_x$ catalyst fitted to a Vauxhall Cavalier 2l SRi using 95RON 'street' grade fuel gave >95% conversion of HC and 65% conversion of NO$_x$ driving an ECE+EUDC cycle followed by a second EUDC under stoichiometric conditions. NO$_x$ conversions increased to 75% when the monolith temperature reached 660°C after a 135km/h drive. Light-off for the catalyst was similar to that observed in the catalyst test rig, viz 359^{0}C.

Lean burn analysis was performed on a Golf gti 1.6l fuel injection engine fitted with Signal gas analysing equipment. Increasing the oxygen partial pressure, from stoichiometric ratio over the catalyst monolith situated in the exhaust gas flow, shows that for a conventional $Pt/Rh/Al_2O_3$ catalyst, the NO_x emissions increase as the CO emissions decrease when moving to lean burn conditions at an oxygen partial pressure of 151.88 torr [Figure 1]. This increase in NO_x is consistent with the disproportionation reaction between adsorbed CO and NO_x where a reduction in p_{CO} leads to a corresponding increase in p_{NO_x} (to 1700 ppm). On the other hand the $Pd-Zn/Ce-ZrO_x$ catalyst exhibits a fall in CO and in NO_x. The latter falls from 1980ppm at p_{O2} of 142.31 torr to 480ppm at p_{O2} of 155.15 torr. The results indicate that removal of NO_x does not depend on the disproportionation reaction between NO_x and CO. Studies to elucidate the surface mechanisms are presently underway and will be reported elsewhere.

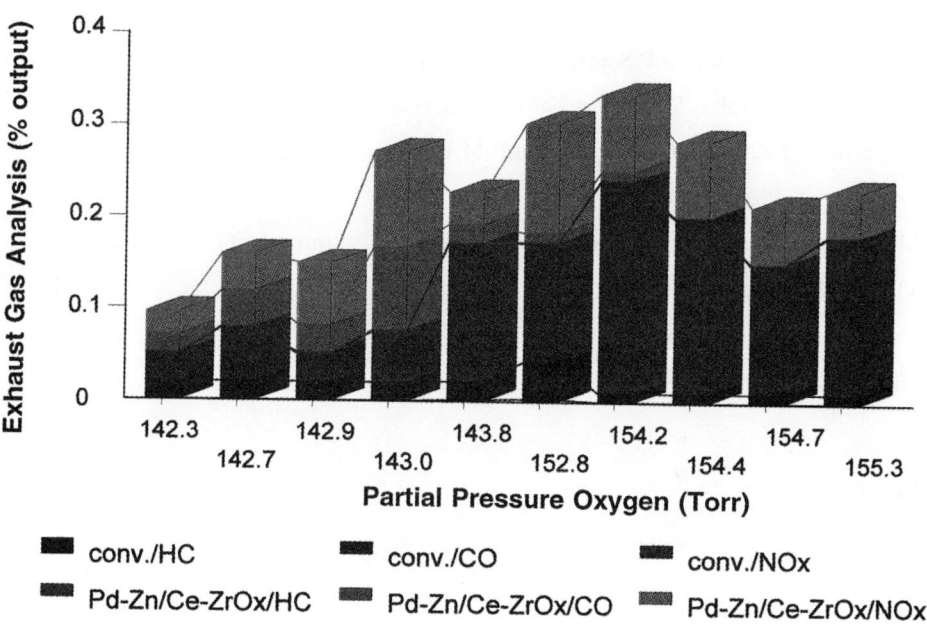

Figure 1 *Engine Trial under Varying Partial Pressure of Oxygen*

Acknowledgements: The authors acknowledge the assistance of Mr J.D.Paton and S.P.Scott who obtained the XRD data and Miss K.Baptiste who obtained the DSC data. The authors also wish to thank Harcros Chemicals, Durham, England and MEL Chemicals, Swinton, Manchester, England for their financial support.

References:

1 K.C.Taylor, *Studies in Surface Sci. and Catal.*, **Vol 30**, 97

2. W.Groenendaal, *Studies in Surface Sci and Catal,* **Vol 30,** 81

3. L.L.Hegedus, J.C.Summers, J.C.Schlatter, K.Baron, *J.Catal.,* 1979, **54**, 321

4. B.J.Cooper, *Platinum Metal Reviews,* 1983, **27,** 146

5. M.Ivamoto, *Proc. Meeting of Catalytic Technology for Removal of Nitrogen Monoxide,*   Tokyo 1990, 17

6. J.A.Cairns, A.C.Hourd, J.Thomson *Intl. Patent Pending,* PCT/GB96/00239 (1996)

7. J.A.Cairns, A.C.Hourd, S.P.Scott, J.Thomson, H.Bradshaw, P.Goulding, C.Norman I.McAlpine,and P.Moles, *SAE Technical Paper Series,Detroit,* 1997, **Paper No. 970468**

8. R.J.Farruato,J.K.Lampert, M.C.Hobson,E.M.Waterman, *J.Applied Catal.,B. Environmental,* 1995, **6**, 263

9. S.P.Scott, M.Sweetman, J.Thomson, A.G.Fitzgerald, E.Sturrock, *J.Catal.,* (paper no. JCAT-N11 in press)

Control of Organic Reactions by Use of Solids: Orientations and Reactivities of 6-Aminocaproic Acid and Glycine Adsorbed on Silica Gel Observed by IR Measurement and Dehydration

H. Ogawa

Department of Chemistry, Tokyo Gakugei University, Koganei, Tokyo 184, Japan

The relationship between adsorption states and reactivities of 6-aminocaproic acid (6-ACA) and Glycine (Gly) on the surface of silica gel (SiO_2) was ascertained by IR measurement with the diffuse reflection technique and by dehydration of amino acids. A decrease in amounts of 6-ACA and Gly on SiO_2 led the selective dehydration of the amino acids to giving ε-caprolactam (ε-CL) and glycylglycine (Gly- Gly), respectively, with depression of polymerization, presumably suppressing intermolecular interactions between amino acid molecules. A decrease in the amount of 6-ACA and Gly on SiO_2 ($\theta < 1.0$) caused the IR spectra to change from the spectra having NH_3^+ and CO_2^- groups like a solid state of the acids to the spectra having NH_2 and C=O groups with the hydrogen bonding.

1 INTRODUCTION

The application of adsorbents such as alumina and silica gel (SiO_2) in organic synthesis affords a new procedure for selective reactions,[1-3] where substrates are adsorbed and orientated on the surface of adsorbents with suppression of translational movement. We have reported the selective monomethyl esterification of dicarboxylic acids[4] and of long chain acids in the presence of acids having shorter carbon chains adsorbed on alumina,[5] orientation of terephthalic acid on alumina,[6] and the selective dehydration of 6-aminocaproic acid (6-ACA) adsorbed on SiO_2.[7] The characteristics of selectively promoted dehydration are attributed to the orientation of the amino acids on the surface of SiO_2. This explains why the reaction occurs selectively on the surface of SiO_2. This paper reports an investigation of the relationship between the orientation of the amino acids and the selective dehydration on the basis of IR measurements and the dehydration.

2 EXPERIMENTAL

2.1 Preparation of adsorption sample

An adsorption sample of amino acid was prepared as follows. The SiO_2 powder (C-200, Wako Chemicals) was added to a DMF solution or an aqueous solution of a certain amount of the acid. The mixture was allowed to stand for periods up to 8 hours at room temperature with occasional shaking. The solvent water was then slowly removed under reduced pressure. Water-additive adsorption sample was prepared as follows. A certain amount of water was added into a glass tube containing the adsorption sample, and the tube was sealed and allowed to stand for a day at room temperature.

2.2 IR measurement

The infrared spectra were recorded on a FT-IR (Jasco. model FT/IR 7000) equipped with DR-81 module for the diffuse reflection method. The adsorption sample was measured as a powder. Spectra were measured with 100 scans at 8 cm^{-1} resolution and obtained by rationing the background spectra of adsorbent SiO_2 or water-additive SiO_2 to those of adsorption samples.

3   RESULTS AND DISCUSSION

3.1 6-Aminocaproic acid (6-ACA)

3.1.1 Selective dehydration of 6-ACA to ε -Caprolactam (ε -CL) on SiO$_2$.
Dehydration of 6-ACA to ε -CL readily occurred by use of the adsorption sample of SiO_2 containing a certain amount of 6-ACA (Figure 1). The smaller the amount of 6-ACA contained in the adsorption sample, the higher the yield and selectivity for ε -CL become.
 ε -CL was obtained in *ca.* 90 % yield and in 99.8 % selectivity using the adsorption sample containing 0.46 x 10^{-4} mol g^{-1} of 6-ACA, while ε -CL was obtained in a 23 % yield in 1.0 x 10^{-3} M solution of 6-ACA in DMF. In addition, this adsorption method is moderately insensitive to water; even on addition of 30 wt% water, ε -CL was obtained in a 68% yield and in 92% selectivity.[7]
3.1.2   IR spectra on various loadings.   IR spectra of 6-ACA on SiO_2 are really complicated and show ambiguous absorption bands (Figure 2). However, they exhibited

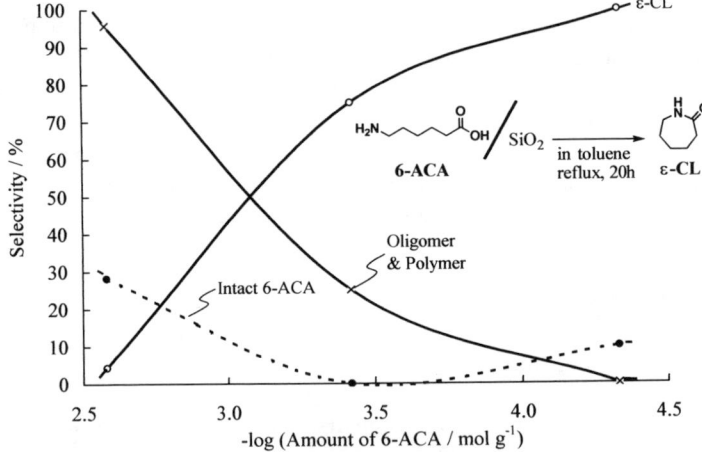

Figure 1 *Selective dehydration of 6-ACA to ε -CL on SiO$_2$*

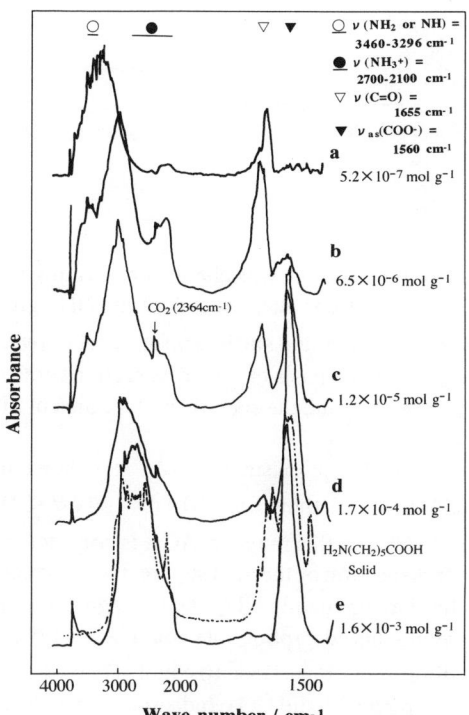

Figure 2 *IR spectra of 6-ACA on various loadings*

several characteristic absorption bands, *i.e.* at 3460-3296 cm-1 due to the stretching vibration of NH$_2$ { ν (NH$_2$)}, in the 2700-2100 cm-1 region due to multiple combination bands of NH$_3^+$ being the most prominent band at near 2200 cm-1, at 1655 cm-1 due to ν (C=O) from the hydrogen bonding, and at 1560 cm-1 due to ν $_{as}$(CO$_2^-$). The characteristic bands were assigned according to the IR absorption of authentic samples and references.[8,9] At fairly high loading (1.6 x 10-3 mol g-1), the spectrum (e) of the adsorption sample was similar to that of neat 6-ACA in the solid state, *i.e.* both of the spectra exhibited the characteristic bands at 2700-2100 cm-1 due to NH$_3^+$ and at 1560 cm-1 due to CO$_2^-$. On decreasing the amount of 6-ACA on SiO$_2$ (less than 1.7 x 10-4 mol g-1), changes in the IR spectra were observed, *i.e.* new bands appeared (i) at 1655 cm-1 due to ν (C=O) from the hydrogen bonding accompanying a disappearance of the characteristic band of carboxylate anion at 1560 cm-1 due to ν $_{as}$(CO$_2^-$) and (ii) at 3460-3296 cm-1 due to ν (NH$_2$) accompanying a disappearance of the bands at 2700-2100 cm-1 due to NH$_3^+$. Selective dehydration of 6-ACA to ε -CL occurred in this region of the loading less than 1.7 x 10-4 mol g-1.

IR spectra of water-additive adsorption samples {1.0 wt% and 10 wt% to the adsorption sample (4.6 x 10-5 mol g-1)} were similar to that of non-additive adsorption sample having IR absorptions attributed to C=O, CO$_2^-$, NH$_2$, and NH$_3^+$ groups.

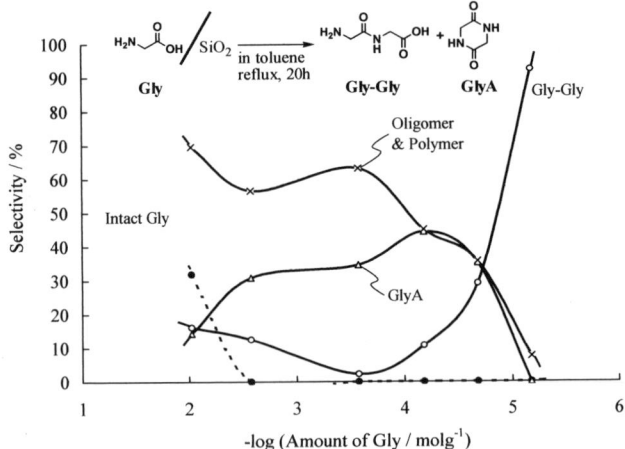

a

$H_3\overset{\oplus}{N}-(CH_2)_5-\overset{\ominus}{CO_2}$

$H_3\overset{\oplus}{N}-(CH_2)_5-\overset{\ominus}{CO_2}$ $H_3\overset{\oplus}{N}-(CH_2)_5-\overset{\ominus}{CO_2}$

Loadings $> 1.6 \times 10^{-3}$ mol g^{-1}-SiO$_2$ (θ =1.0)

b (CH$_2$)$_5$

$\overset{\oplus}{NH_3}$ $\overset{\ominus}{O_2C}$

c (CH$_2$)$_5$

Loadings $< 3.8 \times 10^{-4}$ mol g^{-1}-SiO$_2$ (θ =0.2)

Figure 3 *Adsorption spcices of 6-ACA*

The spectrum of the adsorption sample was changed by heating at 149 °C, showing a increase in intensities of the absorptions attributed C=O and NH$_2$ groups.

3.1.3 Adsorption state.   The amount sufficient for surface coverage (θ = 1) could be estimated at 1.6×10^{-3} mol g^{-1} on the basis of the cross section for 6-ACA (*ca.* 0.39 nm^2 molecule^{-1} by molecular modelling) and the specific surface area of SiO$_2$ (371 m^2 g^{-1}) by BET measurement.

A schematic presentation of adsorption state of 6-ACA is shown in Figure 3. Loadings of 6-ACA on SiO$_2$ are quite low, less than 3.8×10^{-4} mol g^{-1} ($\theta < 0.28$), for the selective dehydration of 6-ACA to ε -CL. In this range, 6-ACA is considered to be adsorbed on the surface of SiO$_2$ with easier dispersion of the acid suppressed intermolecular interactions between the acid molecules, having mainly NH$_2$ and C=O groups (adsorption state **c**). Dehydration of 6-ACA having this adsorption sate could be readily promoted. On the other hand, in the range of loadings more than 1.6×10^{-3} mol g^{-1} ($\theta > 1.0$), 6-ACA might be adsorbed on the surface having NH$_3$$^+$ and CO$_2$$^-$ (adsorption state **a** and slight quantity of **b**) with intermolecular interaction between acids taking into account the low yield of ε -CL by the dehydration.

3.2 Glycine (Gly)

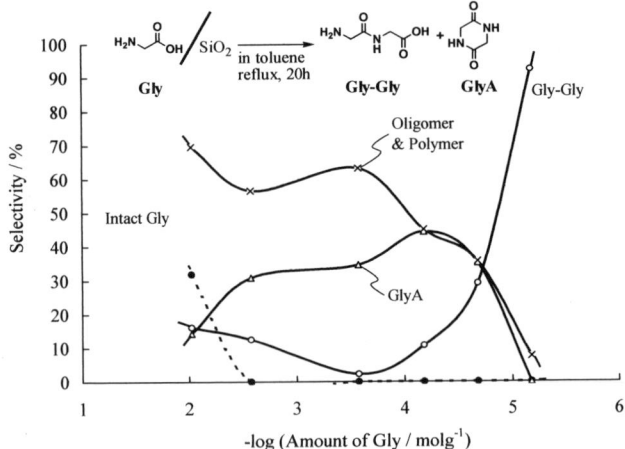

Figure 4 *Selective dehydration of Gly to Gly-Gly on SiO$_2$*

3.2.1 Selective dehydration of Gly to glycylglycine (Gly-Gly) on SiO₂. Selective dehydration readily occurred by use of the adsorption sample of SiO_2 containing a certain amount of 6-ACA (Figure 4), *i.e.* Gly-Gly was obtained in *ca.* 92% yield using the adsorption sample containing 0.66×10^{-5} mol g^{-1} of Gly, while Gly-Gly was obtained in *ca.* 14 % yield in 5.0×10^{-2} M solution of Gly in ethylene glycol accompanying a slight formation of glycine anhydride (Gly A) in 4 % yield. The smaller the amount of Gly contained in the adsorption sample, the higher the yield for Gly-Gly becomes.

3.2.2 IR spectra on various loadings. IR absorptions of various loadings of Gly on SiO_2 were measured and are illustrated in Figure 5. Spectra are really complicated and show ambiguous absorption bands. However, they give to us some information about adsorption state of Gly on SiO_2. On decreasing the amount of Gly on SiO_2 (less than 6.7×10^{-6} mol g^{-1}), changes in the IR spectra were observed, *i.e.* new bands appeared (i) at 1700-1630 cm^{-1} due to ν (C=O) accompanying a disappearance of the characteristic bands of carboxylate anion at 1600-1560 cm^{-1} due to $\nu_{as}(CO_2^-)$ and (ii) at 3400-3200 cm^{-1} due to ν (NH₂) accompanying a disappearance of the bands at 2900-2500 cm^{-1} due to NH₃⁺. Selective dehydration of Gly to Gly-Gly occurred on the loading of 6.7×10^{-6} mol g^{-1} ($\theta = 0.0025$).

3.2.3 Adsorption state. The amount sufficient for surface coverage ($\theta = 1$) could be estimated at 2.7×10^{-3} mol g^{-1} by molecular modelling. Loadings of Gly on SiO_2 are

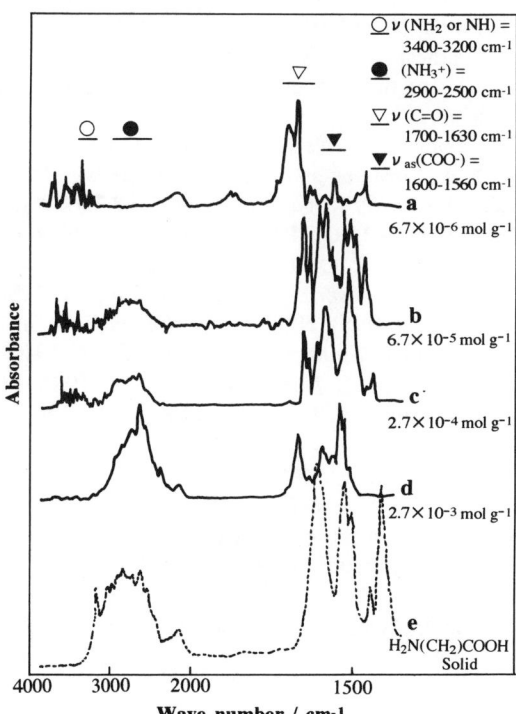

Figure 5 *IR spectra of Gly on various loadings*

quite low, less than 6.7×10^{-6} mol g^{-1} ($\theta = 0.0025$), for the selective dehydration to Gly-Gly. Gly is considered to be adsorbed on the surface of SiO_2 with easier dispersion of the acid suppressed intermolecular interactions between the acid molecules, having mainly hydrogen-bonded NH_2 and C=O groups. On the other hand, in the range of loadings more than 6.7×10^{-5} mol g^{-1} ($\theta > 0.025$), Gly might be adsorbed on the surface having NH_3^+ and carboxylate anion CO_2^- with intermolecular interaction between acids taking into account the low yield of Gly-Gly by the dehydration.

4 CONCLUSION

The selective dehydration of 6-ACA to ε -CL and Gly to Gly-Gly could be achieved by the adsorption of the acids on SiO_2 with the aid of dispersion of the acids, presumably suppressing intermolecular interactions between acid molecules, accompanying the changes of adsorption species from the species having NH_3^+ and CO_2^- groups like a solid state of the acids to the species having NH_2 and C=O groups. Dehydration of the acids of this species could be readily promoted. SiO_2 is considered to play a dual role as a reaction medium like an organic solvent and a catalyst for dehydration.

References

1. J.H.Clark, A.P.Kybett, and D.J.Macquarrie,"Supported Reagents: Preparation, Analysis, and Applications" VCH Publishers, N.Y.(1992); "Solid Supports and Catalysis in Organic Synthesis", ed. By K. Smith, Prentice Hall, West Sussex (1992); "Preparative Chemistry Using Supported Reagents", ed. by P. Laszlo, Academic Press, San Diego (1987).
2. A. McKillop and D.W. Young, *Synthesis*, **1979**, 401; G.H. Posner, *Angew. Chem. Int. Ed.*, **17**, 487(1978); A. Cornelis and P. Laszlo, *Synthesis*, **1985**, 909.
3. S. Hojo and R. Masuda, *Yukigosei Kagaku Kyoukaishi*, **37**, 689(1979); S. Hojo, *Yukigosei Kagaku Kyoukaishi*, **42**, 635(1984); T. Ando, J. Ichihara, and A. Hanahusa, *Kagaku Sosetsu*, **47**, 166(1985); A. Onaka and Y. Izumi, *Shokubai*, **34**, 159(1992); T. Okuhara and M. Misono, *Yukigosei Kagaku Kyoukaishi*, **51**, 128(1993); H. Hattori and H. Tsuji, *Shokubai*, **37**, 2(1995); H. Ogawa, *Hyomen Kagaku*, **11**, 124(1990); H. Ogawa, M. Kodomari, and T. Chihara, *PETROTECH*, **19**, 404(1996).
4. H. Ogawa, T. Chihara, ans K. Taya, *J. Am. Chem. Soc.*, 1985, **107**, 1365.
5. H. Ogawa, N. Hiraga, T. Chihara, S. Teratani, and K. Taya, *Bull. Chem. Soc. Jpn.*, **61**, 2383(1988).
6. H. Ogawa, *J. Phys. Org. Chem.*, **4**, 346(1991).
7. H. Ogawa, K. Nozawa, and P. Ahn, *J. Chem. Soc., Chem. Commun.*, **1993**, 1393.
8. Robert M. Silverstein, G. Clayton Bassler, and Terence C. Morrill," Spectrometric Identification of Organic Compounds, 4th Ed." John Wiley & Sons, N.Y., (1981), p. 126.
9. S. Sternhell and J. R. Kalman, "Organic Structures from Spectra" John Wiley & Sons, N.Y., (1987), p. 13.

ZEOLITE-SUPPORTED OXIDISED COPPER AND RHODIUM SPECIES: COMPARISONS OF ACTIVITIES FOR NO REDUCTION BY C_2H_4 WITH/WITHOUT O_2

Joseph Cunningham and James A. Sullivan

Department of Chemistry,
University College Cork,
Cork City,
Ireland

1 INTRODUCTION

Literature reports of quite different effects of excess gas phase oxygen upon the efficiencies of Cu-ZSM-5 or Rhodium-containing materials in catalysing the reduction of nitric oxide by hydrocarbons or other reductants provided the stimulus for the comparative studies upon Rh/Cu-ZSM-5 and Cu-ZSM-5 described in this paper. In respect of rhodium-containing catalysts, the very extensive literature concerning three-way catalytic (TWC) converters for nitric oxide removal from automotive exhaust gases had amply demonstrated the severe loss of efficiency of rhodium for nitric oxide removal when excess O_2 is present in the exhaust gas (eg. from operation at air-to-fuel ratios > 14.9)[1]. On the other hand the extensive literature concerning catalytic reduction of nitric oxide to N_2 by hydrocarbons or ammonia over Cu-ZSM-5 materials[2] had established the beneficial effect of added O_2 upon efficiency of such conversions,.

Those literature observations led to our interest in the possibility that Cu-ZSM-5 materials into which rhodium was subsequently incorporated as a second cationic promoter might maintain catalytic activity for nitric-oxide reduction by hydrocarbons, not only in O_2-free reactant streams (due principally to the rhodium content) but also in reactant streams with moderate O_2 content (due principally to the Cu-ZSM-5 content).

2 MATERIALS PREPARATION AND CHARACTERISATION

Cu-ZSM-5: A series of Cu ZSM-5 catalysts was prepared by shaking aliquots of H ZSM-5 (Si/A1 = 26, Degussa) overnight in 0.5M Cupric Acetate (Aldrich > 99.99% pure). Each such ion-exchanged sample was then filtered and repeatedly washed in deionised water. Increased copper loadings were achieved by repeating the procedure with fresh solutions of cupric acetate. The resultant solids were vacuum dried overnight at 383K, prior to being calcined in a flow of pure O_2 for 2 hours at 773K. The % copper was determined by atomic absorption following acid digestion.

Rh ZSM-5: Two samples of Rh ZSM-5 were prepared, differing in preparation technique and rhodium loadings, the first via a conventional ion-exchange procedure using a solution of rhodium acetate (Aldrich). This is designated Rh ZSM(ie), and contained only 0.0015% rhodium. The second, designated Rh ZSM(wi), contained 0.025% rhodium deposited onto ZSM-5 by wet impregnation from an acetonitrile solution of rhodium acetate after which it was vacuum dried overnight and calcined in flowing O_2 for two hours at 773K.

Cu-Rh/ZSM-5: The two-cation catalysts were prepared by wet impregnating a previously prepared Cu ZSM-5 material containing 2.4% copper with appropriate amounts of a saturated methanolic solution of rhodium acetate (Aldrich), after which the resulting powders were again vacuum dried overnight and calcined in an O_2 flow at 773K for two hours. In addition to 2.4% copper, the two-cation catalysts contained rhodium (presumably in the form of Rh_2O_3 clusters or crystallites after calcination) at levels of 0.5 and 0.05% Rh and are designated as Cu/Rh ZSM-5 (0.5) and Cu/Rh ZSM-5 (0.05).

FTIR: Difference-spectra, sensitive only to additional infra-red absorbances generated by chemisorption of nitric oxide onto vacuum outgassed discs of Cu-ZSM-5 or RhCu/ZSM-5 were measured in efforts to identify the surface species induced thereon by exposure to 7 torr NO. The difference-spectra illustrated in fig. 1, which resulted from stepwise exposures of a disc of 2.4% Cu/ZSM-5 to NO for 5, 15 and 30 mins (after each of which the gas phase NO was evacuated), typify the results obtained, especially in respect of time-dependent changes in the relative intensities of the IR features peaking at 1811 and 1733 or at 1912 and 1635 cm^{-1}. The former pair may be assigned to $ON\text{-}Cu^+$ and $(ON)_2$ $-Cu^+$, and their decrease with increased NO exposure understood on the basis that chemisorbed NO progressively reoxidised to Cu^{2+} some Cu^+/ZSM-5 sites formed during prior vacuum-outgassing at 773K. Increases in the peak due to $Cu^{2+}\text{-}NO$ at 1912 cm^{-1} resulted from such reoxidation. Accompanying growth of the feature at 1635 cm^{-1} implied slow conversion of some chemisorbed NO to surface – NO_2 species. The main features of difference-spectra obtained from a disc of Rh Cu/ZSM-5 containing 0.05% rhodium and 2.4% Cu were qualitatively similar to those from Cu/ZSM-5 alone. However, comparisons of values for the ratio, Area of $Cu^{2+}\text{-}NO$ @ 1912: Area of $Cu+\text{-}$ NO @ 1811 cm^{-1}, after 5, 15 and 30 mins exposure to NO indicated that rate of reoxidation of Cu^+ in this overexchanged sample was markedly enhanced by the 0.05% Rh – suggesting that Cu^{2+} would greatly outnumber any Cu^{+1} under a continuous flow of NO.

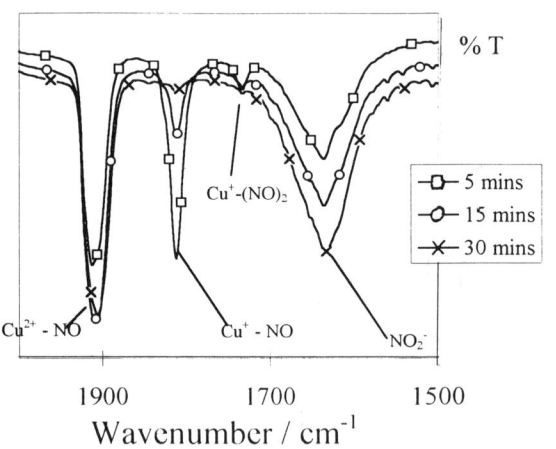

FIGURE 1:FTIR-based 'difference-spectra' illustrating intensity variations in IR bands caused by increased duration of exposure to 7 torr nitric oxide

3 CATALYTIC ACTIVITY

The catalytic runs were carried out on a flow system connected to a gas chromatograph (Pye Unicam Series 104) fitted with molecular sieve or porapak columns and a thermal conductivity detector. Four "Tylan" mass flow controllers were used to blend appropriate gas mixtures from cylinders of 3% NO/He, 3% C_2H_4/He, 10% O_2/He and pure He. The most frequently used mixture featured 0.6% NO, 0.6% C_2H_4 and 5% O_2 in a helium flow totalling 30 mls per minute. When 25mg of catalyst were used this flow corresponded to a nominal GHSV of 36,000 h^{-1}. Studies with varying partial pressures of oxygen were also carried out by using the O_2 and He flow-controllers to achieve different $P(O_2)$ values while keeping the overall flow rate constant.

The SCR-HC (0.6% NO, 0.6% C_2H_4 and 5% O_2) and NO-Red (0.6% NO, 0.6% C_2H_4, 0% O_2) reactions were separately studied over samples of Cu ZSM-5 having differing amounts of copper. Results are presented in Figures 2a and 2b respectively as histograms showing % conversion to nitrogen as a function of temperature and of % Copper within the samples. Prior to measuring the indicated values of conversion, the catalyst aliquot had been maintained in the reactant flow and the temperature kept constant for one hour at each temperature or until stable GC readings were obtained. As may be seen in fig. 2a, the Cu ZSM-2.4% catalyst was the most active material for SCR-HC at 623K, being slightly more active than Cu ZSM-2.1%, whilst Cu ZSM-1.8% was less active and H ZSM very much less active showing less than 10% conversion to N_2. However, at 673K, Cu ZSM-1.8% gave the largest conversion to N_2, followed by the 2.4% sample which again was slightly more active than the 2.1% catalyst. Apparently the presence of over-exchanged copper in the Cu ZSM-2.4% was important for enhancing SCR-HC activity at 623K, whereas at higher temperatures this feature did not predominate in determining such activity. Data in fig 2a taken at temperatures higher than 673K show the well known[3] decrease in % conversion to N_2 attributable to onset of hydrocarbon combustion, $C_2H_4 + 3O_2 = 2CO_2 + 2H_2O$, in competition with C_2H_4 involvement in the selective reduction of NO. However, over undoped H-ZSM conversion to N_2 increased to 40% at 873K thus reproducing well known activity of the parent zeolite for the SCR-HC reaction.

Results from parallel studies of the NO-Red reaction over the same four catalysts are depicted in Figure 2b where it may be noted that none gave over 10% conversion at 573K. At 623K the 1.8% Cu ZSM sample yielded 40% conversion, in contrast with the H ZSM or the 2.1 and 2.4% Cu ZSM samples each of which yielded conversions ca. 10%. At 673K all the samples became more active, that with 1.8% copper again being the most active and yielding 60% conversion. Next most active was H-ZSM, showing a conversion of 40% whereas the 2.1% and 2.4% catalysts yielded lower conversions of 30% and 15% respectively. Thus for temperatures $\leq$ 673K with C_2H_4 as reductant, an inverse correlation seemed to exist in respect of relative activities for SCR-HC and NO-Red conversions: catalysts most active for SCR-HC being least active for the conversion of NO to N_2 in the absence of oxygen and vice versa. These inverse copper-concentration dependences of NO-red and SCR-HC activities for N_2 production at temperatures $\leq$ 673K require an alternative explanation to those based upon the competition for hydrocarbon oxidation between O_2 and NO mentioned above in explanation of decreases in SCR-HC observed at higher temperatures. Such an alternative is developed after figures 2 and 3.

Figures 2 and 3:
Comparative profiles versus T_{rx} for % NO reduction to N_2 by C_2H_4 in SCR-HC conditions (upper pair of figures) and in NO-red conditions (lower pair of figures)

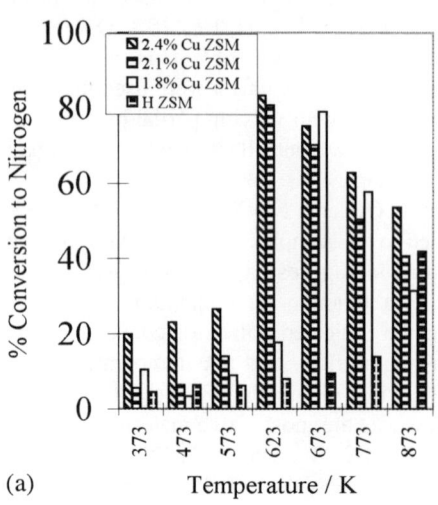

(a)

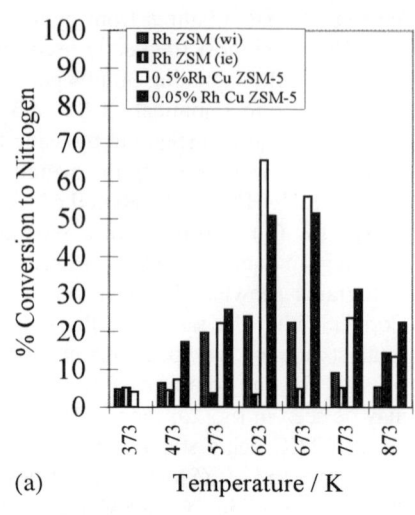

(a)

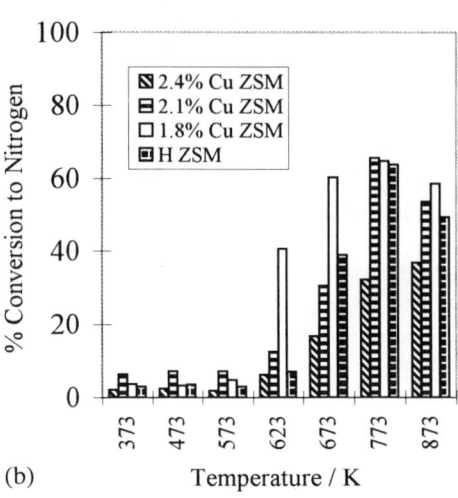

(b)

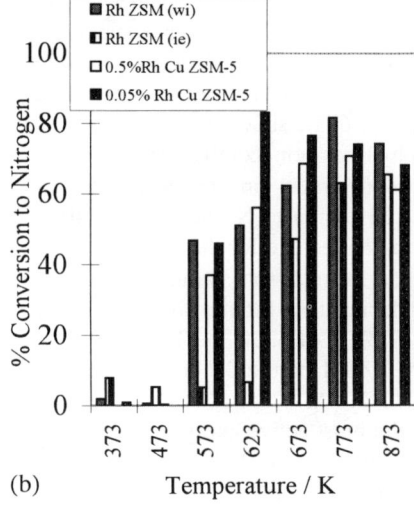

(b)

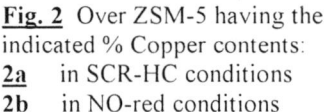

Fig. 2 Over ZSM-5 having the indicated % Copper contents:
2a   in SCR-HC conditions
2b   in NO-red conditions

Fig. 3 Over ZSM-5 having the indicated Rhodium contents:
3a   in SCR-HC conditions
3b   in NO-red conditions

In the NO-red process the catalysts exist under a stream of NO plus a hydrocarbon in the absence of oxygen. Interaction of hydrocarbons and the acid form of zeolites is well known, e.g. the very recent work of Hayes et al[4] concerning interaction of propene with Cu-ZSM-5, and proceeds via formation of carbocations capable of reaction in well-defined acid-catalysed pathways. Facile formation of such reactant carbocation intermediates upon acidic sites in H-ZSM-5 and Cu-ZSM may in turn promote onward conversion to N_2 of NO species adsorbed upon copper species ion-exchanged into the zeolitic framework[4]. Present observations that the presence of copper at 1.8% within the ZSM-5 lattice results in an increase in activity for the NO-Red reaction at $\leq$ 673K (as seen by comparing the histograms for 1.8% Cu ZSM and H ZSM in Figure 2b) are consistent with operation of such a bifunctional mechanism involving cooperative action between NO upon copper at ion-exchanged sites and carbocations from ethylene upon residual acidic sites of the ZSM-5 lattice. This effect is lessened upon introduction of some excess copper species in 2.1% Cu-ZSM-5 and is reversed at the highest copper loading (2.4%). However, it is notable that similar conversions to N_2 could be achieved over the Cu-ZSM-1.8% material from either the NO-Red reaction or the SCR-HC reaction at temperatures ca.673K. Higher temperature data in fig. 2b show that at 773K similar ($65 \pm 2\%$) conversion to N_2 resulted from the NO-Red process over H ZSM-5, 1.8% Cu ZSM and 2.1% Cu ZSM-5, whereas only 30% conversion was achieved over the Cu ZSM 2.4% catalyst.

Details of the ways in which SCR-HC and NO-red activities were modified by incorporating rhodium species into/onto ZSM-5 may be seen by comparing fig. 3a with fig. 2a (SCR-HC activity) or fig. 3b with 2b (NO-red activity). In contrast to the Rh-ZSM (i.e.) material, which was practically inactive until 873K, measurable SCR-HC was found for H-ZSM or 2.4% Cu-ZSM wet impregnated with ionic rhodium species. A maximum of only 25% was reached at 673K over such Rh-ZSM which decreased rapidly at higher T_{rx}. However, maxima of ca. 60% were reached at T_{rx} 623 and 673K over the Rh-Cu ZSM materials wet-impregnated with 0.5 or 0.05% rhodium allied to much more gradual declines at higher-temperatures. Detailed comparison of fig. 3a with data in fig. 2a from 2.4% Cu-ZSM indicate that moderate inhibitory effects upon SCR-HC activity of the latter resulted from rhodium wet-impregnated thereon. On the other hand, detailed comparison of data in fig. 3b with those in 2b demonstrate that NO-red activity of the 0.5% Rh Cu-ZSM material at $T_{rx} \geq 573$ K was substantially improved by such wet impregnation. Indeed the NO-red conversion of ca. 45% evidenced at 573K over those materials significantly exceeded conversions in other plots of figs. 2 or 3 for that temperature.

Sequences of catalytic measurements under different fixed partial pressures of O_2, $P(O_2)$ (viz, 0, 0.83, 2.16, 3.93 and 5%) admixed with NO and C_2H_4 in the reactant flow, were carried out to span the transition from NO-red to SCR-HC conditions and to identify which $P(O_2)$ maximised NO conversion to N_2 over catalysts used in the present study. Results obtained from such sequences of measurements at T_{rx} = 673K and the indicated $P(O_2)$ over rhodium-containing catalysts are summarised in fig. 4. The following similarities/differences emerged from detailed comparison of those results with others obtained from similar measurement–sequences over Cu-ZSM materials: (a) similar highest efficiency conversions to N_2 at 673K were achieved with 2.16% O_2 in the reactant-flow over Cu-ZSM having only 2.1 or 2.4% copper as were attained over the latter material having 0.5% or 0.05% rhodium wet-impregnated thereon: (b) however,

Supported Reagents and Catalysts in Chemistry

Figure 4. Promoting effect of wet-impregnated Rhodia in extending activity of Rh-ZSM and Rh-Cu-ZSM for N_2 production from NO + C_2H_4 at 673K to low and zero $P(O_2)$

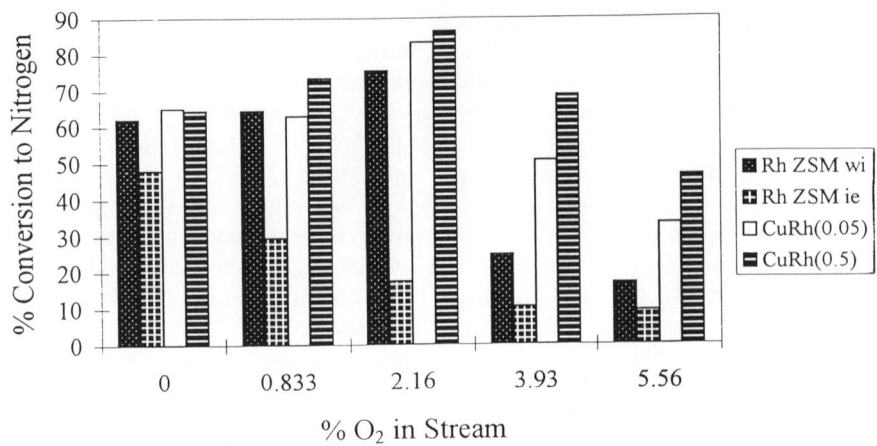

% O_2 in Stream

the steep drops in efficiency noted over those Cu-only zeolites under $P(O_2)$ = 0.82 or zero (to 40% or ca. 25% respectively) did not occur over the 0.5 Rh-Cu-ZSM on 0.05% Rh-Cu-ZSM (which retained activities > 60% in such conditions). Point (b) confirms the capability of wet-impregnated rhodium to compensate for the decline in activity usually found for Cu-ZSM alone at low or zero % O_2.

4 INTERIM CONCLUSIONS AND POINTS FOR FUTURE STUDY

Improvements here demonstrated in the catalytic activity of over-exchanged Cu-ZSM-5 for N_2 production from NO + C_2H_4, as a result of wet-impregnation with rhodium species thereon, include: (i) lowering to 573 K the T_{rx} required for 50% conversion by NO-red; (ii) at T_{rx} = 673K, retention of activity at low and zero $P(O_2)$. It will be of interest in future work to compare those findings with effects of rhodium wet-impregnation upon activities of 1.8% Cu-ZSM-5 having copper species only at ion-exchange sites, since these already evidenced here substantial SCR-HC and NO-red activities at T_{rx} ≥ 673K. Further work is also needed to quantify the effect of wet-impregnated rhodium upon selectivity towards N_2O especially at moderate T_{rx} in NO-red conditions.

5 REFERENCES

1. K.C. Taylor *Catal. Rev. Sci Eng.* 1993, **35**, 457 and references therein.
2. M. Iwamoto Studies *in Surface Science and Catalysis*, 1991, **54**, 121 and refs therein
3. S.Sato, H. Hirabay, H. Mizune and M. Iwamoto, *Catal. Lett.*, 1992, **12**, 193.
4. N.W. Hayes, R.W. Joyner, E.S. Shpiro, *Appl. Catal. B. Environ*, 1996, **158**, 301.

Acknowledgement: This work was in part supported under EC Contract EV-CT-92-0234

ESTERIFICATION OF ACRYLIC ACID WITH BUT-1-ENE OVER SOLID ACID CATALYSTS BASED ON SULFATED ZIRCONIA

N.Essayem, V. Martin, A. Riondel*, G.Coudurier, J.C.Védrine

Institut de Recherches sur la Catalyse, associé à l'UCB Lyon I, CNRS, 2 avenue Albert Einstein, F- 69626 Villeurbanne, France. Fax : (33) 4 72 44 54 13
* Centre De Recherches et de Developpement de l'Est, Elf Atochem, F-57501 Saint Avold, France

1 INTRODUCTION

Esters of acrylic acid are of great importance for the chemistry of polyacrylates which are used as bases for resins, plastics or paints. Among industrial acrylates processes, esterification of acrylic acid with alcohols is the most generally used. For example, for butyl acrylates production, the reaction is carried out in liquid phase at moderate temperature (323 to 363 K) and moderate pressure (20 to 100 kPa) in presence of H_2SO_4 or heteropolyacids[1] as acid catalyst. Replacement of soluble acids for solid catalysts such as acid resins, heteropolyacids supported on activated carbons[2] or sulfated zirconia[3,4] was shown very promising due to an easier separation from the products and regeneration facilities.

However, esterification of acrylic acid with olefins is even more attractive by two ways : use of olefins, rather cheap reactants, and absence of water formation which makes easier the separation of acrylates. The reaction conditions are the same as for esterification with methanol and the industrial catalysts are also H_2SO_4 or acid resins as Amberlysts[5] which are characterized by high activity and stability. Sulfated zirconia presents a potential interest since it has been shown to be as active and more selective than the resins[6]. Various sulfation agents have been used to prepare sulfated zirconia from freshly precipitated zirconium hydroxide and it is generally known that the best results are obtained by percolation or impregnation methods with H_2SO_4 or $(NH_4)_2SO_4$ and that the calcination conditions strongly influence the generation of acid sites.

An attractive method for preparation at the industrial scale consists to grind a mechanical mixture of zirconium hydroxide and ammonium sulfate before calcination. Such a procedure eliminates the energy demanding step of evaporation following the impregnation.

We have already described the acid and catalytic properties of sulfated zirconia [7,8] and mixed tin oxide-zirconia[9] obtained by H_2SO_4 impregnation. The present work deals with the study of the activity and of the stability of two sulfated zirconia, prepared by impregnation with H_2SO_4 or by grinding with ammonium sulfate, in the reaction of acid acrylic esterification with but-1-ene.

2 EXPERIMENTAL

2.1 Catalysts preparation

Zirconium hydroxide $Zr(OH)_4$ was prepared as usual by precipitation of a solution of $ZrOCl_2$. $8H_2O$ by NH_4OH at a pH value of 9, followed by careful washing and dryness at 373 K for 20h under a vacuum of 5.10^3 Pa.

Sulfation was carried out following two procedures. The usual one consists of impregnating zirconium hydroxide by sulfuric acid solution. 20g were suspended in 300 cm^3 of 0.2 N sulfuric acid solution for 20 min. The solid was then filtred and dried at 393K for 24 h in an oven. The material was calcined under air flow running through the bed at 823 K for 2h. The samples will be referred to by the symbolism SZ H. The second procedure consists in grinding a mechanical mixture of zirconium hydroxide with a given amount of ammonium sulfate (15 wt %). Both constituents (Zr(OH)$_4$ and (NH$_4$)$_2$SO$_4$) were also ground before mechanical mixture. The solids were also calcined as previously under air flow at temperatures of 823, 873 and 923 K for 2h. The samples will be referred to by the symbolism SZ A.

2.2 Physico-chemical techniques

DTA-TGA analyses were carried out with a Setaram A92-12 equipment and UV spectra were recorded with a Perkin Elmer Lambda 9 spectrometer in diffuse reflectance mode and BaSO$_4$ as reference. IR spectroscopy measurements were done on KBr pellets using a Vector 22 Bruker instrument.

2.3 Catalytic testing

Two reactions were studied, n-butane isomerization in gas phase and acrylic acid esterification with but-1-ene in liquid phase. For n-butane isomerization reaction a flow microreactor was used with gas chromotography analysis on line.
Acrylic acid esterification with but-1-ene was performed in a batch reactor operating in liquid phase under the following conditions : 11g catalyst, 1 mol. acrylic acid, 0.9 mol. but-1-ene, 700 ppm of polymerization inhibitor (EMHQ : monomethyl ether hydroquinone). The reaction temperature was 343 K and the pressure reached 9.10^5 Pa at the beginning of the reaction. The analysis was carried out after 4 h reaction by potentiometry for the conversion of acrylic acid and by gas chromatography for all products.

3 RESULTS AND DISCUSSION

3.1 Physico-chemical characterizations of the samples

3.1.1 SZ A sample : Effect of the Temperature of Calcination   The relative proportion of tetragonal to monoclinic phases can be represented by the ratio of intensities of the 30.14 and 28.25° (2θ) XRD peaks (see *infra* in figure 2 some spectra). From the data reported in table 1 it was observed that the tetragonal phase was favoured at low temperature of calcination in agreement with the stability of this phase. In addition, the sulfur content and the BET surface area decreased as the calcination temperature increased (table 1).

Table 1 *Evolution of the Sulfur Content , Texture and Surface Area as a Function of the Temperature of Calcination*

T calcination	I 30°/I 28°	S Content	S BET	Porous Surface Area (m^2.g^{-1})	
(K)	(2θ)	(wt %)	(m^2.g^{-1})	1.8<Ø<3.6 nm	Ø > 3.6nm
823	3.9	3.2	100	40	60
873	3.5	2.6	94	15	79
923	1.4	1.6	89	0	89

If the zirconium hydroxide presented exclusively micropores with pore diameter between 0.8 and 1.2 nm, it was important to underline that the calcined sulfated zirconia was mainly mesoporous and the pore apertures increased as the temperature of calcination increased. These textural evolutions have been previously described for the calcination of $Zr(OH)_4$ [10].

3.1.2 Comparison of SZ H and SZ A samples.   TG-DT analyses were performed for the two sulfated hydroxides before the calcination step. The results were quite different. The TG-DTA profiles for SZ H samples were similar to those already described[9,11]. The first weight loss accompanied by two endothermal peaks at 423 and 473 K was due to the departure of physically adsorbed water and to water released by the dehydroxylation reaction (fig 1-a). The second weight loss above 893 K corresponded to the evolution of SO_2 due to the sulfate group decomposition. The exothermal peak at 893 K characterized the crystallization of ZrO_2, the tetragonal phase being stabilized by the presence of sulfate groups. It was observed that the crystallization of unsulfated $Zr(OH)_4$ took place near 693 K. In that case the monoclinic phase was obtained. Let us notice that on pure $Zr(OH)_4$ the weight loss was complete at 873K (fig 1-c).

On the other hand for SZ A samples several different weight losses below 673 K occurred simultaneously with distinct endothermal phenomena (fig 1-b). Presumably, at temperature lower than 473 K the weight losses were due also to the departure of physically adsorbed water and to dehydroxylation. Between 523 K and 673 K the weight losses also accompanied by endothermal peaks were attributed to the departure of compounds produced by the thermal decomposition of the ammonium sulfate, in agreement with the TG-DTA profiles observed for bulk ammonium sulfate (fig 1-d). Unlike SZ H sample the exothermal peak was observed at 686 K, similarly to the crystallisation of unsulfated $Zr(OH)_4$, suggesting that the crystallisation of $Zr(OH)_4$ was not modified by the presence of sulfates as if there was no interaction between the sulfate and ZrO_2.

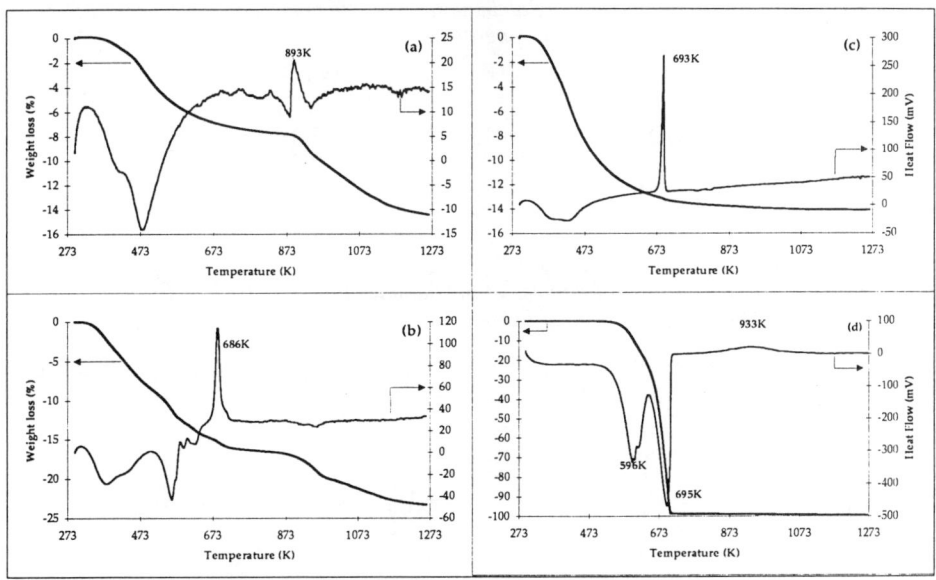

Figure 1 *TGA-DTA curves of the samples before calcination a) SZ H sample b) SZ A sample, c) unsulfated $Zr(OH)_4$ d) bulk $(NH_4)_2SO_4$*

However, a weight loss was observed at temperatures higher than 873 K, while for unsulfated $Zr(OH)_4$ and for bulk ammonium sulfate the weight losses were complete before 873 K. This result suggests that some sulfate groups formed by the decomposition of $(NH_4)_2SO_4$ may be trapped within the zirconium oxide inducing an interaction detected by a weight loss at high temperature.

3.1.3 Comparative Characterization of the two Kinds of Sulfated zirconia. The main characteristics of the SZ H and SZ A samples are summarized in tables 2 and 3 together with the acidic properties estimated in the test reaction of n-butane isomerization. The calcination temperature was chosen for each preparation as corresponding to the highest activity in n-butane isomerization, namely 823 K for SZ H sample[8] and 873 K for SZ A sample.

Labelling 1 and 2 as subscripts indicates two different batches prepared independently.

The sulfur content and the textural properties of the two kinds of sulfated zirconia were comparable, particularly the average pore diameters were similar although the surface area of the SZ A sample was lower that is explained by its higher calcination temperature. SZ H samples presented mainly the tetragonal phase while SZ A samples never presented purely tetragonal phase and the proportion tetragonal/monoclinic phase was badly reproducible.

The activities in n-butane isomerization of the samples SZ H and SZ A were of the same order of magnitude. It is interesting to observe that SZ A_2 presented high acidic properties although it contained highest proportion of monoclinic phase.

Table 2 *Sulfation process, Calcination Conditions, Sulfur Content and Structural Features of the Various Sulfated Zirconias Used in this Study*

Sample	Sulfation agent	Calcination Temperature	Sulfur Content	*I 30° / I 28°*
		(K)	(wt %)	(2θ)
SZ H_1	H_2SO_4	823	3.1	∞
SZ H_2	"	"	2.1	4.8
SZ A_1	$(NH_4)_2SO_4$	873	2.6	3.1
SZ A_2	"	"	-	1.1

Table 3 *Structural and Textural Features of the Samples and Activity in n-Butane Isomerization*

Samples	iC_4 Formation* (10^{-8} mol.g^{-1}.s^{-1})	S BET (m^2.g^{-1})	Average Pore Diameter (nm)
$Zr(OH)_4$	0	190	<1.2
SZ H_1	58	100	5
SZ H_2	63	99	-
SZ A_1	50	71	5
SZ A_2	62	-	-

* Pretreatment conditions : 2h under flowing air at 673 K. Reaction conditions : 3.4 % nC_4 in N_2, total flow rate : 1.3 dm^3.h^{-1} (298 K), T reaction : 473 K, m catalyst = 200 mg.
- : means that the experiment was not performed

3.2 Liquid Phase Esterification : Reactivity and Deactivation Studies

3.2.1 Comparison of SZ H and SZ A samples.   All samples, characterized by a high activity in n-butane isomerization, exhibited also a high activity in the reaction of esterification (table 4). The selectivities in sec-butyl acrylate as regards to the but-1-ene reached 85% and as regards to the acrylic acid reached 95 %. Other products were also formed by but-1-ene or acrylate self condensations.

On the first cycle the SZ A samples exhibited low activity while the SZ H samples still provided an activity after one and even two cycles. After two cycles the conversion was reduced by a factor of 4.

3.2.2 Physico-chemical Characterizations of the Used Catalysts after the Esterification in Liquid Phase.   XRD spectra recorded on SZ A samples after one cycle showed a decrease of the proportion of the tetragonal phase as regards to the monoclinic one (figure 2). On the other hand, XRD spectra recorded on SZ H samples after 1, 2 or 3 cycles remained unchanged as regards to the fresh catalyst. Although similar observation for the sulfated zirconia deactivation was already mentioned in the literature[12], but after gas phase n-butane isomerization as we did also, this structural modification observed on SZ A samples was quite surprising since the reaction temperature was only 343 K.

As it can be seen in table 5, the sulfur content of the catalysts remained constant after several cycles. These data are the proof that the possible deactivation of the catalyst by sulfate leaching by the liquid phase did not take place. In addition these data show that the modification of the catalyst structure was not a consequence of the loss of sulfur.

For the two kinds of sulfated zirconia a moderate decrease of the BET surface area was observed, while the pore size distribution remained unchanged. Note that the extent of the catalyst deactivation was not at all of the same order of the magnitude than the decrease of the catalyst surface area.

Table 4 *Activity of SZ A and SZ H Samples in Liquid Phase Esterification*

Samples	Acrylic acid conversion fresh samples (%)	Acrylic acid conversion after one cycle (%)	Acrylic acid conversion after two cycles (%)
SZ H_1	47	15	-
SZ H_2	58	22	14
SZ A_1	59	4	-
SZ A_2	60	-	-

Table 5 *Sulfur Content (wt %) of the Fresh and Used Samples*

Samples	Fresh	After one cycle	After 2 cycles	After 3 cycles
SZ H_2	2.1	2.2	2.2	2.3
SZ A_1	2.6	2.4	2.5	-

Table 6 *Modifications of the S BET ($m^2.g^{-1}$) after Esterification Reaction*

Samples	Fresh	After one cycle	After two cycles
SZ H_1	100	96	72
SZ A_1	90	87	83

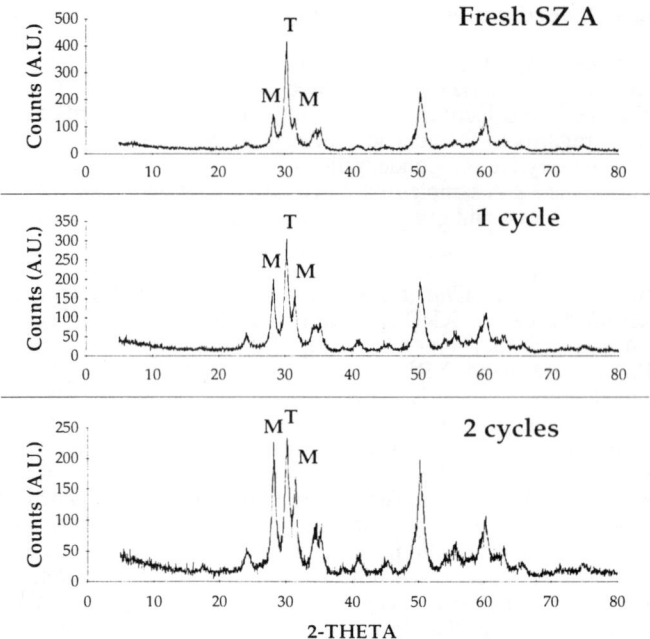

Figure 2 *XRD spectra for used SZ A sample*

3.2.3 Acidic Properties of the Used Catalysts.   SZ A samples were observed to present a strong decrease of activity in esterification after one cycle and negligible conversion after a second cycle. The measure of the acidic properties by means of the reaction of n-butane isomerization showed also a drastic decrease after the first esterification run (table 7).

For SZ H samples a more progressive decrease of the acidic properties was observed after each cycle, which is in agreement with the progressive loss of activity observed in esterification.

More generally, these results show the strongest decrease of activity of SZ A samples as regards to SZ H ones with time on stream.

In conclusion, it appears that deactivation in esterification follows very well the decrease in n-butane isomerization activity and thus corresponds to a loss of acidity. This is probably due to a poisoning of the acid sites by some adsorbed compounds (reactants, intermediaries or products).

Table 7 *Activities in n-Butane Isomerization at 473 K of Fresh and Used* SZ Samples*

		Rate of iC_4 Formation (10^{-8}.mol.g^{-1}.s^{-1})		
Samples	*Fresh*	*After one cycle*	*After two cycles*	*Regenerated ***
SZ H_1	58	29	16	52
SZ H_2	63	-	20	51
SZ A_1	50	4.4	1.0	18
SZ A_2	62	6.5	-	-

* used means used in esterification reaction
** regeneration conditions : 2h under air flow at 773 K

3.2.4 Nature of the Products Adsorbed on Used Catalysts. Thermal analyses of the aged samples were performed under air flow, up to 873 K, in order to characterize the carbonaceous deposits, quantitatively (weight losses) and qualitatively (desorption or combustion temperatures). The thermal curves recorded on SZ A after one esterification cycle are presented in figure 3 together with the thermal curves recorded on SZ H sample after two esterification cycles.The two samples presented several weight losses. Below 423 K the weight loss must correspond to the departure of slightly adsorbed compounds and the second loss near 473 K to the release of more strongly adsorbed poisoning compounds.

For temperatures higher than 500 K the samples exhibited weight losses accompanied with one or two separate exothermal peaks. The exothermal peaks were observed at 528 K and at 641 K on SZ A samples while on SZ H samples, only the exothermal peak at 630K appeared clearly. These exothermal peaks should correspond to the combustion of strongly adsorbed species.

In order to investigate the nature of the poisoning compounds, selective adsorptions on the catalysts of each reactant (but-1-ene and acrylic acid) and of the product (sec-butyl acrylate) were performed under the reaction conditions at 343 K for 4 h and the resulting solid was then studied by DTA-TGA, UV and IR spectroscopies. This study was performed on a SZ A sample and EMHQ was added to prevent polymerisation of acrylic acid or acrylate in the same conditions as for the reaction. The resulting solids are abbreviated SZ+AA (acrylic acid adsorption), SZ+sec.BA (sec-butyle acrylate adsorption) and SZ+BU (but-1-ene adsorption). Thermal analyses were performed on the samples, under air flow, up to 873 K. Curves recorded on SZ+sec.BA or on SZ+BU were characterized by two weight losses. The first one occurring at temperatures lower than 423 K together with endothermal peaks, corresponds to the desorption of weakly adsorbed compounds, the second one at higher temperature, accompanied by a broad and weak exothermal peak from 473 to 873 K, is assigned to a combustion process (Total weight losses were 6 and 6.5 % for the SZ+sec.BA and SZ+BU samples respectively).

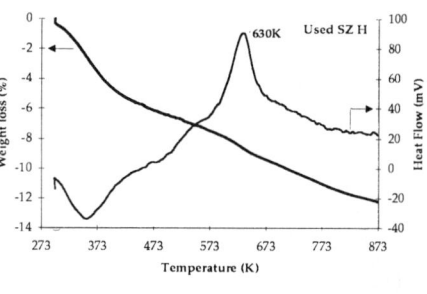

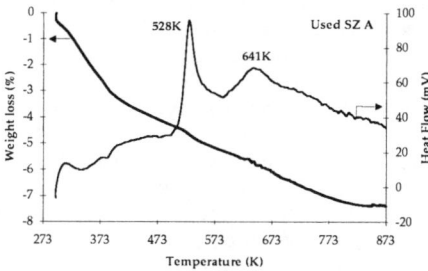

Figure 3 *TGA-DTA for used SZ A and used SZ H samples*

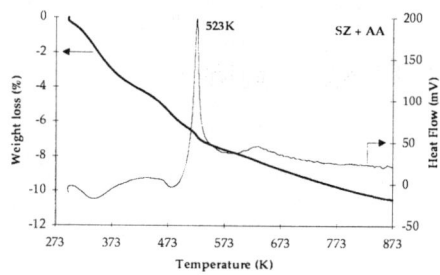

Figure 4 *TGA-DTA for SA+AA sample*

On the other hand, thermal analysis on SZ+AA was different, the third weight loss coincided with a strong and narrow exothermal peak at 523 K and the total weight loss was about 10 % indicating the presence of carbonaceous residues in higher amount (figure 4).

For used SZ A, the thermal curves, described previously, presented this narrow exothermic peak at 528 K, characteristic of the adsorption of acrylic acid in addition to a large exothermic phenomenon at higher temperature, which can be rather ascribed to the strong adsorption of sec-butyl acrylate and/or but-1-ene without possible distinction.

Upon but-1-ene adsorption on SZ A sample, a band appeared at 295 nm in the UV-vis spectrum (figure 5), besides the band at 230 nm attributed to the $Zr^{4+} \leftarrow O^{2-}$ charge transfer. This band has already been observed upon olefin adsorption on acidic catalysts[13-16] and its assignment to monoalkenyl cations has been well established [15,17]. The IR spectrum agreed well with this attribution. (In figure 6A, the spectra of

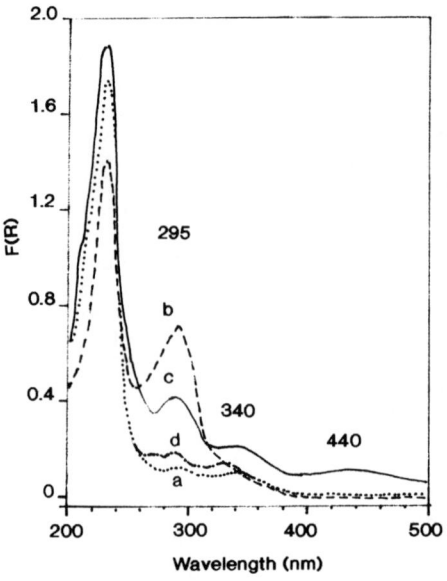

Figure 5 *UV spectra at room temperature of a) SZ+sec.BA, b) SZ+BU, c) SZ+AA samples and d) SZ A sample after one esterification cycle*

the adsorbed species are reported and in figure 6B, those corresponding to the subtraction of SZ A sample which eliminates the contribution of δH_2O at 1620 cm^{-1}.) Only vCH and δCH vibrations appeared at 2960, 2870, 1385 cm^{-1} (CH_3 groups) 2935, 2850, 1460 cm^{-1}, (CH_2 groups) (spectrum not shown in the vCH range). The absence of the $vC=C$ vibration, the red shift of the vCH vibrations and the blue shift of the δCH vibrations as compared to but-1-ene vibrations account for electron delocalization over three or more carbon atoms. The mechanism of formation of such species, which are precursors of polyenes, may be described as follows :

$$CH_2=CH-CH_2-CH_3 + H^+ \rightarrow CH_3\text{-}^+CH\text{-}CH_2\text{-}CH_3 \rightarrow (CH_3)_3\text{-}C^+$$

$$(CH_3)_3\text{-}C^+ + 2\ CH_2=CH\text{-}CH_2\text{-}CH_3 \rightarrow (CH_3)_3\text{-}C\text{-}CH_2\text{-}CH=CH\text{-}^+CH_2$$
$$+ CH_3\text{-}CH_2\text{-}CH_2\text{-}CH_3$$

Upon sec-butylacrylate adsorption, both IR and UV spectra were of low intensity. Curiously no band due to $vC=O$ vibration was observed. By gas chromatography an important conversion of sec.butylacrylate in acrylic acid and C_8H_{16} was observed. So it may be concluded that the sec-butyl acrylate was decomposed on the sulfated zirconia and that the olefinic fragments underwent the condensation as described above.

Upon adsorption of acrylic acid, UV-vis spectrum became more complex, three bands were observed at 290, 340 and 440 nm. In the IR spectrum, bands appeared at 1720, 1568, 1450, 1415, 1380 cm^{-1} and at 1325 and 1308 cm^{-1} with very low intensities (figure 6 A). In the vCH range, bands were observed at 2953 and 2850 cm^{-1} (CH_2 groups) and at 2890 cm^{-1} (CH groups). The relative higher intensity of these two spectra suggests that the amount of adsorbed species was higher than for adsorbed but-1-ene. The intense band at 1720 cm^{-1} in the IR spectra accounts for the presence of $vC=O$ vibration of carboxylic

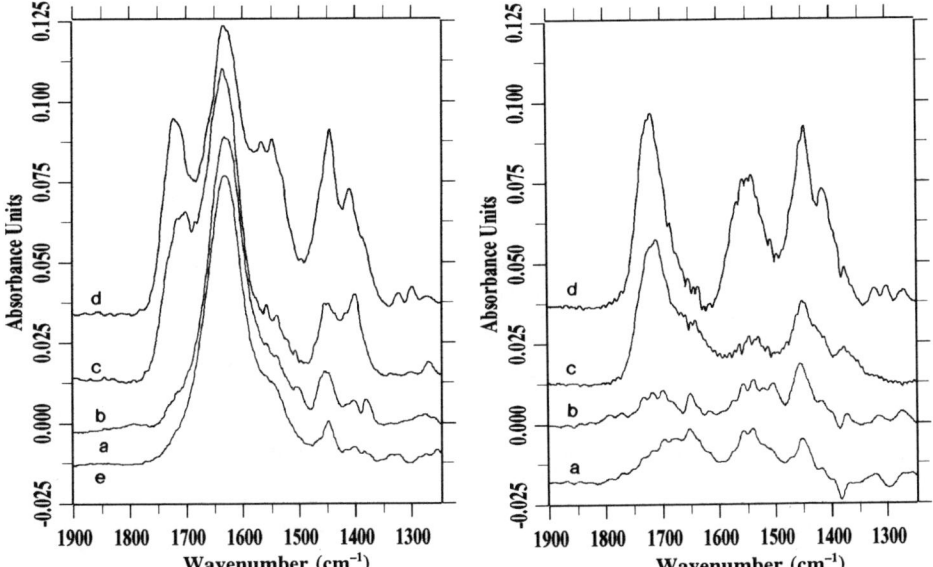

Figure 6 A *IR spectra at room temperature a) SZ+sec.BA, b) SZ+BU, c) SZ+AA samples, d) SZ A sample after one cycle and e) SZ A sample before adsorption*

Figure 6 B *IR spectra obtained by subtraction of the SZ A spectrum before adsorption a) SZ+sec.BA, b) SZ+BU, c) SZ+AA samples and d) SZA sample after one cycle*

groups while the band near 1568 and 1450 cm^{-1} may be due to asymmetric and symmetric vibrations of a carboxylate species. Moreover it is to be noted the presence of bands at 1415 and 1325 cm^{-1} attributed to δCH vibrations of vinyl groups. The presence of intense bands in the UV spectra indicates the presence of conjugated double bonds and may not be attributed to acrylic acid (200 nm) or to the acylium cation $CH_2=CH-^+CO$ (235 nm) or to a polymer of the acrylic acid (no conjugated double bonds). So we propose a more condensed species resulting from the addition of the acylium cation to a second acrylic molecule. Such species may be precursors of unsaturated polymers.

$$CH_2=CH-COOH + H^+ \rightarrow CH_2=CH-^+CO + H_2O$$

$$CH_2=CH-^+CO + CH_2=CH-COOH \rightarrow \ ^+CH_2-CH-COOH$$
$$| $$
$$C- CH=CH_2$$
$$\| $$
$$O$$

After the esterification reaction the IR spectrum looked like that of SZ+AA sample except the bands due to carboxylate species while the UV spectrum resembled that of SZ+sec.BU which was shown above to correspond to both AA and BU adsorbed. Thus, it may be concluded that the deactivation was due to accumulation of unsaturated species issued primarily from the acrylic acid.

3.2.5 Regeneration . Attempts were performed in order to regenerate the SZ catalysts through calcination under air flow at 773 K for 2h. Thermal analysis of used samples, previously described, showed that at this temperature at least part of the poisoning compounds were released; this temperature being low enough in order to avoid the sulfate decomposition. In the case of SZ H catalysts, such regeneration process restored almost the initial activity while on SZ A samples only part of the initial activity was restored (table 7). It is important to recall that SZ A catalyst presented a structural modification during the esterification reaction in liquid phase unlike SZ H sample. It may be the reason why SZ H solid can be regenerated through a calcination unlike SZ A solids. The structural change of SZ A sample must be irreversible.

4 CONCLUSION

Some general conclusions may be drawn from the present work. Firstly the two different preparation procedures, H_2SO_4 solution impregnation and $Zr(OH)_4 + (NH_4)_2SO_4$ mixing and grinding, resulted in active catalysts for liquid phase acrylic acid esterification with but-1-ene at 343 K. However the physico-chemical charateristics and the deactivation process of both types of samples were different, the impregnated sample being less deactivated under reaction conditions and more reproducible.

The strong adsorption of acrylic acid and but-1-ene was shown to be the main reason of the deactivation for SZ H samples prepared with H_2SO_4 impregnation while for SZ A samples prepared by grinding its structural unstability was an additional reason for its fast deactivation. The nature of the adsorbed species as reactants (acrylic acid, but-1-ene) or product (ester) has been studied using TG-DTA and IR, UV spectroscopies and were tentatively assigned to unsaturated cations, precursors of polymers.

Acknowledgement:The zirconium hydroxide was kindly prepared by Mr A.Levray from Elf-Atochem. The authors would like thank Mrs M.Roullet for her help in establishing chromatographic analyses method.

References

1. E. Paumard, Elf Atochem, French patent FR 007 368, 1990.
2. P. Dupont, J.C. Védrine, E. Paumard, G. Hecquet and F. Lefebvre, Appl. Catal., 1995, **129**, 217.
3. M. Hino and A. Arata, Appl. Catal., 1985, **18**, 401.
4. T.S. Thorat, V.M. Yadav and G.D. Yadav, Appl. Catal., 1992, **90**, 73.
5. CEPSA, Elf Atochem, European patent, EP 445 859, 1991
6. A. Riondel, Elf Atochem, European patent, EP 677,506, 1994.
7. F.R. Chen, G. Coudurier, J-F. Joly and J.C. Védrine, J. Catal., 1993, **143**, 616.
8. P. Nascimento, C. Akratapoulou, M. Oszagyan, G. Coudurier, C. Travers, J.F. Joly and J.C. Védrine, "New Frontiers in Catalysis", Proc. 10th International Congress on Catalysis, L. Guczi et al (ed), Elsevier, Amsterdam, 1993, p.1185.
9. A. Patel, G. Coudurier, N. Essayem and J.C. Védrine, J. Chem. Soc, Faraday Trans., 1997, **93**, 347.
10. P.D.L. Mercera, J.G. Van Ommen, E.B.M.Doesburg, A.J.Burggraaf and J.R.H. Ross, Appl. Catal., 1990, **57**, 127.
11. R. Srinivasan, R.A. Keogh, D.R. Milburn and B.H. Davis, J. Catal., 1995, **153**, 123.
12. C. Li and P.C. Stair, Catal. Lett., 1996, **36**, 119.
13. H.P. Leftin, in "Carbonium ions", G.Olah and P.R.Schleyer (ed), Interscience, New York, 1978, vol.1.
14. E.D. Garbowski and H. Praliaud, J.Chim.Phys.,1979, **76**, 687.
15. G. Coudurier, T. Decamp and H. Praliaud, J. Chem. Soc., Faraday Trans. I, 1982, **78**, 2661.
16. D.Spielbauer, G.A.H.Mekhemer, E.Bosch and H.Knözinger, Catal.Lett., 1996, **36**, 59.
17. T.S.Sorensen, J.Am.Chem.Soc., 1965, **87**, 5075.

SCOPE AND LIMITATIONS OF HETEROGENEOUS ENANTIOSELECTIVE CATALYSTS

Hans-Ulrich Blaser and Benoit Pugin

Catalysis & Synthesis Services, Novartis Services AG, R 1055.6
CH-4002 Basel, Switzerland.

1 SUMMARY

Scope and limitations as well as the state of the art for the preparation and synthetic application of the following types of enantioselective heterogeneous catalysts are described: (Supported) metals and metal oxides modified by chiral compounds; chiral organic polymers; and chiral metal complexes immobilized by covalent attachment or by adsorption on organic polymers or metal oxide surfaces. Emphasis is placed on the description of the most useful synthetic applications, of important system parameters and of the problems associated with the preparation and application of these catalysts. At the present time, only two types of modified metallic catalysts, two types of polymers and a few immobilized metal complexes have been developed to a point where synthetic applications on a technical level can be considered to be feasible.

2 INTRODUCTION

There is no doubt that the trends towards the application of single enantiomers of chiral compounds is increasing. This is especially the case for the pharmaceutical industry but is also true for agrochemicals, and for the preparation of flavors and fragrances.[1] Among the various methods to selectively produce one single enantiomer of a chiral compound, enantioselective catalysis is arguably the most attractive method. With a minute quantity of a (usually expensive) chiral auxiliary, large amounts of the desired product can be produced. At this time, homogeneous metal complexes with chiral ligands are the most widely used and versatile enantioselective catalysts.[2,3] From an industrial point of view, heterogeneous catalysts are of interest for a number of reasons, and a comparison of the advantages and problems of homogeneous and heterogeneous catalysts is made in Table 1. Several recent reviews give a digest of the most relevant aspects of heterogeneous enantioselective catalysts.[4-7] In this short overview we will give a summary of the state of the art for the various types of chiral heterogeneous catalysts based on metals, metal oxides and organic polymers as well as for immobilized chiral metal complexes. Special emphasis is placed on the problems a synthetic chemist might have with their application for the preparation of organic molecules and on interesting new developments reported in the last few years.

Table 1 *Strong and weak points of homogeneous and heterogeneous catalysts*

	Homogeneous	*Heterogeneous*
Strong points	defined on molecular level scope, variability (design?) availability	separation, recovery handling stability, re-use
Weak points	sensitivity activity, productivity (separation)	characterization reproducibility availability, preparation narrow scope

3 CHIRALLY-MODIFIED METALLIC HYDROGENATION CATALYSTS

Investigation of this class of heterogeneous chiral catalysts started in the late fifties in Japan and has seen a renaissance in the last few years. In spite of many efforts, only two classes of modified catalyst systems have been found that are of synthetic use at this time. However, efforts continue at several laboratories to expand the scope of this interesting and potentially very versatile class of chiral catalysts.

3.1 Tartrate-modified Nickel Catalysts

The development and successful applications of the nickel-tartrate system have been reviewed by Tai and Harada[6]. The preferred catalyst is freshly prepared Raney nickel; the only useful modifier is tartaric acid with bromide as co-modifier. The impregnation procedure (tartaric acid and NaBr concentrations, pH, T, time, ultra sound[8]) as well as the reaction conditions (solvent, p, T, acid co-modifiers) are crucial for getting good results. In general, the catalysts have relatively low activity and high pressures and temperatures are usually necessary.[6] The Ni/SiO$_2$ - tartrate system has recently been investigated in great detail (effect of impregnation and reaction parameters, catalyst characterization) and it looks as if modification by tartaric acid increases the specific activity by a factor of two.[9] Careful optimization of the impregnation and reaction conditions have improved the enantioselectivity of the Raney nickel system for β-keto esters[10] as well as for 2- and 3-alkanones[11,12] (see Figure 1).

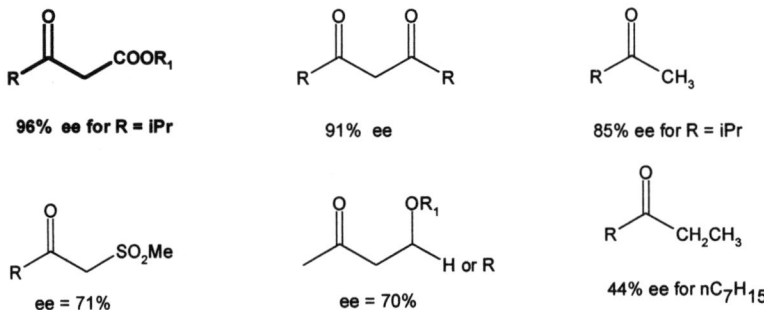

Figure 1 *Best optical yields reported for various ketones using Ni/tartrate/Br catalysts*

Synthetic applications of the Ni-tartrate catalyst have been described for sex pheromones of the pine sawfly;[13,14] for biologically active C_{10} - C_{16}-3-hydroxyacids;[15] for a diphosphine ligand;[6,16] and for an intermediate of tetrahydrolipstatin, a pancreatic lipase inhibitor (ee 90-92%, 6-100 kg scale).[17] There is no doubt that this well studied catalyst has the potential to be useful for producing several types of chiral alcohols.

3.2 Cinchona modified Pt and Pd and Related Catalysts[18]

The cinchona-modified platinum catalysts are at the moment among the most selective catalytic systems known for the hydrogenation of α-keto acid derivatives and for other activated keto groups, whereas some Pd - cinchonidine catalysts give reasonable ee s for α,β-unsaturated acids (see Figure 2). The best catalysts are 5% Pt/Al₂O₃ modified with cinchona derivatives. Carriers like SiO_2 or $BaCO_3$ and carbon supports, and also some new types of modifiers[19] give good optical yields as well (see Figure 2). The catalyst has to be pretreated in hydrogen at 300-400 °C before the reaction. The modifier can be directly added to the reaction solution but more elaborate modification procedures have also been reported.[5] The reaction conditions (solvent, modifier concentration, p, T) have a strong effect on rate and optical yield. The modified catalyst is 10-100 times more active than the unmodified Pt/ Al₂O₃ (modifier accelerated catalysis).[20]

The hydrogenation of ethyl 2-oxo-4-phenylbutyrate, an intermediate for the ACE inhibitor benazepril, has been developed and scaled-up into a production process (10-200 kg scale, chemical yield >98%, ee 79-82%).[21]

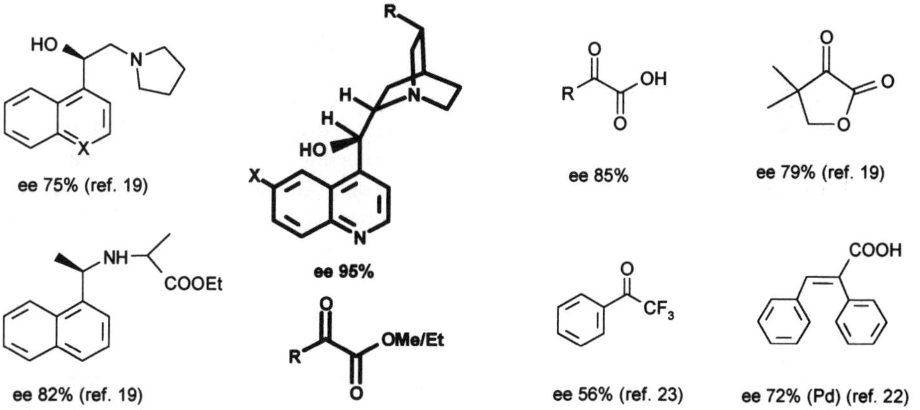

Figure 2 *Most selective modifiers and best optical yields for various types of substrates in Pt (Pd) catalyzed hydrogenation reactions*

3.3 Miscellaneous Enantioselective Hydrogenation Reactions

Anionic Pt and Ru carbonyl clusters anchored on a polystyrene functionalized with cinchonidine and ephedrine were able to hydrogenate methyl pyruvate with ee s up to 80%.[24] Rh colloids stabilized and modified by chiral amines gave low but significant ee s of 3-6% for the hydrogenation of o-substituted toluenes.[25] New results were reported for the hydrogenation of pyruvic acid oxime (Pd - ephedrine, ee 26%),[26] for an α,β-unsaturated ketone (Pd - ephedrine, ee 36%),[27] and for a pyrazine derivative (Pd - camphor-10-sulfonic acid, ee 50%).[28] However, in all cases 0.5 to more than 1 equivalents of modifier were necessary to give good results suggesting that 1:1 modifier-substrate adducts were the reactive species.

4 MODIFIED METAL OXIDES

Solid acids and bases are increasingly used for the catalytic synthesis of fine chemicals. However, chirally-modified versions have not yet been developed to the point where their synthetic application seems feasible. Therefore only a cursory overview is given. A zeolite β, partially enriched in polymorph A, was reported to give a low but significant ee of 5% for the ring opening of trans-stilbine epoxide.[29] Titanium-pillared montmorillonite (Ti-PILC) modified with tartrates were described as heterogeneous Sharpless epoxidation catalysts[30] as well as for the oxidation of aromatic sulfides[31]. At the moment it is not clear whether the preparation of the catalysts is reproducible. Metal oxides modified with histamine showed modest efficiencies for the kinetic resolution of activated amino acid esters ($k_R/k_S \approx 2$).[32] Silica or alumina treated with diethyl aluminium chloride and menthol catalyzed the Diels-Alder reaction between cylopentadiene and methacrolein with modest enantioselectivities of up to 31% ee.[33] Zeolite HY, modified with chiral sulfoxides had remarkable selectivities for the kinetic resolution of 2-butanol ($k_S/k_R = 39$) but unfortunately the catalyst is not very stable.[34] Clearly, this class of chiral catalysts, although of potential interest because of its variability, is not ready for synthetic applications.

5 CHIRAL POLYMERS AS CATALYSTS OR SUPPORTS

Because nature has provided us with so many chiral polymeric materials, their use for enantioselective synthesis is obvious. Indeed, some of the first successful attempts at enantioselective catalysis were carried out using amino cellulose (Bredig, 1932), metals on quartz (Schwab, 1932) and Pd supported on silk fibroin (Akabori, 1956).[4] However, the catalysts were very difficult to reproduce and other strategies were pursued more successfully. In the meantime, impressive progress has been made for the catalytic application of polypeptides and of dipeptide gels.

5.1 Polypeptides as Epoxidation Catalysts

Synthetic poly(amino acid) derivatives are highly selective catalysts for the asymmetric epoxidation of electron deficient olefins with NaOH/H₂O₂ in a two-phase reaction system.[35] Parameters of importance for the catalytic performance are the type of amino acid, the degree of polymerization, the substituent at the terminal amino group, pre-treatments and the organic solvent. From these and other observations it was inferred that supramolecular interactions inside the polymer aggregate might be important for catalysis and stereocontrol. Poly(amino acids) grafted onto 2% cross-linked polystyrene catalyzed the epoxidation of several substituted chalcones with optical yields up to 99%; re-use was possible.[36] Poly-L-alanine was used to prepare chiral intermediates for a leukotriene receptor antagonist with very high optical yields.[37] Recently, the scope of the reaction was somewhat broadened (see Figure 3) and extended to other oxidants, bases and solvents.[38] In general, the catalysts seem relatively easy to prepare, the reactions are reproducible but the catalytic activities are very low.

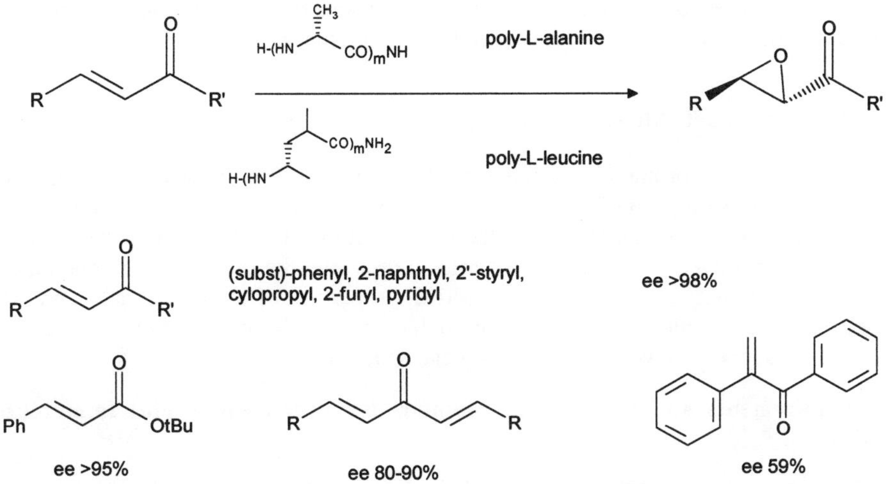

Figure 3 *General reaction scheme and optical yields for the epoxidation of various activated ketones using poly peptide catalysts[38]*

5.2 Cyclic Dipeptides as Catalysts for Cyanohydrin Formation

Cyclic dipeptides, especially cyclo[(S)-phenylalanyl-(S)-histidyl], are efficient catalysts for the hydrocyanation of aromatic aldehydes (see Figure 4) if the reaction is carried out in a "clear gel", in the heterogeneous state (best ee 92%).[39] Even higher optical yields were claimed for a catalyst adsorbed on a non-ionic polymer resin (ee >98%).[40] Application to various aromatic aldehydes is possible but the pattern and the nature of substitution is important.[45] A recent study confirmed the importance of the aggregate formation and reported a second order rate dependence on the concentration of the cyclic dipeptide.[42]

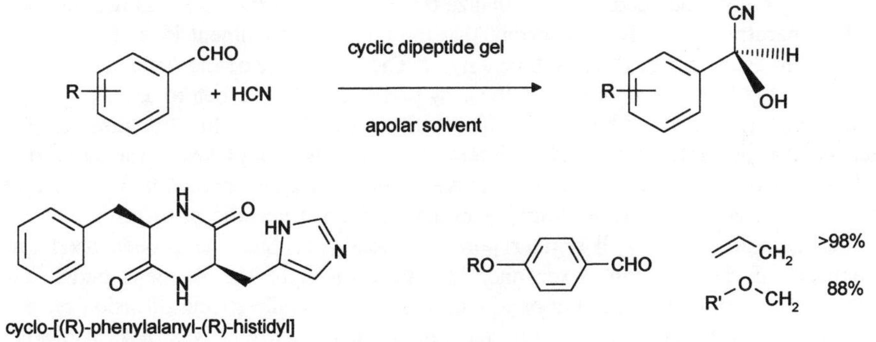

cyclo-[(R)-phenylalanyl-(R)-histidyl]

Figure 4 *General reaction scheme and some recent results[41] for the dipeptide catalyzed hydrocyanin formation*

These findings indicate that there is a higher order in the gel state and maybe also when the dipeptide is adsorbed on the resin. It is therefore plausible that the catalyst preparation and purification methods strongly affect the catalytic performance.

6 IMMOBILIZED METAL COMPLEXES

The heterogenization or immobilization of active metal complexes on insoluble supports or carriers seems to be a promising strategy to combine the best properties of homogeneous and heterogeneous catalysts. However, the results published in the last two decades have made it clear that even though the idea works in principle, the reality is more complex. In effect, one can not simply add the positive properties of two materials and as a consequence, the method of immobilization has to be adapted carefully to the metal complex, the support as well as the reaction that should be catalyzed.[7]

6.1 Requirements and Preparation Methods for Practically Useful Immobilized Catalysts[7]

Several requirements should be met in order to make immobilized enantioselective catalysts practically useful: the *preparation methods* should be generally applicable, simple and efficient; the *catalytic performance* should be comparable or better than that of the corresponding free catalysts; *separation* should be achieved by a simple filtration, and at least 95% of the catalyst should be recovered; *metal leaching* should be minimal and re-use of the catalyst should be possible; the *supports* must be chemically and mechanically stable, solvent compatible and commercially available in reproducible quality (texture, purity etc.).

Different immobilization approaches are possible and most have their advantages and disadvantages. *Heterogenization by covalent attachment* of the metal complex via the ligand and a suitable bifunctional linker or tether is by far the most important and versatile strategy. Both organic polymers and inorganic solids are suitable but support properties (e.g. chemical composition, degree of crosslinking, porosity etc.) can strongly influence the catalytic performance and separation of the catalyst. The complexes can be attached to the support by *grafting* or by *co-polymerization* of ligands functionalized with appropriate functional groups. The need to functionalize the ligands is a drawback and usually requires a high preparative effort. It is important that the point of attachment is far from the chiral active site in order not to disturb the catalysis. The advantage of the *heterogenization via adsorption and ion pair formation* is the easy preparation of the heterogenized catalyst by a simple adsorption procedure very often without the need to functionalize the ligand. Since electrostatic forces or lipophilic interactions bind the complexes to the support these catalysts are usually quite sensitive to solvent effects. *Heterogenization via entrapment* relies on the size of the metal complex rather than on a specific adsorptive interaction. Various materials with small regular pores, especially zeolites, have been used for the entrapment of metallic complexes and this approach is at the moment under intense investigation. A drawback of entrapped catalysts is the usually strong diffusion resistance. Recently, the successful use of *supported liquid-phase catalysts* was described where a solution of the metal complex is placed on the surface of a porous support. The results were similar to those obtained with two-phase systems and similar solvent restrictions were observed.

6.2 Synthetically Useful Immobilized Complexes

In Table 2 we give results for selected catalyst systems that we consider to be useful for synthetic purposes, i.e., they fulfill most of the requirements defined above. Our selection is by no means comprehensive and the inclusion of a catalyst does unfortunately not mean that it can be applied routinely like any organic reagent. Especially the preparation and characterization of some of these solid organic or inorganic materials will be difficult for many organic chemists and none of the catalysts are at this time available commercially. The given ee values are the best reported in the literature, turnover frequencies (tof) are average values for full conversion.

Table 2 *Immobilized catalysts with outstanding catalytic performance (max. ee and tof)*

Reaction, Type of substrate	Metal	Ligand	Attachment and Support	ee (%)	tof (h^{-1})	Ref.
Hydrogenation of enamides	Rh	pyrphos	grafted on SiO_2	100		43
		various	grafted on SiO_2	95	1500	44
		various	grafted on soluble polymers	95	2000	45
		various	grafted on crosslinked polymers	97	200	45
		glup	adsorbed on SiO_2/ Ph-SO_3H	95	≈500	46
Hydrogenation of N-phenyl-imines	Ir	ferrocenyl-diphosphines	grafted on SiO_2	79	10'000	47
			grafted on polymers	78	1300	47
Hydrogenation of C=C-COOH	Ru	binap-4-SO_3Na	supported liquid-phase	96	54	48
Dihydroxylation of alkenes	Os	(DHQD)$_2$PHAL	co-polymerized with acrylates	99	6	49
			grafted on SiO_2	99	6	50

6.3 Future Trends

There is no doubt that immobilized catalysts are much more complex than their homogeneous counterparts. The effect of important factors is not well understood and, as a consequence, the effect of immobilization is usually unpredictable. Nevertheless, the results summarized in Table 2 are proof that with enough effort and experience very selective and, as the immobilized Ir-ferrocenyldiphoshine catalysts demonstrate, also very active catalysts can be prepared. There is a strong current trend towards the development of bi-phasic catalysis, where the substrate and the catalyst are in different, non-miscible solvents. Of special interest are water-soluble catalysts[51], a technology that is used on a technical level, e.g. for the oxo process of Ruhrchemie / Rhône-Poulenc. The main advantage is that the catalysts remain homogeneous but there is a restricted choice of suitable pairs of solvents and mass transport between the two phases can be a problem.

References

1   For a discussion of the industrial problems see R.A. Sheldon, 'Chirotechnology', Marcel Decker, Inc., New York, 1993.

2   R. Noyori, 'Asymmetric Catalysis in Organic Synthesis', John Wiley & Sons,
     Inc., Chichester, 1994.
3   I. Ojima (Ed.), 'Catalytic Asymmetric Synthesis', VCH , Weinheim, 1993.
4   H.U. Blaser and M. Müller, *Stud. Surf. Sci. Catal.,* 1991, **59**,73.
5   G. Webb and P.B. Wells, *Catal. Today,* 1992, **12**, 319.
6   A. Tai and T. Harada, in Y. Iwasawa (Ed.), 'Tailored Metal Catalysts', D.
     Reidel, Dordrecht, 1986, p. 265.
7   A more detailed assessment of immobilization concepts is presented in H.U.
     Blaser and B. Pugin, in G. Jannes and V. Dubois, Eds., 'Chiral Reactions in
     Heterogeneous Catalysis', Plenum Press, New York, 1995, p. 33.
8   A. Tai, T. Kikukawa, T. Sugimura, Y. Inoue, S. Abe, T. Osawa and T.
     Harada, *J. Chem. Soc., Chem. Commun.,* 1991, 795.
9   M. Keane, *Langmuir,* 1997, **13**, 41 and references cited therein
10   T. Sugimura, T. Osawa, S. Nakagawa, T. Harada and A. Tai, *Stud. Surf. Sci.
     Catal.,* 1996, **101**, 132.
11   T. Osawa, O. Takayasu and I. Matsuura, Abstracts of the 15th International
     Conference on Catalysis, The Taniguchi Foundation, November 1996, Sanda,
     Hyogo, Japan.
12   T. Osawa, T. Harada, A. Tai, O. Takayasu and I. Matsuura, *Stud. Surf. Sci.
     Catal.,* 1997, **108**, 199.
13   A. Tai, M. Imaida, T. Oda and H. Watanabe, *Chem. Lett.,* 1978, 61.
14   A. Tai, N. Morimoto, M. Yoshikawa, K. Uehara, T. Sugimura and T.
     Kikukawa, *Agric. Biol. Chem,.* 1990, **54**, 1753.
15   M. Nakahata, M. Imaida, H. Ozaki, T. Harada and A. Tai, *Bull. Chem. Soc.
     Jpn.,* 1982, **55**, 2186.
16   J. Bakos, I. Toth and L. Marko, *J. Org. Chem,* 1981, **46**, 5427.
17   E. Broger, Hoffmann-LaRoche, Basel, personal communication.
18   For a comprehensive review see H.U. Blaser, H.P. Jalett, M. Müller and M.
     Studer, *Catalysis Today,* 1997, **37**, 441.
19   A. Baiker, *J. Mol. Catal. A: Chemical,* 1997, **118**, 473
20   H.U. Blaser, M. Garland and H.P. Jalett, *J. Catal.,* 1993, **144**, 139.
21   G.H. Sedelmeier, H.U. Blaser and H.P. Jalett; EP 206'993, 1986; assigned to
     Ciba-Geigy AG.
22   Y. Nitta and K. Kobiro, *Chem. Lett.,* 1996, 897.
23   T. Mallat, M. Bodmer and A. Baiker, *Catal. Lett.,* 1997, **44**, 95
24   S. Bhaduri, V.S. Darshane, K. Sharma and D. Mukesh, *J. Chem. Soc., Chem.
     Commun.,* 1992, 1738.
25   K. Nasar, F. Fache, M. Lemaire; J-C. Beziat, M. Besson and P. Gallezot, *J.
     Mol. Catal.,* 1994, **87**, 107.
26   K. Borszeky, T. Mallat, R. Aeschiman, W.B. Schweizer and A. Baiker, *J.
     Catal.,* 1996, **161**, 451
27   C. Thorey, F. Henin and J. Muzart, *Tetrahedron: Asymmetry* 1996, **7**, 975.
28   K. Kikegawa, JP 07291943-A, 1994, assigned to Koei Chem. Ind. Co Ltd.;
     Chem. Abstr., 1996, **124**, 176155
29   M.E. Davis, *Acc. Chem. Res.,* 1993, **26**, 111.
30   B.M. Choudary, V.L.K. Valli and A. Durga Prasad, *J. Chem. Soc., Chem.
     Commun.,* 1990, 1186.

31 B.M. Choudary, S. Shobha Rani and N. Narender, *Catal. Lett.*, 1993, **19**, 299.

32 T. Moriguchi, Y. G. Guo, S. Yamamoto, Y. Matsubara, M. Yoshihara and T. Maeshima, *Chem. Express,* 1992, **7**, 625.

33 J.M. Fraile, J.I. Garcia, J.A. Mayoral and A.J. Royo, *Tetrahedron: Asymmetry,* 1996, 7, 2263

34 S. Feast, M. Rafiq, H. Siddiqui, R.P.K. Wells, D.J. Willock, F. King, C.H. Rochester, D. Bethell, P.C. Bulman Page and G.J. Hutchings *J. Catal.*, 1997, **167,** 533 and references cited therein.

35 For a review see M. Aglietto, E. Chiellini, S. D'Antone, G. Ruggeri and R. Solaro, *Pure & Appl. Chem.*, 1988, **60**, 415.

36 S. Itsuno, M. Sakakura and K. Ito, *J. Org. Chem.,* 1990, **55**, 6047.

37 J.R. Flisak, K.J. Gombatz, M.H. Holmes, A.A. Jarmas, I. Lantos, W.L. Mendelson, V.J. Novack, J.J. Remich and L. Snyder, *J. Org. Chem.,* 1993, **58**, 6247.

38 M.E. Lasterra-Sanchez, U. Felfer, P. Mayon, S.M. Roberts, S.R. Thornton and C.J. Todd, *J. Chem. Soc., Perkin Trans. I*, 1996, 343; W. Kroutil, M.E. Lasterra-Sanchez, S.J. Maddrell, P. Mayon, P. Morgan, S.M. Roberts, S.R. Thornton, C.J. Todd and M. Tüter, *J. Chem. Soc., Perkin Trans. I,* 1996, 2837; P.A. Bentley, S. Bergeron, M.W. Cappi, D.E. Hibbs, M.B. Hursthouse, T.C. Nugent, R. Pulido, S.M. Roberts and L.E. Wu, *Chem. Commun.*, 1997, 739.

39 K. Tanaka, A. Mori and S. Inoue, *J. Org. Chem.*, 1990, **55**, 181; H. Danda, *Synlett,* 1991, 263.

41 H.J. Kim and W.R. Jackson, *Tetrahedron: Asymmetry*, 1994, **5**, 1541

42 Y. Shvo, M. Gal, Y. Becker and A. Elgavi, *Tetrahedron: Asymmetry*, 1996, 7, 91.

43 U. Nagel and E. Kinzel, *J. Chem. Soc., Chem. Commun.*, 1986, 1098.

44 B. Pugin and M. Müller, *Stud. Surf. Sci. and Catal.*, 1993, **78**, 107. B. Pugin, J. Mol Catal., 1996, **107**, 273.

45 B. Pugin, EP 728768 A2 (1996), assigned to Ciba-Geigy AG.

46 R. Selke and M. Capka, *J. Mol. Catal.,* 1990, **63**, 319.

47 B. Pugin, EP 729969 A1, 1996; WO 9632400 A1, 1996; WO 9702232 A1, 1997, assigned to Ciba-Geigy AG.

48 K.T. Wan and M.E. Davis, *J. Catal,.* 1995, **152**, 25.

49 C. E. Song, J.W. Yang, H-J. Ha and S. Lee, *Tetrahedron: Asymmetry*, 1997, **8**, 841.

50 C.E. Song, J.W. Yang and H-J. Ha, *Tetrahedron: Asymmetry*, 1996, **7**, 645.

51 H.N. Flach, I. Grassert, G. Oehme and M. Capka, *Colloid Polym. Sci.*, 1996, **274**, 261 and references cited therein.

SILICA-SUPPORTED 1,5,7-TRIAZABICYCLODECENE AS A CATALYST FOR MICHAEL ADDITIONS

Y.V. Subba Rao, D.E. De Vos and P.A. Jacobs*

Centre for Surface Science and Catalysis
Katholieke Universiteit Leuven
Kardinaal Mercierlaan 92, 3001 Heverlee, Belgium

1 INTRODUCTION

The Michael addition is a familiar methodology to create new C-C bonds, and is of tremendous use in synthetic organic chemistry.[1] Michael reactions proceed through intervention of a base, as deprotonation of the acidic donor substrate is required before its addition to the Michael receptor. Solid oxide base catalysts are an attractive alternative to stoichiometric base reagents, because of the ease in handling and the potential re-use. However, the real challenge is to adapt the properties of these bulk solid materials to specific reactions by well-chosen catalyst tailoring.[2] Such modifications can be achieved by exchange or impregnation with alkali metals or alkali metal ions,[3-5] via nitridation[6] or by immobilisation of organic molecules.[7]

Strong nitrogen bases such as guanidines are particularly attractive to immobilise. Both cyclic and acyclic guanidines are available, and the geometric constraints imposed by the carbon skeleton result in a varying steric hindrance at the protonation site, and in different pK_{BH^+} values.[8] Moreover, the anchoring of a single reactive species should ensure that the basicity of the solid falls in a well-controlled domain. Traditional immobilisation strategies start from aminopropyl or halopropyl functionalised silica,[7,9] and involve the formation of new covalent links via amide formation or via substitution at the haloalkyl sites. Particularly the latter technique is unsuitable for base catalyst preparation, as the liberated strong acids might be difficult to remove from the base sites. We have recently explored an alternative methodology, employing surface glycidylation.[10] For instance, the 1,4,7-triazacyclononane ligand is easily attached to a surface which is modified with glycidyloxypropyl groups.

In this contribution we study the heterogenisation of 1,5,7-triaza[4,4,0]bicyclodec-5-ene (TBD) on a glycidylated silica surface. The resulting heterogeneous strong base is used in a series of Michael additions. The effect of solvents on the base-catalysed reactions is studied, and the performance of the hybrid silica material is compared to that of a mixed oxide.

2 RESULTS AND DISCUSSION

The immobilisation route is summarised in the Scheme below. The surface silanol groups of the starting silica material are first reacted with the glycidyloxypropyltrimethoxysilane. We have previously demonstrated via CP-MAS ^{13}C NMR that the oxirane ring withstands the reaction with the silica surface.[11] Next the secondary amine function in TBD is reacted with the oxirane group. This results in formation of an immobilised tertiary amine:

1. *3-Glycidyloxypropyltrimethoxysilane, toluene, reflux*
2. *1,5,7-triazabicyclodec-5-ene, toluene, 60° C*

The course of these reactions was characterised *i.a.* with FTIR spectroscopy. The original silica material has a sharp band at 3745 cm^{-1}, due to surface silanol groups. This band disappears completely after glycidylation, and ν_{C-H} vibrations appear. In the spectroscopic window between 1600 and 1200 cm^{-1}, very weak vibrations with mostly δ_{C-H} character can be observed, *e.g.* at 1479, 1458, 1442, 1413 and 1342 cm^{-1}. The bands in this domain become much stronger after anchoring of TBD. For instance, the bands at 1526, 1446, 1380 and 1322 cm^{-1} for the silica-anchored TBD match well those of the pure TBD precursor. Via thermogravimetric analysis, an average loading of 0.3 mmol TBD per g of silica support was determined.

Inspection of Table 1 reveals that the silica-immobilised triazabicyclodecene (S-TBD) is a good catalyst for Michael additions under mild conditions. Depending on the reactivity of the donor-receptor couple, reaction temperatures as low as 313 K are sufficient to allow a close to quantitative reaction, for instance with acrolein and diethylmalonate (DEM). These conditions are much less severe than those reported for *e.g.* Cs^{+} exchanged MCM-41.[5] The reactivity order is primarily determined by electronic factors. Thus ethylcyanoacetate (ECA) gives higher yields than diethylmalonate. On the receptor side acrolein and methylvinylketone are most reactive, while more drastic conditions are required for reaction of the cyclic enones.

Different solvents can be applied, at the condition that they are sufficiently polar. In the reaction of cyclopentenone with ethylcyanoacetate, considerable activities were obtained in isopropanol, dioxane and acetonitrile (Figure 1). At contrast, no product was obtained at all in the apolar medium cyclohexane.

In general, the selectivities for the Michael reactions with the S-TBD catalyst are excellent. For several reactions, the regioselectivity of the nucleophilic attack was checked by ^{1}H NMR. It was established that the formation of 1,4 adducts is predominant, *e.g.* in the reaction of methylvinylketone and diethylmalonate (see Experimental). In the reaction of cyclopentenone or cyclohexenone with ethylcyanoacetate, a yellow (λ_{max}=432 nm) or red colour (λ_{max}=510 nm) developed in the initial stage of the reaction. This colour is ascribed to formation of a small amount of 1,2 Knoevenagel condensation compounds, which have an extended conjugated system. Figure 2 shows the GC-MS trace for the reaction of cyclohexenone and ethylcyanoacetate. Based on such data, the fraction of Knoevenagel products in the total of mono-addition compounds is maximally 10 %.

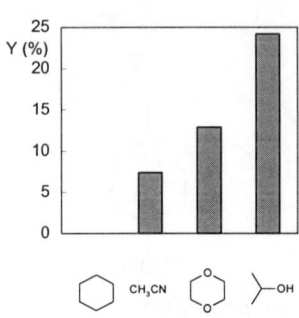

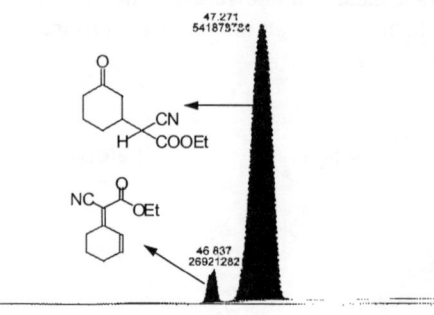

Figure 1 *Yields (Y, %) of mono-adduct in the reaction of c-pentenone with ethylcyanoacetate in different solvents (9 h, 353 K, 20 mg S-TBD)*

Figure 2 *GC-MS trace of the reaction of cyclohexenone and ethylcyanoacetate, with Knoevenagel and Michael products*

The first addition of the donor on the unsaturated carbonyl compound is in some cases followed by a second addition. This indicates that the catalyst can not only deprotonate the initial donors ethylcyanoacetate or diethylmalonate but also the mono-addition products:

Contributions from competing reaction types are small. When methylvinylketone is used as the receptor enone, a minor fraction of the enone dimerises. This condensation is essentially a thermal, non-catalysed reaction. Because of the strong base nature of the S-TBD catalyst, the reaction can be conducted at a moderate temperature (338 K), and the enone loss to the thermal reaction is small. Polymerisation of methylvinylketone or acrolein was not observed.

Table 1 *Michael Additions Catalysed by the Solid Bases S-TBD and Calcined LDH* [a]

Receptor	Donor [b]	T, t (K, h)	Main Product	Molar Yields [c] for mono (M) or di-adducts [d] (D) (%)	
				S-TBD	LDH
	DEM	353 K, 24 h		16 M	3 M
	ECA	353 K, 5 h		28 M	15 M
	DEM	353 K, 24 h		13 M	2.6 M
	ECA	353 K, 2 h		22 M	3.1 M
	DEM	338 K, 1 h		30 M	18 M
	ECA	338 K, 1 h		14 M, 71 D	25 M, 21 D
CHO	DEM	313 K, 1 h		66 M, 32 D	92 M, 5 D
	ECA	313 K, 1 h		61 M, 39 D	71 M, 24 D

[a] Conditions: 1 mmol receptor, 1 mmol donor, 20 mg catalyst, 2 ml CH_3CN, N_2 atmosphere. [b] DEM = diethylmalonate, ECA = ethylcyanoacetate. [c] Yields are based on the receptor. [d] In some cases (*e.g.* DEM + acrolein), the di-addition product underwent cyclisation and dehydration in an intramolecular aldol reaction.

Finally the activity of the S-TBD catalyst was compared to that of a mixed Mg-Al oxide, prepared by calcination of a layered double hydroxide (LDH).[12] Calcined LDHs are among the stronger highly dispersed base oxide catalysts, and are for instance much more active than AlPO-nitrides or alkali-exchanged materials.[13] Comparison between the last 2 columns in Table 1 shows that the activity of the LDH per catalyst weight is clearly lower than that of the S-TBD material. This is remarkable, as the number of catalytically active sites on S-TBD ($\cong$ 0.3 meq/g) is probably much lower than on the LDH. Another striking feature is the product distribution over mono- and di-addition compounds. The S-TBD catalyst systematically produces more di-addition than the LDH. This indicates that the base sites of the LDH material are not sufficiently strong to perform the second addition; with S-TBD, the deprotonation of the mono-addition product is much easier, illustrating the exceptionally high proton affinity of the base sites.

3 CONCLUSIONS

The glycidylation is an easy route to heterogenisation of the strong guanidine base TBD. Even if the number of active sites is relatively limited, Michael reactions can be performed in mild conditions. The strong base nature of the reactive sites becomes most apparent in the comparison with the mixed oxide catalyst prepared by calcination of an LDH. Not only are the yields higher with S-TBD; the changing selectivities (*e.g.*, more double addition with S-TBD) give supplementary evidence that the performance of the heterogeneous TBD catalyst is superior to that of many known oxide catalysts.

4 EXPERIMENTAL

Vacuum-dried commercial silica gel (Fluka, 70 to 235 mesh, 5 g) was suspended in dried and degassed toluene (100 ml) under nitrogen atmosphere. 0.95 g of 3-glycidyl-oxypropyltrimethoxysilane (4 mmol) was added and the mixture was refluxed for 24 h. The resulting modified silica was washed thoroughly with toluene, subjected to overnight soxhlet extraction to remove any adsorbed reagent, and stored under vacuum.

For the anchoring of the TBD, 1 g of the glycidylated silica was suspended in toluene, and 0.14 g of 1,5,7-triazabicyclo[4.4.0]dec-5-ene (TBD, Fluka, 1 mmol) was added. After 20 h stirring at 60 °C, the material was isolated and purified via similar extractions as in the previous step.

A mixed oxide catalyst was prepared by precipitation of a mixture of Mg and Al nitrate with Na_2CO_3 / NaOH at pH 10. The isolated layered double hydroxide (LDH) material was dried at 60°C and calcined at 450 °C.

The procedure for the Michael reactions was as follows: a 10 ml Schlenck tube was charged sequentially with 20 mg of catalyst, 1.0 mmol of donor, 2.0 ml of dry solvent and 1.0 mmol of receptor. Reactions were conducted under inert atmosphere, and the progress was monitored by GC. Quantification was based on the sensitivity factors as measured for the pure compounds. For non-isolated addition products, sensitivity factors were estimated by interpolation from those of receptor and donor.

Product identification was based on MS and ^{1}H-NMR spectroscopy. This allowed to distinguish between 1,4 and 1,2 addition. For instance, following NMR spectrum was obtained for the 1,4 mono-adduct of diethylmalonate and methylvinylketone: δ = 4.22 (q, 4H), 3.38 (t, 1H), 2.58 (t, 2H), 2.14 (s, 3H), 2.14 (m, 2H), 1.29 (t, 6H).

Acknowledgements

This work was supported by the Belgian Government in the frame of an I.U.A.P project. YVSR and DDV are grateful for post-doctoral fellowships to K.U.Leuven and F.W.O. (Belgium) respectively. We thank Bert Sels and Dirk Hoegaerts for experimental assistance.

References

1. J. March, 'Advanced Organic Chemistry: Reactions, Mechanisms and Structure', John Wiley & Sons, 1992.
2. H. Hattori, Chem. Rev., 1995, **95**, 537.
3. G. Suzukamo, M. Fukao and M. Minobe, Chem. Lett., 1987, 585.
4. L. R. M. Martens, P. J. Grobet and P. A. Jacobs, Nature, 1985, **315**, 568.
5. K. R. Kloetstra and H. van Bekkum, Chem. Commun., 1995, 1005.
6. M. J. Climent, A. Corma, V. Fornes, A. Frau, R. Guil-Lopez, S. Iborra and J. Primo, J. Catal., 1996, **163**, 392.
7. A. Cauvel, G. Renard and D. Brunel, J. Org. Chem., 1997, **62**, 749.
8. R. Schwesinger, J. Willaredt, H. Schlemper, M. Keller, D. Schmitt and H. Fritz, Chem. Ber., 1994, **127**, 2435.
9. D. J. Macquarrie, Chem. Commun., 1997, 601.
10. Y. V. Subba Rao, D.E. De Vos and P.A. Jacobs, Chem. Commun, 1997, 355.
11. Y. V. Subba Rao, D. De Vos and P.A. Jacobs, Stud. Surf. Sci. Catal., 1997 (in press).
12. W. T. Reichle, J. Catal., 1985, **94**, 547.
13. M. J. Climent, A. Corma, S. Iborra and J. Primo, J. Catal., 1995, **151**, 60.

TRANSITION METAL CATALYSTS AND IMMOBILISED BIOCATALYSTS IN ENANTIOSELECTIVE SYNTHESIS

A.R. Maguire* and L.L. Kelleher

Department of Chemistry,
University College Cork,
Cork,
Ireland

1 INTRODUCTION

Development of new methodology for asymmetric synthesis is currently an extremely active area of research, principally due to the recognition that the two enantiomers of a compound may display significantly different biological activity. A number of strategies for asymmetric synthesis are possible: *e.g.* use of the 'chiral pool', application of chiral auxiliaries or chiral reagents, but perhaps the most elegant approach involves the use of asymmetric catalysts. Two of the main strategies employed for catalytic enantioselective synthesis are the use of transition metal catalysts and the use of enzymatic systems as biocatalysts. These two approaches have usually been explored independently as the techniques involved differ substantially; however, there have been a few isolated reports of combination of the two complementary approaches in recent years.[1]

$$ R\diagup\diagdown\diagup CO_2H \longrightarrow \longrightarrow R\diagup\underset{X}{\overset{\gamma}{\diagdown}}\diagup CO_2H \qquad (1) $$

X = alkyl group *or* functionalised substituent

Despite significant advances in asymmetric synthesis, and especially in enantioselective catalysis, there remain many synthetic transformations for which efficient asymmetric methods are yet to be developed. For example, enantioselective functionalisation at positions α or β to a carboxylic acid moiety can be readily achieved by existing methods, the former by enantioselective enolate alkylation and the latter by asymmetric conjugate addition, among other methods. However, functionalisation at the γ–position in an enantioselective manner as illustrated in equation (1) is not readily achieved using existing synthetic methods. Our objective was to establish a general method for enantioselective functionalisation at the unactivated γ-position of carboxylic acids. Ideally, the new methodology should allow access to each of the two complementary enantiomeric series and should be applicable to a wide range of carboxylic acids. Furthermore, development of methodology which would allow introduction of either simple alkyl groups or functionalised groups for exploitation in subsequent transformations would be advantageous.

2 REGIOSPECIFIC ACTIVATION *VIA* TRANSITION METAL CATALYSIS

The first challenge was to regiospecifically activate an unactivated C-H bond at the γ-position. This was achieved under mild conditions by rhodium catalysed carbenoid C-H insertion[2] as outlined in Scheme 1. Transformation of the carboxylic acids **1**, under standard conditions, *via* esters to β-keto sulfones **2**, followed by diazo transfer[3] led to the α-diazo-β-keto sulfones **3** which are carbenoid precursors. Rhodium(II) acetate catalysed decomposition resulted in essentially quantitative C-H insertion to form the α-benzenesulfonylcyclopentanones **4**. The carbenoid insertion is complete in just a few hours in refluxing dichloromethane. While the crude product may contain a mixture of *cis* and *trans* isomers of **4**, equilibration to form exclusively the thermodynamically more stable *trans* isomer is readily achieved.

Scheme 1 R = Me, Et, Pr, Bu, C_5H_{11}, Ph, CH_2Ph

This step, which results in activation of an unactivated C-H bond at the γ-carbon, has many of the advantages typically associated with transition metal catalysed processes: mild reaction conditions *e.g.* neutral pH, highly selective transformation, complete regiocontrol, and readily allows access to a racemic series of cyclopentanones **4**.

3 KINETIC RESOLUTION

With the racemic cyclopentanone derivatives in hand, the next challenge was to develop a method to obtain these compounds in enantiomerically pure (or enriched) form. Application of biocatalysis leading to a kinetic resolution of the enantiomers of **4** was identified as an appropriate strategy. Biotransformation, *i.e.* the application of enzyme catalysed processes in organic synthesis, is now widely recognised as an extremely valuable tool in enantioselective synthesis.[4] Either isolated enzymes or whole cell systems can be employed as biocatalysts. The advantages of biocatalysis include the following:
*   reactions can be achieved under very mild conditions of temperature, pH and pressure
*   in some instances, transformations can be effected which are difficult using traditional organic reactions
*   biotransformations can be conducted in aqueous media or in organic solvents

• most importantly, enzyme catalysed processes are frequently extremely enantioselective and therefore lead to products with very high enantiopurity which can be exploited in asymmetric synthesis.

One of the microorganisms which has been widely investigated in organic synthesis is baker's yeast, *Saccharomyces cerevisiae*, largely due to its ready availability, ease of use and the wide range of substrates which it accepts.[5] While enantioselective reduction of β-keto esters with baker's yeast is an extremely successful transformation which has been widely exploited in asymmetric synthesis, reduction of β-keto sulfones with baker's yeast has been less successful.[6] Short chain derivatives were successfully reduced with yeast but extension of the alkyl chain resulted in dramatic decreases in both the efficiency of the yeast mediated reduction and the enantioselectivity of the process.

Despite this disappointing precedent, baker's yeast mediated reduction of the α-benzenesulfonylcyclopentanone derivatives **4** was attempted and, as illustrated in Scheme 2, resulted in extremely efficient kinetic resolution.[7] The 2R-enantiomer of each of the cyclopentanone derivatives **4** was selectively reduced by the enzymes present in the yeast, while the 2S-enantiomer was essentially left unchanged. The diastereoselectivity of the reduction was very good; the 1S,2R-diastereomers of the cyclopentanols **5** were formed as the principal product of the yeast reduction in each case.[8] A minor diastereomer of the cyclopentanol, assigned as 1S,2S-**6**, was detected in the crude product in some reactions. Thus the 2S-enantiomer of the cyclopentanones **4** can undergo baker's yeast mediated reduction less efficiently than the 2R-enantiomers. By variation of the conditions employed for the yeast reduction[9] (*e.g.* concentration, reaction time, yeast immobilisation, use of organic solvents, additives) the efficiency, enantio- and diastereoselectivity of the transformation were optimised for each of the cyclopentanone derivatives **4** so that it is possible to obtain samples of each of the cyclopentane derivatives 2S-**4** and 1S,2R-**5** in highly enantioenriched form (> 95% ee); formation of the minor diastereomer **6** can be avoided by appropriate choice of conditions.

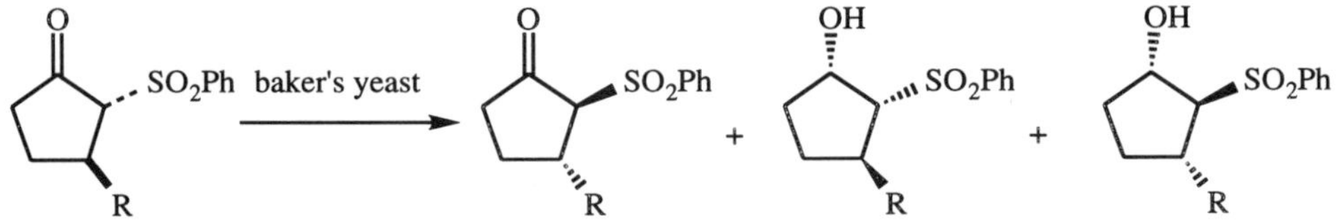

	2S,3R-**4**	1S,2R,3S-**5**	1S,2S,3R-**6**
e.g. R = Et	> 95 % ee	> 95 % ee	
	47 %	41 %	not detected

Scheme 2

Use of baker's yeast immobilised on alginate[9] as the biocatalyst was particularly convenient; the reduction could be conducted in hexane or in water, product isolation is much easier than with free yeast as the solution can be simply decanted from the immobilised yeast, and the enantioselectivity and diastereoselectivity of the reduction using the immobilised yeast were very satisfactory.

The relative stereochemistry of the cyclopentanol derivatives **5** was established by X-ray crystallography on the compound with R = Bu, and extended to the other derivatives by analogy.[7] The absolute stereochemistry of the cyclopentanones 2S-**4** (R = Me, Et, Ph) was determined by desulfonylation to the known 3-methyl / ethyl / phenylcyclopentanones respectively and comparison of specific rotation values with those recorded in the literature. The absolute stereochemistry of the cyclopentanols **5** was determined by Dess-Martin oxidation to form the complementary enantiomeric series of the cyclopentanones 2R-**4** followed by desulfonylation. The enantiomeric purity of each of the compounds isolated from the yeast mediated reductions and kinetic resolution was determined by ^{1}H NMR spectroscopy in the presence of Eu(hfc)$_3$ as chiral shift reagent. Chiral HPLC analysis[7] was also conducted to confirm the enantiomeric purity of some samples of **4** (R = Me, Pr) and **5** (R = Me, Pr) and chiral GC analysis[7] was used to confirm the enantiomeric purity of some samples of **4** (R = Et).

Evidently the efficient kinetic resolution allowed access to the cyclopentanones 2S-**4** and the cyclopentanols 1S,2R-**5** essentially as single enantiomers *i.e.* the minor enantiomer could not be detected by chiral shift NMR experiments. However, the complementary series of these cyclopentane derivatives could also be readily prepared; Dess-Martin oxidation of the cyclopentanols **5** recovered from the yeast reductions gave the 2R-**4** series, while sodium borohydride reduction of the recovered cyclopentanones **4** furnished the 1R,2S-**5** series.

4 BASE INDUCED CLEAVAGE OF β-KETO SULFONES

Treatment of β-keto sulfones with aqueous base leads to cleavage in a retro-Claisen type process.[10] Each of the two complementary enantiomeric series of α-benzenesulfonylcyclopentanones **4** was treated with aqueous sodium hydroxide resulting in efficient ring opening to form the corresponding carboxylic acid derivatives **7** as illustrated in Scheme 3. The enantiomeric purities of the carboxylic acids **7** isolated were determined by chiral HPLC analysis confirming that the stereochemical integrity at C-4 of the carboxylic acids **7** was, as expected, unaffected under these reaction conditions.

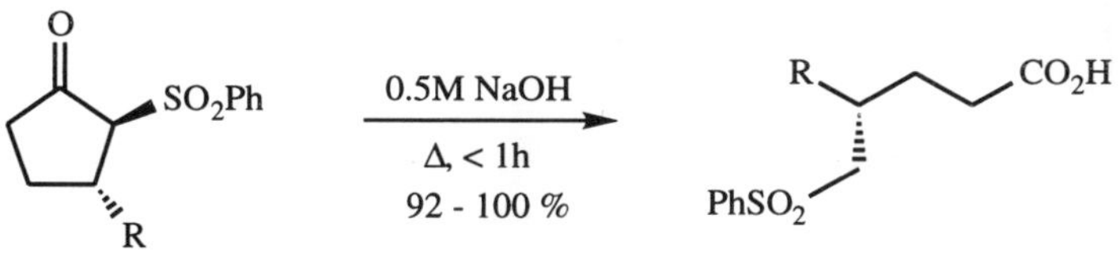

2S,3R-**4** R = Me, Et, Pr, Bu, C$_5$H$_{11}$ 4R-**7** R = Me, Et, Pr, Bu, C$_5$H$_{11}$

2S,3S-**4** R = Ph   4S-**7** R = Ph

The complementary enantiomeric series of **4** was similarly transformed to **7**.

Scheme 3

5 ENANTIOSELECTIVE FUNCTIONALISATION AT A REMOTE UNACTIVATED CARBON ATOM

The processes described above are summarised in Scheme 4. Starting from the carboxylic acid derivatives **1**, conversion to the racemic cyclopentanones **4** is achieved in four steps, the key step involving a rhodium catalysed C-H insertion process which regiospecifically activates an unactivated C-H bond under mild reaction conditions, and is followed by kinetic resolution employing baker's yeast as the biocatalyst. Ring cleavage under basic conditions produces efficiently the 4-(benzenesulfonylmethyl)carboxylic acids **7**. The overall transformation resulting from this reaction sequence is the enantioselective introduction of a benzenesulfonylmethyl substituent at the γ-position of the carboxylic acids **1**. Most significantly, access to the two complementary series is possible *via* this methodology. As the reaction conditions involved are quite mild, the sequence is envisaged to be widely applicable. The benzenesulfonylmethyl substituent is a versatile group which can either be reductively desulfonylated to leave the methyl substituted compounds, or alternatively the sulfone can be used for further synthetic transformations leading to a range of valuable chiral synthons.

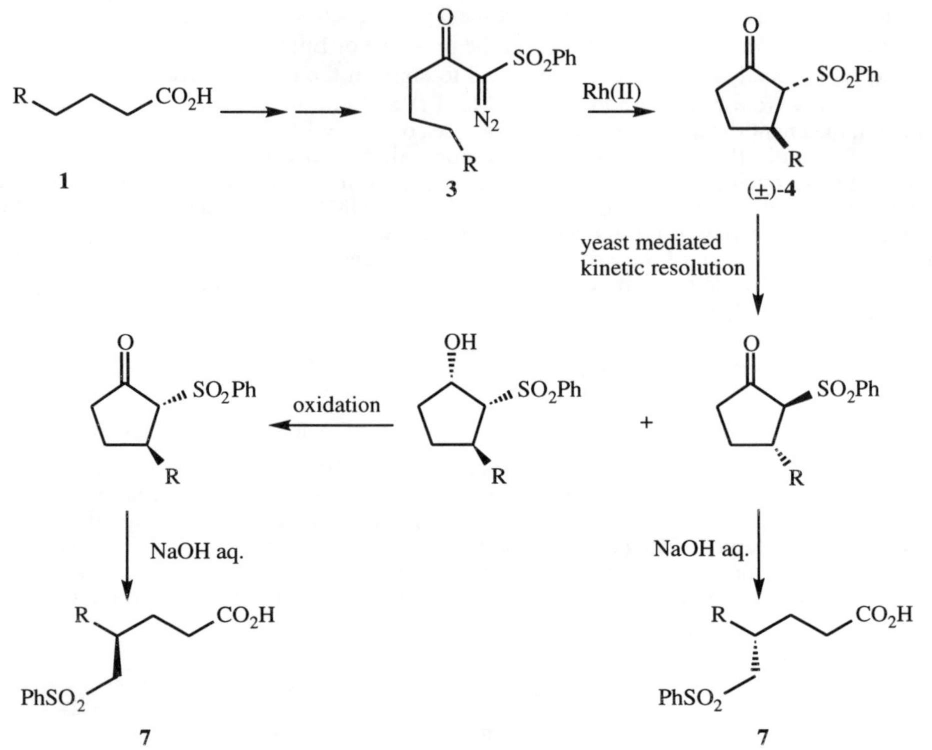

Scheme 4

6 ONE POT PROCESS

As the yeast reduction can be efficiently conducted using immobilised yeast in an organic solvent, the synthetic utility and efficiency of this sequence has been further enhanced by incorporating the key transformations into a one-pot process. While the rhodium catalysed C-H insertion is usually conducted in dichloromethane, this transformation can instead be effected in refluxing hexane; addition of immobilised baker's yeast directly to the cooled hexane solution results in efficient kinetic resolution to form the cyclopentanones 2S-4 and the cyclopentanols 1S,2R-5. Critically, the yeast mediated reduction is unaffected by the presence of the rhodium acetate catalyst from the C-H insertion step. The hexane solution is simply decanted from the immobilised yeast; exposure of this solution to aqueous base results in cleavage of the cyclopentanones 2S-4 to the enantiomerically enriched (typically > 95 %ee) carboxylic acids 7. The carboxylic acids 7 are recovered from the basic aqueous phase, while the cyclopentanols remain in the hexane layer. Thus, the transition metal catalysed process and the biotransformation can be conducted in a single pot, then a chemospecific transformation of one component, followed by phase separation of the two components results in a rapid, efficient, one pot enantioselective transformation of the α-diazo-β-keto sulfones 3 (derived from carboxylic acids 1 *via* three standard reaction steps) into the γ-substituted derivatives 7. To illustrate the synthetic utility of this process, α-diazo-β-keto sulfone 3 (R = Et) was subjected to the one pot sequence and, following neutralisation of the aqueous layer, 34 % yield of the carboxylic acid 4R-7 (R = Et) was isolated. As a kinetic resolution is involved, the maximum possible yield is 50 %; evidently the tandem sequence is remarkably efficient.

7 APPLICATION TO THE SYNTHESIS OF INSECT PHEROMONES

8 **9** **10**

The chirality of many insect pheromones is due to the presence of a branched methyl group on a hydrocarbon chain,[11] for example R-6-methyl-3-octanone **8**, the alarm pheromone of *Crematogaster* and *Myrmecine* ants, S-14-methyloctadecene **9**, the pheromone of the peach leafminer moth, and R-4-methylnonanol **10**, isolated from the yellow mealworm. The utility of our novel methodology for asymmetric synthesis is illustrated by application to the synthesis of compounds **9** and **10**, where the key step is the enantioselective carbon-carbon bond formation introducing the benzenesulfonylmethyl group to an achiral backbone, followed by reductive desulfonylation to reveal the unsubstituted methyl substituent.

Acknowledgements The authors wish to acknowledge support for this research from Forbairt, the Irish Science and Technology Agency, the Royal Irish Academy, and University College Cork, especially the President's Research Fund. X-ray crystallographic studies conducted by Prof. G. Ferguson (Guelph) are gratefully acknowleged. We also wish to thank Prof. I.E. Markó and V.T. Trieu (Louvain-la-Neuve) for chiral HPLC analysis, and Prof. W. König (Hamburg) for chiral GC analysis. An ACE award from Ciba (to ARM) and a loan of rhodium acetate from Johnson Matthey are gratefully acknowledged.

References and notes
1. J.V. Allen and J.M.J. Williams, *Tetrahedron Lett.*, 1996, **37**, 1859; P.M. Dinh, J.A. Howarth, A.R. Hudnott, J.M.J. Williams and W. Harris, *Tetrahedron Lett.*, 1996, **37**, 7623; J.-E. Backvall, *Acta Chem. Scand.*, 1996, **50**, 661.
2. Rhodium catalysed C-H insertion reactions of α-diazo-β-keto esters to form cyclopentanone derivatives were originally developed by Taber, D.F. Taber and E.H. Petty, *J. Org. Chem.*, 1982, **47**, 4808, and later extended to α-diazo-β-keto sulfones, H.J. Monteiro, *Tetrahedron Lett.*, 1987, **28**, 3459; M. Kennedy, M.A. McKervey, A.R. Maguire and G.H.P. Roos, *J. Chem. Soc., Chem. Commun.*, 1990, 361.
3. M. Regitz and G. Maas, *Diazo Compounds, Properties and Synthesis*, Academic Press, New York, 1986.
4. L. Poppe and L. Novak, *Selective Biocatalysis*, VCH, Weinheim, 1992; H.G. Davies, R.H. Green, D.R. Kelly and S.M. Roberts, *Biotransformations in Preparative Organic Chemistry*, Academic Press, London, 1989; H.L. Holland, *Organic Synthesis with Oxidative Enzymes*, VCH, New York , 1992.
5. For reviews see S. Servi, *Synthesis*, 1990, 1; R. Csuk and B.I. Glanzer, *Chem. Rev.*, 1991, **91**, 49; T. Sato and T. Fujisawa, *Biocatalysis*, 1990, **3**, 1.
6. R.L. Crumbie, B.S. Deol, J.E. Nemorin and D.D. Ridley, *Aust. J. Chem.*, 1978, **31**, 1965; K. Nakamura, K. Ushio, S. Oka, A. Ohno and S. Yasui, *Tetrahedron Lett.*, 1984, **25**, 3979; S. Robin, F. Huet, A. Fauve and H. Veschambre, *Tetrahedron: Asymmetry*, 1993, **4**, 239; R. Tanikaga, K. Hosoya and A. Kaji, *J. Chem. Soc., Perkin Trans. 1*, 1987, 1799; A.P. Kozikowski, B.B. Mugrage, C.S. Li and L. Felder, *Tetrahedron Lett.*, 1986, **27**, 4817; A. Svatos, Z. Hunkova, V. Kren, M. Hoskovec, D. Saman, I. Valterova, J. Vrkoc and B. Koutek, *Tetrahedron: Asymm.*, 1996, **7**, 1285; also references in the reviews in ref. 5.
7. A.R. Maguire, L.L. Kelleher and G. Ferguson, *J. Mol. Catalysis B: Enzymatic*, 1996, **1**, 115; A.R. Maguire and D.G. Lowney, *J. Chem. Soc., Perkin Trans. 1*, 1997, 235.
8. Note the stereochemical assignment of the stereocentre at C-3 in **4** and **5** depends on the substituent R as the priority of the groups at C-3 changes. With the alkyl substitued derivatives 2S,3R-**4** and 1S,2R,3S-**5** are isolated, whereas with R = phenyl or benzyl the stereochemical course of the reduction is the same producing 2S,3S-**4** and 1S,2R,3R-**5**.
9. K. Nakamura, *Microbial Reagents in Organic Synthesis*, ed. S. Servi, 1992, 389 - 398, Kluwer Academic Publishers, Dordrecht; K. Nakamura, Y. Kawai and A. Ohno, *Tetrahedron Lett.*, 1991, **32**, 2927; K. Nakamura, Y. Kawai, N. Nakajima and A. Ohno, *J. Org. Chem.*, 1991, **56**, 4778.
10. J. Ficini and G. Stork, *Bull. Soc. Chim. Fr.*, 1964, 723; H.J. Monteiro, *Synlett*, 1992, 990; R. Ballini, G. Bosica and T. Mecozzi, *Tetrahedron*, 1997, **53**, 7341.
11. K. Mori, *Tetrahedron*, 1989, **45**, 3233; S. Sankaranarayanan, A. Sharma, B.A. Kulkarni and S. Chattopadhyay, *J. Org. Chem.*, 1995, **60**, 4251; M. Kato and K. Mori, *Agric. Biol. Chem.*, 1985, **49**, 2479; T. Kitahara and S.-H. Kang, *Proc. Japan Acad., Ser. B*, 1994, **70**, 181.

THE REGIOCHEMISTRY AND STEREOCHEMISTRY OF THE [2+2] PHOTOCYCLOADDITION OF A 2-ALLYL-2(1H)-NAPHTHALENONE ON THE SURFACE OF SILICA. AN EXPERIMENTAL AND MOLECULAR MODELLING STUDY

N. W. A. Geraghty and M. J. Monaghan

Chemistry Department
University College Galway
Ireland

1 INTRODUCTION

The stereoselectivity of catalytic hydrogenation is considered to be due to the control exercised by the surface of the catalyst on the adsorption of the substrate and on the preferred direction of approach of the hydrogen atoms. The high levels of stereochemical control which are obtainable in this classic synthetic process would encourage the reasonable belief that one of the principal advantages of using a supported reagent should be the stereochemical and regiochemical control resulting from the involvement of the surface. The alkylation of ambient anions on alumina,[1] a variety of electrophilic aromatic substitutions on silica[2] and intermolecular [2+2] cycloaddition reactions in clays[3] are among the processes that have been considered from the regiochemical point of view. Stereochemically, the effect of surface adsorption on the *exo/endo* selectivity of Diels-Alder reactions[4] and on the face selectivity of intermolecular [2+2] cycloaddition reactions[5] has been considered. In general, however, the degree of control that has been achieved in this way is disappointing.

The study described here is concerned with correlating the changes in regiochemistry and stereochemistry that occur in a photochemical intramolecular [2+2] cycloaddition reaction due to the effect of surface adsorption, and in developing a simple molecular modelling approach by which the system can be explored. In general terms the use of a photochemical reaction and a "dry" support in this context is particularly attractive as there are no substrate – reagent interactions to be considered, the reagent (a photon) being delivered directly to the substrate which is unambiguously adsorbed on the surface.

1.1 Photochemical [2+2] Cycloaddition Reactions on Solid Supports

Until recently[6] it was considered that the regiochemistry of photochemical [2+2] cycloaddition reactions was determined by electronic interactions in an excited state complex (exciplex) formed between a ground state and an excited state alkene. The current belief is that the 1,4-biradical produced by the formation of a single bond between the two reacting alkene systems is the product–determining intermediate in most cases. This is particularly true for the intramolecular variant of the reaction for which exciplex formation is precluded by the geometric constraints inherent in the system. The lifetime of these 1,4-biradicals is of the order of 50ns[6] and thus bond rotation prior to ring closure is possible, leading to the formation of mixtures of stereoisomers. Such lifetimes would also allow conformational relaxation of the biradicals to occur prior to the bond formation which would complete the cyclobutane ring, or the bond cleavage, giving starting material, which is an important source of inefficiency in these reactions (Figure 1). Thus the regiochemistry of these reactions may depend on the relative stability of the possible biradicals, or on their geometry, with those which are product forming having a favourable distance and angular relationship between the two radical centres.

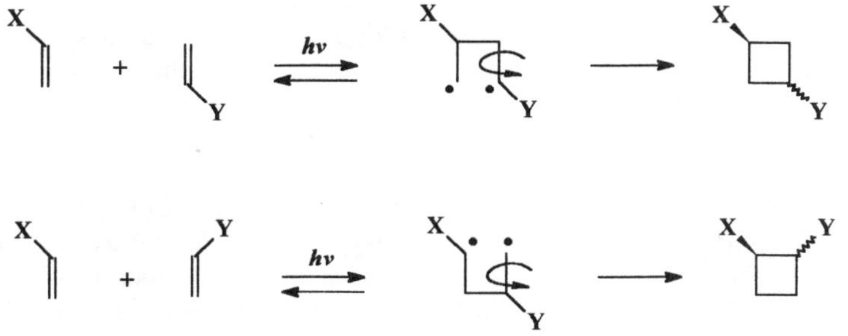

Figure 1 *Biradical involvement in [2+2] cycloaddition reactions*

As these reactions generally give a mixture of regioisomeric and stereoisomeric products, the composition of which is dependent on the structure of the intermediate biradicals, they constitute an excellent model system for assessing the effect of adsorption on the outcome of the reaction. The basic concept underpinning their use is the possibility that the surface conformation of the biradical intermediates would be different to that adopted in solution, thus resulting in a change in the regiochemical and stereochemical outcome of the reaction.

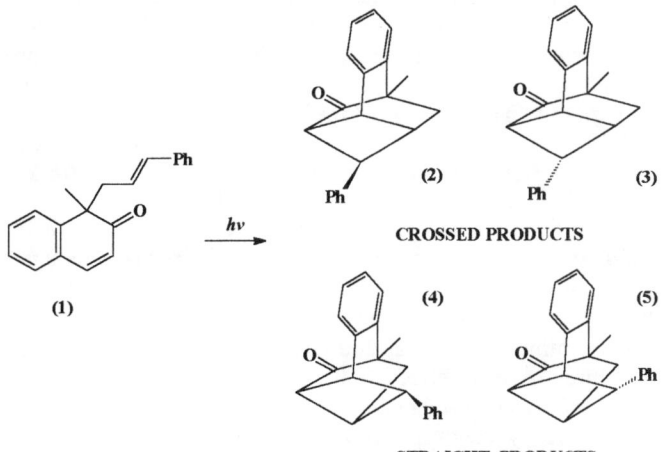

Figure 2 *Photochemistry of 1-cinnamyl-1-methyl-2(1H)-naphthalenone*

2 PHOTOCHEMISTRY OF 1-CINNAMYL-1-METHYL-2(1H)-NAPHTHALENONE

2.1 Solution Photochemistry

The irradiation of 1-cinnamyl-1-methyl-2(1H)-naphthalenone (1), through pyrex and using a medium pressure mercury lamps, results in the formation of all possible photoadducts – both stereoisomers of the two possible regioisomers (Figure 2)[7]. The structures of the adducts were assigned using 1-and 2-D-NMR, IR and, in the case of (5), X-ray crystallography. This naphthalenone system is thus a particularly suitable probe for the effect of surface adsorption on the reaction, facilitating an assessment of both stereochemical and regiochemical effects. The fact that the composition of the product mixture is essentially independent of the solvent used is also an attractive feature of the system, as any differences observed between solution and surface reactions can be attributed to factors other than a simple solvent effect.

The regiochemical outcome of the reaction is remarkable in that the major product is neither that favoured by the "rule of five" [8,9] nor is it that which involves the intermediacy of the dibenzyl biradical, the most stable of the four possible biradicals. However, a biradical involved in the formation of (2) (III, Figure 4) has a minimum energy conformation in which the radical centres have a favourable spatial relationship for bond formation and thus the outcome of the reaction may be under biradical conformation control.

2.2 Photochemistry on Silica

The enone (1) was adsorbed from a hexane solution onto the surface of chromatographic grade silica (Merck 70-230mesh, mean surface area, $550m^2g^{-1}$)[10] by the slow evaporation of the solvent. Sufficient enone was used to give approximately 10% surface coverage, assuming that the naphthalenone ring is adsorbed parallel to the surface. Activation of the silica involved heating for 2h under reduced pressure (1mmHg) and subsequent cooling in a dessicator; silica activated in this way was used immediately. After adsorption of the enone the silica was placed in a pyrex tube which was attached to the drive motor of a rotary evaporator and rotated in front of a medium pressure mercury lamp; both lamp and reaction vessel were mounted horizontally in a cylindrical container lined with aluminium foil. The progress of the reaction was monitored by HPLC.

2.3 Regiochemical and Stereochemical Analysis of Product Ratio Data

An analysis of the results obtained with unactivated silica and with silica which had been activated at a range of temperatures (Table 1), suggests that although there is only a small change in the stereoselectivity of the reaction, there is a substantial change in its regioselectivity, with the "straight" products (4) and (5) being formed in larger amounts on silica. Although activating the silica at 200°does slightly enhance this preference, higher activation temperatures have no effect. The effect of ultrasonication of the silica after adsorption of the enone was investigated in the belief that as adsorption from solution is in all probability a kinetically controlled process, a reorientation of the molecules on the surface might occur. However, no change in the product composition resulted.

Table 1 *Regiochemical and Stereochemical Analysis of Product Ratio Data*

Reaction Medium	Regiochemistry Crossed / Straight Product Ratio[a]; [(2)+(3)] / [(4)+(5)]	Stereochemistry Trans / Cis Product Ratio[a] [(2)+(4)] / [(3)+(5)]
C_6H_{12}	2.7	3.3
CF_3CH_2OH	2.1	3.3
CH_3OH	2.4	2.6
SiO_2	1.2	4.0
SiO_2 (200°)	1.0	3.8
SiO_2 (350°)	1.0	4.6
SiO_2 (350°, ultrasound)	1.0	4.9

[a]Determined by ^1H-NMR

3 MODELLING THE SURFACE REACTION

The change in the regiochemistry of the reaction can be interpreted in terms of a model that involves surface control of biradical formation (Figure 3). Thus the orientation (6), which results in biradicals that subsequently close to give the crossed product, is disfavoured, relative to orientation (7) which leads to the straight product, as a result of surface adsorption. Although this model allows the effect of surface adsorption to be rationalized, it is not in keeping with the current mechanistic thinking in relation to [2+2] cycloaddition reactions which considers the biradicals (Figure 4) to be the key intermediates. Thus a more appropriate explanation for the effect of the surface would lie in its ability to change the geometry of the minimum energy conformation of the biradicals so that their closure / cleavage ratios, and hence the product composition, is altered.

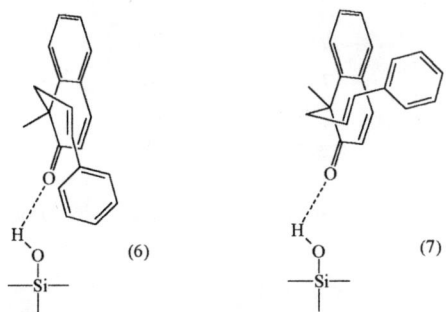

Figure 3 *Surface control of biradical formation*

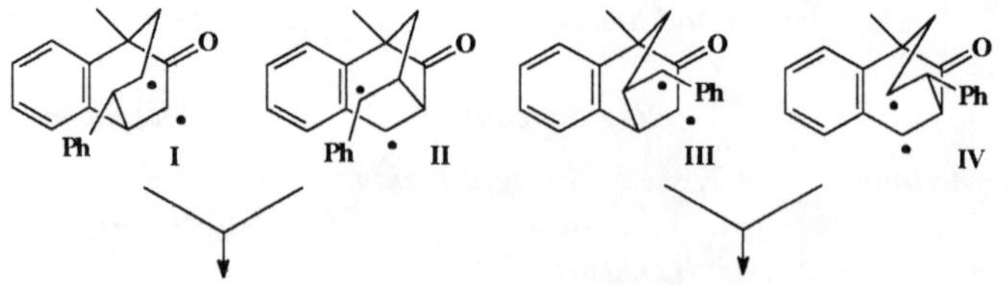

STRAIGHT PRODUCTS: (4),(5)   CROSSED PRODUCTS: (2),(3)

Figure 4 *Biradicals involved in cycloaddition reaction*

As it had already been shown[11] that considerable insight into the behaviour of these biradicals can be obtained from molecular modelling studies, a simple model for a silica adsorbed biradical was developed in an attempt to find a basis for understanding the results of these photochemical experiments. A twelve membered siloxane ring was used as a model for the surface; hydroxyl groups were attached to in a *cis* configuration to each silicon atom and hydrogen atoms were used to complete the structure. After optimising this structure (AM1)[12], six water molecules were placed close to the hydroxylated face of the ring and these were optimised (AM1) keeping the ring conformationally fixed. The biradical was placed close to the surface and was optimised keeping the ring and the water molecules fixed; this involved an initial MM$^+$ optimisation, the minimum energy geometry being further optimised using AM1.

Table 5 *Results of Biradical / Surface Modeling*

Biradical[a]	Dihedral Angle		Inter-radical Distance	
	Isolated	*Adsorbed*	*Isolated*	*Adsorbed*
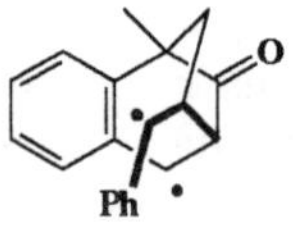 I	171.0°	158.6°	2.86Å	2.92Å
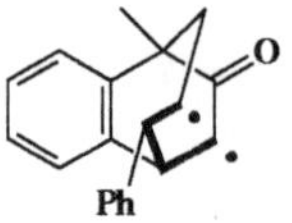 II	73.6°	76.6°	3.10Å	3.12Å
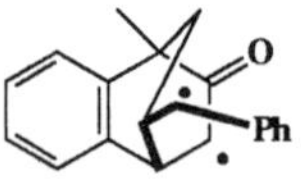 III	78.0°	164.9°	2.99Å	2.96Å
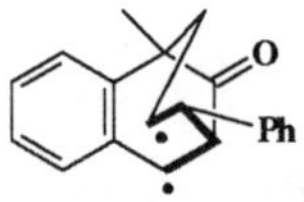 IV	-69.5°	-71.1°	3.07Å	3.09Å

[a]Bonds in bold were used to define the dihedral angle

The results indicate that only in the case of biradical III (Table 5) is there a significant difference between the solution and the surface minimum energy conformations. A consideration of the relative orientation of the radical centres in these conformations (Figure 5), suggests that ring closure of the adsorbed biradical should be more difficult as the radical p-orbitals are mutually orthogonal; in the solution conformation these orbitals are directed towards each other. As ring closure of biradical III results in the formation of a "crossed" product, the reduced amount of these products on silica is as a result understandable.

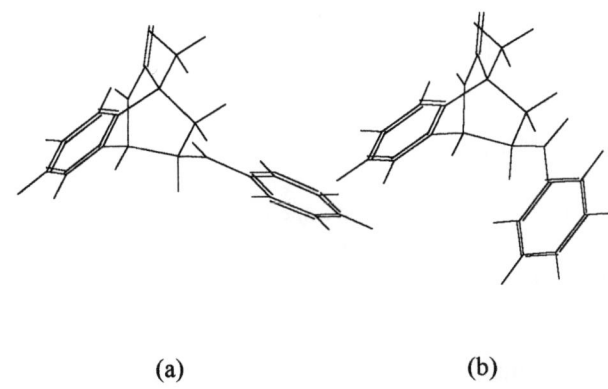

(a)   (b)

Figure 5   (a) *Surface and (b) solution conformation of biradical III*

Thus the outcome of these photochemical cycloadditions on silica is controlled to some extent by the surface, the control being, it is suggested, the result of a surface induced conformational change in a biradical intermediate. It also appears that the simple molecular modelling approach proposed here has a validity in terms of rationalising the effect of the surface on the outcome of the reaction.

References

1. R. G. Benson and N. W. A. Geraghty, *J. Chem. Research(S)*, 1983, 290.
2. K. Smith, D. M. James and I. Matthews, M. R. Bye, *J. Chem. Soc. Perkin Trans. I*, 1992, 1877.
3. T. Shichi, K. Takagi and Y. Sawaki, *Chem. Commun.*, 1996, 2027.
4. C. Cativiela, J. M. Fraile, J. I. Garcia, J. A. Mayoral, E. Pires, A. J. Royo, F. Figueras, L. C. de Menorval, *Tetrahedron*, 1993, **49**, 4073.
5. R. Fawaha, P. de Mayo, J. H. Schauble and Y. C. Toong, *J. Org. Chem.*, 1985, **50**, 245.
6. D. I. Schuster, G. Lem and N. A. Kaprinidis, *Chem. Rev. 1993*, **93**, 3.
7. M. J. Monaghan, PhD Thesis, National University of Ireland, 1988.
8. R. Srinivasan and K. H. Carlough, *J. Am. Chem. Soc.*, 1967, **89**, 4932.
9. R. S. H. Liu and G. S. Hammond, *J. Am. Chem. Soc.*, 1967, **89**, 4936.
10. 'Products for Liquid Chromatography', E. Merck, 1981, p.5.
11. M. Audley and N. W. A. Geraghty, *Tetrahedron Lett.*, 1996, **37**, 1641.
12. Hyperchem®, v4.5, Hypercube Inc., 1995.

ENCAPSULATED BIOCATALYSTS FOR ORGANIC TRANSFORMATIONS

H.L. Holland, J.-X. Gu and D. Riffle

Department of Chemistry
Brock University
St. Catharines
ON L2S 3A1   Canada

E.N. Vulfson and J.A. Khan

Institute of Food Research
Early Gate
Whiteknights Road
Reading RG6 6BZ   UK

1 INTRODUCTION

The recent development of biocatalysis as a method for performing reactions in organic chemistry has been dramatic in both scope and application.[1] The advantages associated with this technique, such as rapid catalytic rates, the regio- and stereoselectivity of reactions, and the ability to carry out transformations under environmentally-friendly conditions of pH and temperature, have earned biocatalysis a secure place in the repertoire of methods available for organic chemistry.

Biocatalytic reactions have traditionally been carried out in aqueous media. While this method has environmental advantages, the technique can suffer from problems associated with factors such as low substrate solubility, limitation of yield by product inhibition or substrate toxicity, or complications in downstream processing arising from the use of large aqueous volumes. A potential solution lies in the use of an organic solvent for biocatalytic reactions and, while the application of this method to many isolated enzyme catalysts has been systematically explored,[2,3] the process has not been extensively applied to the more generally useful range of whole-cell biocatalysts such as bacterial cells, yeast cells, or fungal spores.   For these catalysts a new range of potential problems arises, such as solvent toxicity, poor mass transport of the substrate, and the detrimental effects of a solvent-water interface on living cells.[4,5]
Various methods have been explored in attempts to circumvent these problems while maintaining a satisfactory level of catalyst activity.

1.1 Biotransformations in Two-phase Systems

This method has been successfully applied to several whole cell-catalysed reactions, such as the epoxidation of 1-octene by *Pseudomonas oleovorans* in a water/hydrocarbon solvent system,[6] the oxidation of cholesterol to cholestenone by *Arthrobacter simplex* in a water/carbon tetrachloride mixture,[7] and the conversion of octanoic acid to 2-heptanone by spores of *Penicillium roquefortii* in an aqueous/tetradecane system,[8] but cell death at the solution interface, with resultant loss of enzyme activity, generally predicates against the use of a simple two-phase reaction system.[9]

1.2 Immobilised Cells in Organic Solvents

Several methods have been developed for the protection of a microbial cell biocatalyst from the detrimental effects of an organic solvent. Of these, the one most frequently used is immobilisation of the cell in a matrix such as calcium alginate,[10] polyurethane foam,[11] a photo cross-linked polymer,[12] or a silicone polymer.[13] More complex procedures involve the use immobilised cells in a reverse micelle preparation[14] or a specific interface bioreactor.[15,16] Classical immobilisation techniques such as the formation of alginate or κ-carrageenan beads result in the formation of large (up to 5mm) particles which suffer from mass transfer limitations and low mechanical stability, especially in the presence of an organic solvent.

A specific application of the immobilisation approach for the stabilisation of whole cell catalytic activity in organic solvents is the use of a microencapsulation technique, in which the biocatalyst is protected from the solvent by a thin, semi-permeable membrane.

1.3 Microencapsulation of Biocatalysts[17]

The technology of microcapsule formation by interfacial polymerisation has been modified to permit its application to microbial biocatalysts. This method facilitates the preparation of an encapsulated biocatalyst in an aqueous environment contained within a small (50-300 μm diameter), mechanically stable capsule in a bulk organic phase. The thin capsule wall is formed by interfacial polymerisation surrounding a stabilised micro-droplet of aqueous biocatalyst suspended in an organic solvent of low toxicity, and is produced by reaction of di-acid chloride (typically dodecandioyl chloride) and polyamine (typically polyallylamine) monomer components, present in the organic and aqueous phases, respectively. This procedure results in the formation of mechanically stable capsules and permits easy control over the size of capsule and also, by variation of monomer components, of the membrane permeability.

This technique was originally developed for the encapsulation of baker's yeast cells, and the resulting preparation used for the reduction of 1-phenyl-1,2-propanedione in a variety of solvents.[17] We have now extended the range of encapsulated biocatalysts to include other yeast strains, bacterial cells, and fungal spores, and present herein a preliminary report on the viability and application of these biocatalysts.

2 VIABILITY OF ENCAPSULATED BIOCATALYSTS

2.1 Determination of Viability

The encapsulated biocatalysts used in this study were prepared as described, and contained typically 10^7-10^8 cells or spores per mL of aqueous medium.[17] The resulting microcapsules were stored in octane, and viability of the biocatalyst preparation was determined following transfer of the microcapsules to distilled water and opening of the capsules by irradiation for 2 minutes in an ultrasonic cleaning bath. The percentage of

viable cells in the resulting suspension was determined in two ways: firstly, by the use
of specific viability stains as specified below; and secondly, by counting (in duplicate
experiments) the number of viable colonies resulting from inoculation and incubation of
suitable agar plates with 1 mL of a cell or spore suspension of known concentration,
obtained by serial dilution of the recovered encapsulated material to a total
concentration of 100 cells or spores per mL.   The results obtained by either method
were in good agreement (±5%).   The absence of detrimental effects on cell viability of
the methods used for opening the capsules was confirmed for all examples by the
removal of biocatalyst samples from capsules by microsurgery.   The samples so
obtained expressed viability levels identical within experimental error (±5%) to those
obtained following bulk disintegration of the capsules.   The specific viability stains
used were: for *Pseudomonas putida* and *Rhodococcus rhodochrous*, *Bac*Light™
(Molecular Probes Inc., Eugene, Oregon, USA); for *Saccharomyces cerevisiae, Candida
rugosa,* and *Yarrowia lipolytica, Fungo*light™; and for spores of *Penicillium
roquefortii, Aspergillus niger, Mortierella isabellina,* and *Helminthosporium* species,
fluorescein diacetate/propidium iodide.
Parallel studies, performed by analogous methods, were used to determine the viability
of non-encapsulated biocatalysts stored in a two phase aqueous/octane system.

2.2 Viability of Fungal Spore Preparations

The species used in this study were *Penicillium roquefortii* ATCC 64383, *Aspergillus
niger* ATCC 9142, *Mortierella isabellina* ATCC 42613, and *Helminthosporium* species
NRRL 4671. The results shcwn for *P. roquefortii* spores in Figure 1 are typical.

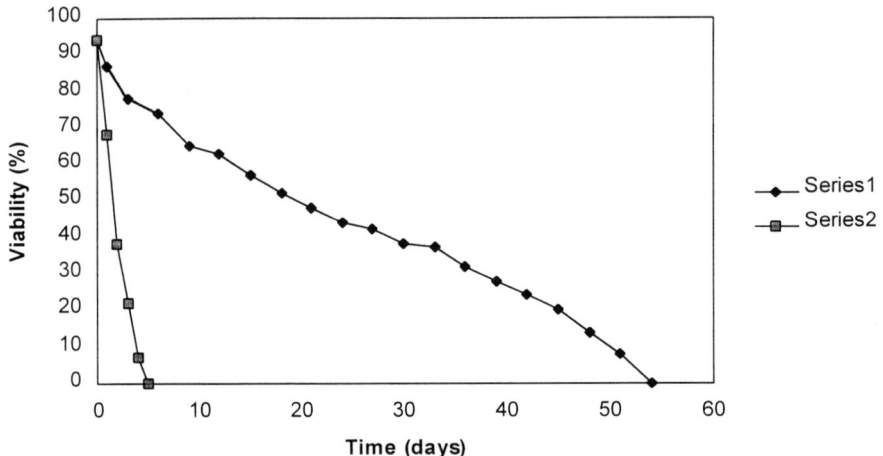

Figure 1 *Viability of encapsulated (Series 1) and non-encapsulated (Series 2) spores of*
P. roquefortii *in octane*

Viability data for spores of other fungal species are summarised in Table 1.

Table 1 *Viability of Encapsulated and Non-encapsulated Fungal Spores in Octane*

Species	Pre-encapsulation Viability (%)	$t\frac{1}{2}$ (days)	
		Encapsulated	Non-encapsulated
P. roquefortii	94	22	1.6
A. niger	93	22	1.5
M. isabellina	92	24	2.2
H. species	84	20	5

2.3 Viability of Bacterial and Yeast Cells

The species studied were *Pseudomonas putida* 39/D, *Rhodococcus rhodochrous* ATCC 17895, *Saccharomyces cerevisiae* (baker's yeast), *Candida rugosa* ATCC 10571, and *Yarrowia lipolytica* Y10. The results for *C. rugosa* are shown in Figure 2, and for this and other species summarised in Table 2.

Figure 2 *Viability of encapsulated (Series 1) and non-encapsulated (Series 2) cells of* C. rugosa *in octane*

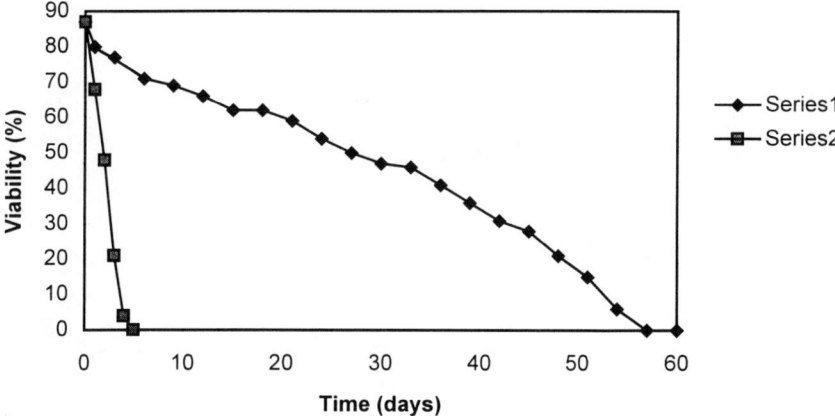

Table 2 *Viability of Encapsulated and Non-encapsulated Bacterial and Yeast Cells in Octane*

Species	Pre-encapsulation Viability (%)	$t\frac{1}{2}$ (days)	
		Encapsulated	Non-encapsulated
P. putida	91	8.5	1.1
R. rhodochrous	85	9.5	1.2
S. cerevisiae	96	18	2
C. rugosa	87	31	2
Y. lipolytica	91	21	1.5

3 REACTIONS CATALYSED BY ENCAPSULATED BIOCATALYSTS

The following reactions (Figure 3) have been carried out in preliminary trials using encapsulated biocatalysts in a bulk organic solvent.

Figure 3 *Reactions catalysed by encapsulated biocatalysts*

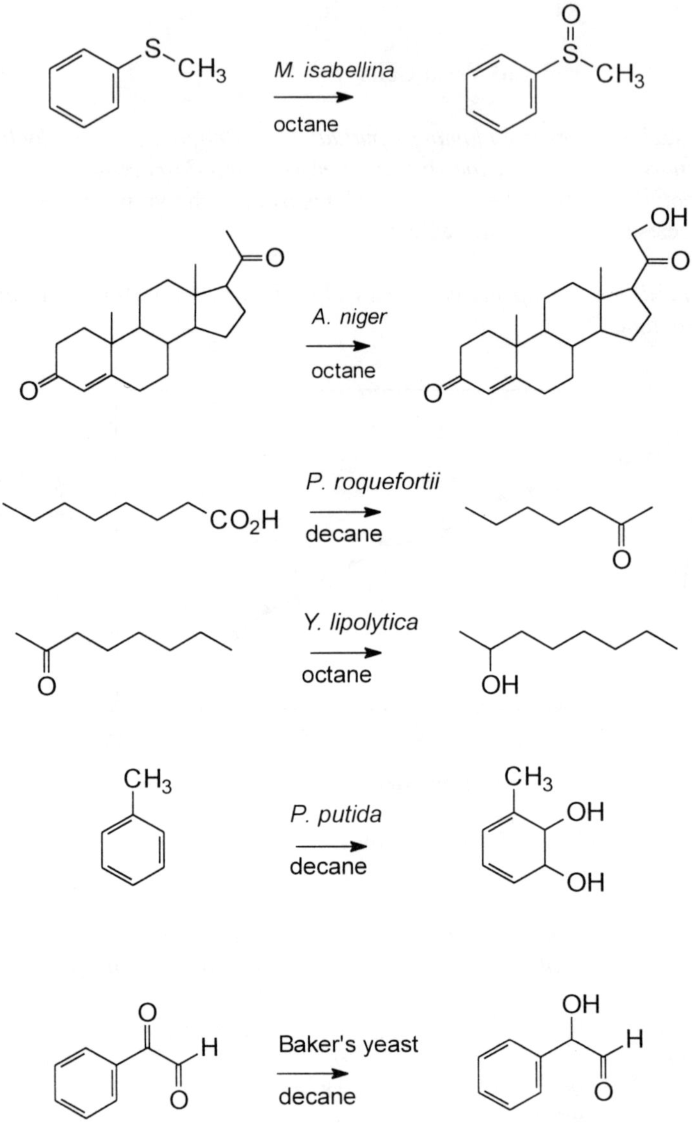

Optimisation of reaction conditions for the effective application of microencapsulated whole cell biocatalysts in organic solvents is currently under way.

Acknowledgement

This work was funded by the Natural Sciences and Engineering Research Council of Canada through a Collaborative Research Grant to HLH and ENV.

References

1. K. Drauz and H. Waldmann, eds., "Enzyme Catalysis in Organic Synthesis", VCH Weinheim, 1995.
2. J.S. Dordick in "Applied Biocatalysis", H.W. Blanch and D.S. Clark, eds., Marcel Dekker, Basel, 1991, p.1.
3. A. Zaks in "Biocatalysts for Industry", J.S. Dordick, ed., Plenum Press, London, 1991, p. 161.
4. G.J. Salter and D.B. Kell, *Crit. Rev. Biotech.*, 1995, **15**, 139.
5. M.H. Vermuë and J. Tramper, *Pure Appl. Chem.*, 1995, **67**, 345.
6. M-J. de Smet, B. Witholt, and H. Wynberg, *J. Org. Chem.*, 1981, **46**, 3128.
7. W.-H. Liu, W.-C. Horng, and M.-S. Tsai, *Enzyme Microb. Technol.*, 1996, **18**, 184.
8. C. Creuly, C. Larroche, and J.-B. Gros, *Enzyme Microb. Technol.*, 1992, **14**, 669.
9. H.M. Van Sonsbeek, H.H. Beeftink, and J. Tramper, *Enzyme Microb. Technol.*, 1993, **15**, 722.
10. Y. Naoshima, Y. Munakata, T. Nishiyama, J. Maeda, M. Kamezawa, T. Haramaki, and H. Tachibana, *World J. Microbiol. Biotechnol.*, 1991, **7**, 219.
11. K. Nakamura, T. Miyai, K. Inoue, S. Kawasaki, S. Oka, and A. Ohno, *Biocatalysis*, 1990, **3**, 17.
12. M. Iida, T. Nakamura, and H. Iizuka, *Nippon Kagaku Kaishi*, 1983, 1393.
13. K. Kawakami, S. Tsuruda, and K. Miyagi, *Biotechnol. Prog.*, 1990, **6**, 357.
14. N.W. Fadnavis, A. Deshpande, S. Chauhan, and U.T. Bhalerao, *J. Chem. Soc. Chem. Commun.*, 1990, 1548.
15. W.J.J. van den Tweel, E.H. Marsman, M.J.A.W. Vorage, J. Tramper, and J.A.M. de Bont in "Bioreactors and Biotransformations", G.W. Moody and P.B. Baker, eds., Elsevier, 1987, p. 231.
16. S. Oda, Y. Inada, A. Kato, N. Matsudomi, and H. Ohta, *J. Ferment. Bioeng.*, 1995, **80**, 559.
17. K.D. Green, I.S. Gill, J.A. Khan, and E.N. Vulfson, *Biotechnol. Bioeng.*, 1996, **49**, 535.

ACID CATALYSIS OF SUPPORTED AND UNSUPPORTED HETEROPOLYACIDS FOR METHYL TERT-BUTYL ETHER (MTBE) SYNTHESIS

Makoto Misono, Toshio Okuhara* and Sawami Shikata

Department of Applied Chemistry
Graduate School of Engineering
The University of Tokyo, Bunkyo-ku, Tokyo 113, Japan

*Graduate School of Environmental Earth Science
 Hokkaido University, Sapporo 060, Japan

1 ABSTRACT

After the introduction of essential concepts that characterize the heterogeneous catalysis of heteropolyacids (HPA), the gas-phase MTBE synthesis catalyzed by supported and unsupported HPA are described. Dawson-type HPA showed a very high catalytic activity which was explained by the formation of active pseudoliquid phase. It was concluded that MTBE synthesis proceeded in the pseudoliquid also in the case of supported HPA.

2 INTRODUCTION

Basic concepts and facts necessary for understanding the heterogeneous catalysis of heteropoly compounds (heteropolyacids (HPA) and their salts), in addition to those for ordinary heterogeneous catalysis, are [1];

(1) The presence of primary, secondary and tertiary structures. Clear distinction between these structures is necessary.

The primary structure is the structure of heteropolyanion (metal oxide cluster); Keggin (e.g., $PW_{12}O_{40}$), Dawson (e.g., $P_2W_{18}O_{62}$), etc. The secondary structure is the ionic crystal (sometimes amorphous solid) consisting of large anions of the primary structure and various kinds of cations. The salts with small cations (H, Na, etc.; group A salts) have in general small surface areas and are soluble in polar solvents. The secondary structures of group A salts are flexible and absorb a large number of polar molecules. Due to this nature, these compounds exhibit "pseudoliquid" behaviour for the reactions of polar molecules. In contrast, the salts of large cations like Cs and NH_4 (group B salts) have large surface areas and rigid secondary structures, and hence exhibit surface type catalysis.

(2) The presence of three classes in heterogeneous catalysis; surface-type, bulk type I (pseudoliquid), and bulk type II. These are illustrated in Figure 1.

The surface type is the ordinary heterogeneous catalysis in which reactant molecules are adsorbed on the surface (on both pore walls and outer surface) and react there. In the bulk-type I catalysis, the reactant molecules are absorbed into the lattice of solid bulk and react there. The solid behaves in a sense like a concentrated solution, so that this is called pseudoliquid catalysis. The bulk-type II catalysis was found in the case of high-temperature dehydrogenation reactions. Due to the rapid diffusion of protons and electrons produced on the surface, the entire solid takes part in catalysis.

In this article, we first describe the evidence for the pseudoliquid behaviour for dehydration of alcohols catalyzed by $H_3PW_{12}O_{40}$. Next, the MTBE synthesis over $H_3PW_{12}O_{40}$ (Keggin) and $H_6P_2W_{18}O_{62}$ (Dawson) is compared with that over these HPA s supported on silica gel. We demonstrate that the pseudoliquid behaviour operates for both supported and unsupported systems. Then the structure and catalysis of Cs salts of HPA particularly of $Cs_xH_{3-x}PW(or\ Mo)_{12}O_{40}$ are described briefly.

3 EXPERIMENTAL

$H_nXW_{12}O_{40}$ (Keggin, X=P, Si, Ge, B, and Co) and $H_6P_2W_{18}O_{62}$ (Dawson) were synthesized according to the methods in the literature [2, 3]. The structures were confirmed by IR and NMR. Silica-supported HPA s were prepared by impregnation of SiO_2 (Davison G-62, 253 m^2/g) with aqueous solutions of HPA. The loading amount (HPA/(HPA + SiO_2)) was varied from 20 to 75 wt %. $Cs_xH_{3-x}PW(or\ Mo)_{12}O_{40}$ s were prepared as in the previous work [1]. H-ZSM-5 and SO_4-ZrO_2 were the same as before [2, 3]. Ion-exchange resin (Amberlyst 15) purchased commercially was pretreated in N_2 at 383 K.

The MTBE synthesis was carried out in a flow reactor at atmospheric pressure in the

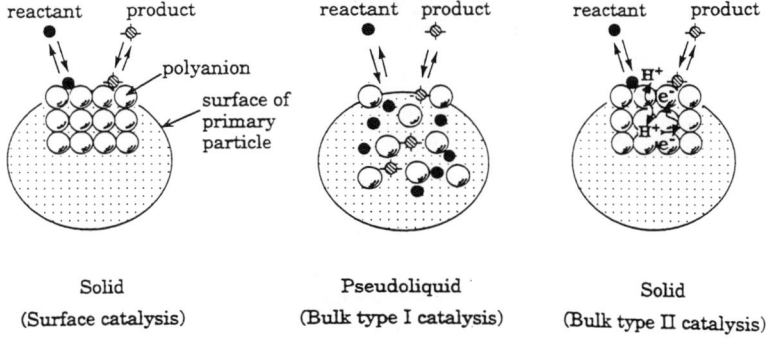

Figure 1. Three types of catalysis by heteropoly compounds.

temperature range from 303 to 383 K. Prior to the reaction, catalysts were treated in a flow of N_2 at a temperature selected for each catalyst as in the previous work unless otherwise stated. In ordinary experiments, the feed gas was a mixture of methanol, isobutylene and N_2 with a volume ratio of 1 : 1 : 3. The yield of MTBE was calculated by the ratio of MTBE at the outlet to methanol at the inlet.

4 RESULTS AND DISCUSSION

4.1 Pseudoliquid behaviour of ethanol dehydration

A large number of ethanol molecules are readily absorbed into the solid bulk of $H_3PW_{12}O_{40}$ almost reversibly [1]. Due to this behaviour, the dehydration of ethanol takes place not only on the surface but also in the solid bulk, resulting in a high catalytic activity and unique selectivity. Figure 2 is a typical result of transient response when the feed of ethanol-d_0 was replaced by ethanol-d_6 at the steady state of the dehydration of ethanol [4]. The result demonstrated that the absorption and desorption were much faster than the dehydration reaction and that the amount of ethanol absorbed in the solid was much larger than the amount of the monolayer adsorption.

Another supporting evidence is the pressure dependency of the rate and selectivity, shown in Figure 3. The unusual variation of the pressure dependency was successfully explained by

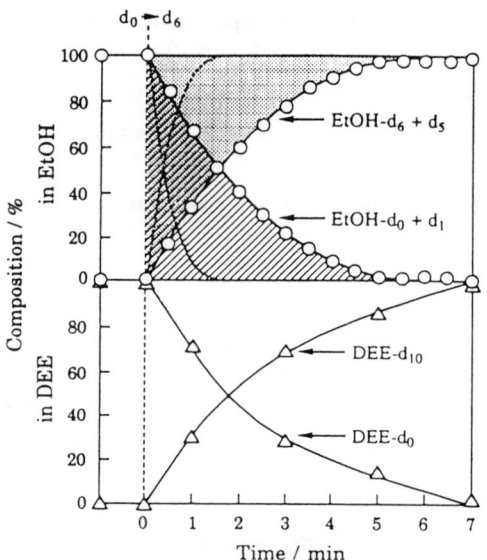

Figure 2. Transient-response at the outlet of the reactor to the replace-ment of the feed from ethanol-d_0 to -d_6 (16.4 kPa) during dehydration over $H_3PW_{12}O_{40}$ (403 K). At the vertical broken line, the feed was changed. See text for the broken and solid curves: top, ethanol (EtOH); bottom, diethyl ether (DEE).

assuming the following reaction scheme in the pseudoliquid phase. We detected directly by NMR the protonated ethanols postulated in this scheme [4], thanks to the uniform, mobile and three-dimensional nature of pseudoliquid.

$$+C_2H_5OH$$
$$C_2H_5OH \text{ (gas)} \rightleftharpoons C_2H_5OH_2^+ \text{ (abs)} \rightleftharpoons (C_2H_5OH)_2H^+ \text{ (abs)}$$
$$\updownarrow \qquad\qquad \updownarrow$$
$$C_2H_4 \text{ (gas)} \qquad\qquad (C_2H_5)_2O \text{ (gas)}$$

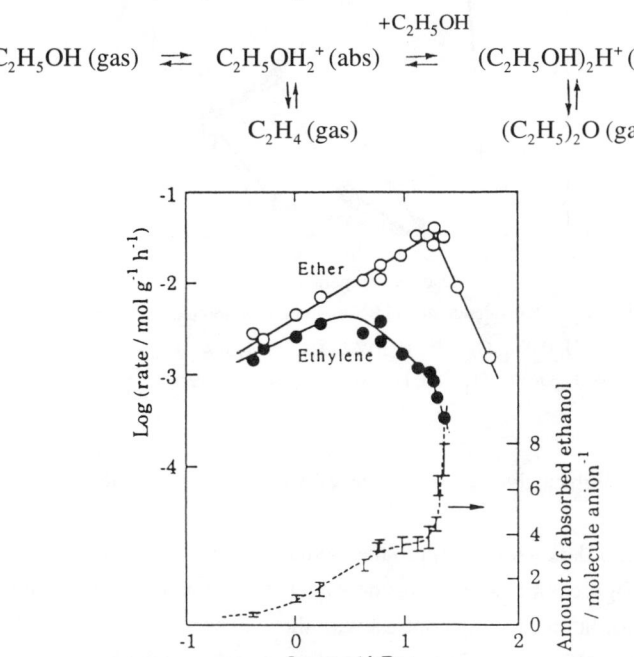

Figure 3. Rates of formation of diethyl ether and ethylene from ethanol over $H_3PW_{12}O_{40}$ (C_2-basis) as well as the amount of absorbed ethanol under the working conditions as a function of the partial pressure of ethanol: reaction temperature, 403 K; W/F (ratio of the catalyst weight to the feed rate), 2–60 g h mol^{-1}.

4.2 MTBE synthesis catalyzed by hydrogen forms, $H_3PW_{12}O_{40}$ and $H_6P_2W_{18}O_{62}$

Figure 4 shows the MTBE yields at the steady states, which were attained within one hour, as a function of reaction temperature for various unsupported HPA [3]. The yields were reproducibly obtained when the reaction temperature was changed up and down. The selectivities (based on both methanol and isobutylene) were close to 100 % below 343 K. It is remarkable that $H_6P_2W_{18}O_{62}$ (Dawson-type) showed a much higher activity than the other Keggin-type HPA s in the lower temperature range that is favourable for the equilibrium of MTBE formation. The decreases of the yield at high temperatures are mainly due to the equilibrium limitation of the reaction.

The very high catalytic activity of $H_6P_2W_{18}O_{62}$ as compared with $H_3PW_{12}O_{40}$, $H_4SiW_{12}O_{40}$, etc. at about 340 K cannot be explained by the acid strength measured by Hammett indicators and by calorimetric measurement of NH_3 absorption. On the other hand, the rates of absorption of methanol and MTBE at 313-323 K were much faster for $H_6P_2W_{18}O_{62}$,

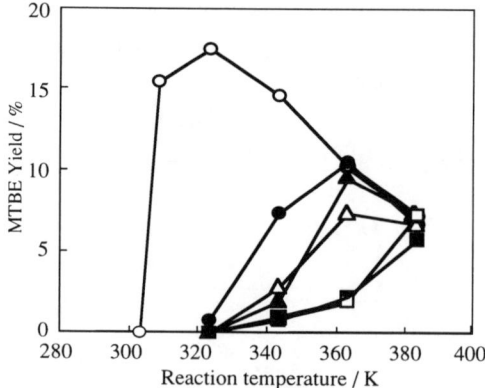

Figure 4. Dependence of MTBE yield on reaction temperature.
$\bigcirc$, $H_6P_2W_{18}O_{62}$, $\bullet$, $H_3PW_{12}O_{40}$, $\triangle$, $H_4SiW_{12}O_{40}$,
$\blacktriangle$, $H_4GeW_{12}O_{40}$, $\square$, $H_5BW_{12}O_{40}$, $\blacksquare$, $H_6CoW_{12}O_{40}$.

corresponding to its high catalytic activity. The dependencies of the rate on methanol pressure are shown in Figure 5.

Unusual reaction orders with respect to the methanol pressure which are somehow similar to those shown in Figure 3 are noted. This dependency is similar to that for an ion-exchange resin (Amberlyst 15), but in striking contrast with those observed for H-ZSM-5, SO_4-ZrO_2 and $Cs_{2.5}H_{0.5}PW_{12}O_{40}$ (Figure 6). These results demonstrate that the MTBE synthesis proceeds in the pseudoliquid phase in the case of HPA.

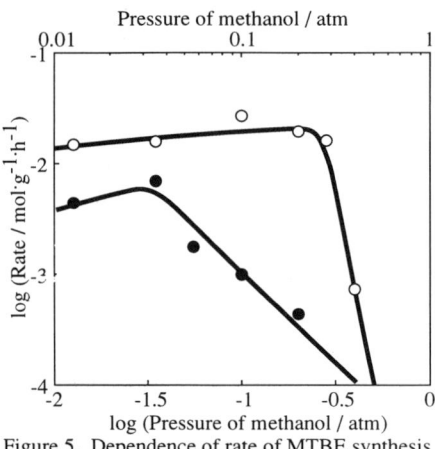

Figure 5. Dependence of rate of MTBE synthesis on methanol pressure at 323 K.
$\bigcirc$, $H_6P_2W_{18}O_{62}$, $\bullet$, $H_3PW_{12}O_{40}$.

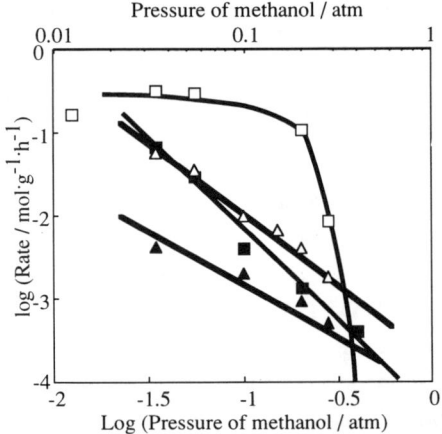

Figure 6. Dependence of rate of MTBE synthesis on methanol pressure.
At 323 K: $\square$, Amberlyst 15; $\blacksquare$, $Cs_{2.5}H_{0.5}PW_{12}O_{40}$.
At 343 K: $\triangle$, SO_4-ZrO_2; $\blacktriangle$, H-ZSM-5.

It was observed for isopropanol dehydration that there were two pseudoliquid phases; one has a much higher activity and a smaller amount of absorbed alcohol than the other [5]. Hence, the fact that the $H_6P_2W_{18}O_{62}$ was much more active at low temperatures is understandable, since the rate of absorption was high and the amount of absorption during the reaction was moderate for $H_6P_2W_{18}O_{62}$.

Table 1 shows a remarkable correspondence between the secondary structures determined by powder XRD (avoiding carefully the exposure to the humidity) and the reaction rates observed for the eight cases; two HPA s with two different pretreatments for the initial and stationary states [6]. High rate was always observed when the secondary structure was amorphous, and low rate for crystalline cubic structure. For example, $H_3PW_{12}O_{40}$ treated at 523 K showed high activity at first, but the activity declined to a very low level at the steady state, accompanying the change of the secondary structure from amorphous to crystalline (cubic), while $H_6P_2W_{18}O_{62}$ always showed a high activity and amorphous XRD pattern. It is likely that the secondary structure is more flexible when it is amorphous and hence the absorption and desorption are easier. It is also probable that the oval shape of the Dawson anion is not suitable for stable crystalline structure and tends to form an amorphous structure, while the spherical shape of Keggin anion favours crystalline cubic structure as in the structure of its hexahydrate.

Table 1. The secondary structures and the catalytic activities for MTBE synthesis over $H_6P_2W_{18}O_{62}$ and $H_3PW_{12}O_{40}$

Primary structure	Pretreatment temp.	Secondary structure		Catalytic activity	
		Before reaction	After reaction	Initial	Stationary
Dawson type ($H_6P_2W_{18}O_{62}$)	423 K 523 K }	amorphous	amorphous	fast	fast
Keggin type ($H_3PW_{12}O_{40}$)	423 K	cubic	cubic + (?)	low	low
	523 K	amorphous	cubic + (?)	fast	low

This correspondence is summarized as follows.

Primary structure ---> Secondary structure ---> Catalytic activity
Keggin (spherical) → crystalline → low
Dawson (oval) → amorphous → high.

It is here assumed that non-polar isobutylene is absorbed into HPA bulk, being assisted by co-existing methanol [2].

In addition, it was found by the in-situ high-temperature XRD study that after dehydration at 423 K the cubic secondary structure of $H_3PW_{12}O_{40}$ shrinked slightly. This shrinkage is reasonable, but this has not been observed previously. Even very short exposure to the air of the dehydrated sample returned the lattice constant back to that of the hexahydrate.

4.3 MTBE synthesis catalyzed by supported $H_3PW_{12}O_{40}$ and $H_6P_2W_{18}O_{62}$

The catalytic activities of the two HPA s increased remarkably (particularly in the case of $H_3PW_{12}O_{40}$) when they were supported on silica gel [3]. The activities of both HPA s were about the same as that of Amberlyst 15 and the selectivities were comparable or better than Amberlyst 15 under the same reaction conditions. The pressure dependencies of the rate are shown in Figure 7. The unusual behaviour similar to that in Figure 5 indicates that the reaction over supported HPA also proceeds in the pseudoliquid (or in the solid bulk of HPA particle).

As the loading level increased, the rate increased. But at high loading levels, the rate decreased. This change is explained by the mass transfer effect in the pseudoliquid phase. XRD and NMR studies have indicated that the particle size of HPA on silica increased with the loading level [7]. With large particles, the mass transfer in the HPA particle becomes more rate-limiting and the effectiveness factor of catalyst decreases.

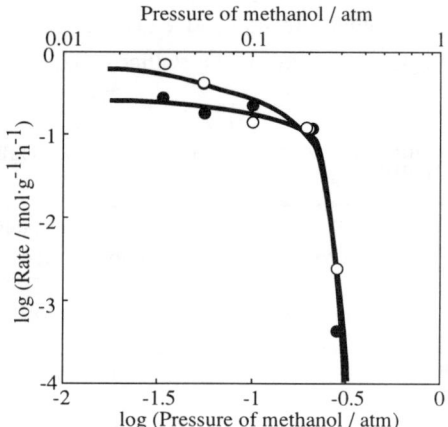

Figure 7. Dependence of rate of MTBE synthesis on methanol pressure at 323 K.
$\bigcirc$, 20wt% $H_6P_2W_{18}O_{62}/SiO_2$;
$\bullet$, 20wt% $H_3PW_{12}O_{40}/SiO_2$.

4.4 Structure and catalysis of $Cs_xH_{3-x}PW_{12}O_{40}$ and $Cs_xH_{3-x}PMo_{12}O_{40}$

It has been suggested that acidic Cs or K salts of $H_3PW(or\ Mo)_{12}O_{40}$ are thin films of acid form covering the surface of stoichiometric Cs salt or mixtures of them [8, 9]. Our NMR study, however, showed that the acidic salts formed nearly uniform solid solutions by the easy diffusion of proton and cesium ion [10]. The high surface acidity and catalytic activity were obtained when x was near 2.5. It was further demonstrated that the pore size can be precisely controlled and shape selective catalysis reveals at $x = 2.1 - 2.7$.

In addition, the Cs salts have higher thermal stabilities. Taking these advantages, it was possible to improve significantly the performance of vanadium-containing heteropolymolybdates for oxidative dehydrogenation of isobutyric acid [12].

REFERENCES

1. E.g., T. Okuhara, N. Mizuno, and M. Misono, *Advan. Catal.*, 1996, **41**, 113.
2. S. Shikata, T. Okuhara, and M. Misono, *J. Mol. Catal.*, 1995, **100**, 49.
3. S. Shikata, S. Nakata, T. Okuhara, and M. Misono, *J. Catal.*, 1997, **166**, 263.
4. K. Y. Lee, T. Arai, S. Nakata, S. Asaoka, T. Okuhara, and M. Misono, *J. Am. Chem. Soc.*, 1992, **114**, 2836.
5. K. Takahashi, T. Okuhara, and M. Misono, *Chem. Lett.*, 1985, 841.
6. S. Shikata, T. Okuhara, and M. Misono, *Shokubai*, 1997, **39**, 174.
7. Y. Izumi, R. Hasebe, and K. Urabe, *J. Catal.*, 1983, **84**, 402; N. Mizuno, T. Watanabe, H. Mori, and M. Misono, *J. Catal.*, 1990, **123**, 157; K. M. Rao, R. Gobetto, A. Iannibello, and A. Zecchina, *J. Catal.*, 1989, **119**, 512.
8. J. B. Black, N. J. Clayden, P. L. Gai, J. D. Scott, E. M. Serwicka and J. B. Goodenough, *J. Catal.*, 1987, **106**, 1.
9. N. Essayem, G. Coudurier, M. Fournier, and J. C. Védrine, *Catal. Lett.*, 1995, **34**, 223.
10. T. Okuhara, T. Nishimura, H. Watanabe, K. Na, and M. Misono, *Stud. Surf. Sci. Catal.*, 1994, **90**, 419.
11. T. Okuhara, T. Nishimura, and M. Misono, *Chem. Lett.*, 1995, 155.
12. K. Y. Lee, S. Oishi, H. Igarashi, and M. Misono, *Catal. Today*, 1997, **33**, 183.

ENANTIOSELECTIVE HYDROGENATIONS ON CHIRALLY MODIFIED PLATINUM. MECHANISTIC ASPECTS AND CATALYST DESIGN

Alfons Baiker

Department of Chemical Engineering and Industrial Chemistry
Swiss Federal Institute of Technology
ETH-Zentrum, CH-8092 Zurich, Switzerland

1 INTRODUCTION

First attempts of heterogeneous enantioselective catalysis used naturally occurring chiral materials as auxiliaries[1] or as supports[2]. A brief description of the historical development of the field has been given in a recent review[3]. Among the various strategies used for designing enantioselective heterogeneous catalysts, the modification of metals by chiral modifiers has been most successful. Judged from the synthetic potential, today the two most relevant types of enantioselective catalysts, based on chirally modified metals, are the nickel-tartrate system[4] for the hydrogenation of β-ketoesters, β-diketones and methyl ketones, and the platinum-cinchona alkaloid system[5] for the hydrogenation of α-ketoesters, α-ketoacids and lactones. Since its discovery in the 1940s, the tartaric acid modified nickel system has been studied intensively and developed further[3,6]. In contrast, deeper insight into the functioning of the Pt-cinchona system has been gained only recently. The progress made till the end of 1995 has been discussed in recent reviews[7,8]. Based on experimental observations and theoretical studies using quantum chemistry techniques, feasible structures of the diastereomeric transition complexes formed between modifier and reactant have been suggested, paving the way to the development of new enantiodifferentiating heterogeneous catalysts[7]. Here we illustrate how the available mechanistic information could be applied in the search of new modifiers and reactants. Emphasis will be given to the most recent developments.

2 PLATINUM MODIFIED BY CINCHONA ALKALOIDS

2.1 General Features

The efficiency of the platinum-cinchona system for the enantioselective hydrogenation of α-ketoesters depends mainly on the structure and concentration of the modifier, the structural properties of the supported platinum, and the solvent used. Although all these properties have to be optimised to achieve high optical yields, the structure of the modifier is most crucial for enantioselection. Already very small quantities of modifier (modifier : $Pt_{surf} \ll 1$) are sufficient to induce both enantiodifferentiation and rate acceleration. The most suitable cinchona alkaloids (scheme 1) used for chirally modifying platinum are cinchonidine (CD) **1a**, 10,11-dihydrocinchonidine **1b**, and 10,11-dihydro-O-methylcinchonidine **1c**. Systematic studies on the relationship between modifier structure and enantiodifferentiation provided information on the structural requirements of modifiers[9]. Changing the absolute configuration at C-8 and C-9 of cinchonidine, i.e. substituting cinchonidine by its near-enantiomer, cinchonine, alters the chirality of the product lactate.

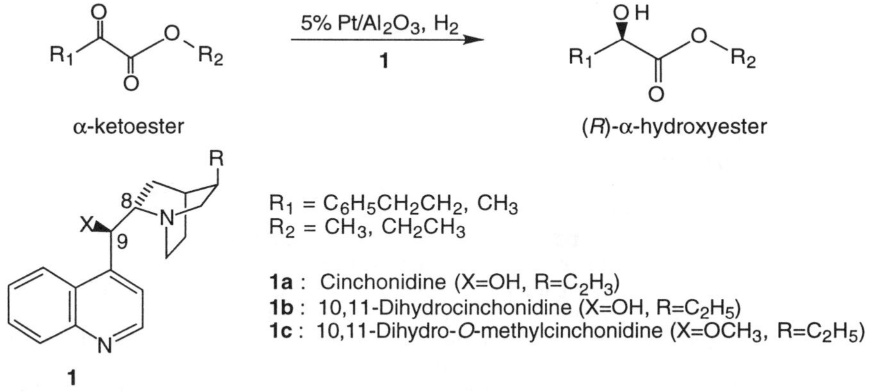

Scheme 1 structures:

α-ketoester → (5% Pt/Al$_2$O$_3$, H$_2$, **1**) → (*R*)-α-hydroxyester

$R_1 = C_6H_5CH_2CH_2$, CH_3
$R_2 = CH_3$, CH_2CH_3

1a : Cinchonidine (X=OH, R=C$_2$H$_3$)
1b : 10,11-Dihydrocinchonidine (X=OH, R=C$_2$H$_5$)
1c : 10,11-Dihydro-*O*-methylcinchonidine (X=OCH$_3$, R=C$_2$H$_5$)

1

Scheme 1

Alkylation of the quinuclidine nitrogen results in a complete loss of enantio-differentiation, which indicates that this centre plays a crucial role in the mechanism of enantioselection. Partial hydrogenation of the quinoline ring causes a drop in enantiomeric excess (e.e.) to below 50%. The selectivity is only marginally influenced by O-methylation, whereas replacing the OH by hydrogen or using the acylated derivative results in a decrease in e.e. by more than 20%. Interestingly, protonation of the quinuclidine nitrogen increases e.e. by 5-15%.

The three structural elements which are crucial for the functioning of the cinchona alkaloids as chiral modifiers are: (i) the *anchoring part*, represented by the flat aromatic ring system (quinoline), which is assumed to be adsorbed parallel to the platinum surface by multicentre π-bonding[10]; (ii) the *stereogenic region*, embracing C-9 and C-8, which determines the chirality of the product; and, (iii) the *basic quinuclidine nitrogen*, which is directly involved in the interaction with the reactant α-ketoester.

As concerns the properties of the platinum catalyst, proper platinum dispersion[11], support material[3] and pore size distribution[9] are important. The solvent can influence the catalytic behaviour due to different solubility of reactants (α-ketoester and hydrogen), and interaction with the modifier, α-ketoester and platinum surface. Solvents with dielectric constants (d.c.) between 2 and 10 were found to be most suitable[12]. Acetic acid (d.c. = 6.2) affords the highest e.e.s (up to 95%) under optimised conditions[13].

2.2 Mechanistic Aspects

2.2.1 Kinetics. Various kinetic studies of the enantioselective hydrogenation of ethyl pyruvate (EP), summarised and discussed in a recent review[8], revealed the dependence of the observed reaction rate and enantiomeric excess (e.e.) on crucial reaction parameters (concentration of reactants and modifier, solvent, temperature) and indicated the role of external and intraparticle mass transfer in this complex reaction system. Although a complete mechanistic model is so far out of reach, the kinetic results can be described reasonably well using models which imply that a modified active site is formed by the adsorption of a cinchona molecule. Both α-ketoester and hydrogen are assumed to be reversibly adsorbed on Pt. The adsorbed modifier interacts with the adsorbed α-ketoester forming a stabilised half-hydrogenated intermediate. The rate acceleration and the enantiodiscrimination are considered to originate from preferential stabilisation of one of the two diastereomeric intermediates formed by the interaction between α-ketoester and cinchona modifier. The question whether the addition of the first or the second hydrogen is rate determining is not conclusively answered from presently available kinetic investigations. A possibility for discriminating these steps could be the study of the dependence of the rate acceleration and enantioselectivity as a function of hydrogen pressure. Very recently a model providing a quantitative rationalisation of the effect of hydrogen concentration on enantioselectivity has been proposed[14].

2.2.2 Structure of Reactant—Modifier Complex. Theoretical studies[7] aimed at rationalising the interaction between the cinchona alkaloid and reactant methyl pyruvate (MP) have been undertaken using quantum chemistry techniques, at both ab initio and semi-empirical levels, and molecular mechanics. The calculation of the structure of these interaction complexes was hoped to provide some insight into the structures of the enantiodifferentiating complexes, which are assumed to resemble the corresponding transition complexes. Figure 1 shows the optimised structure of the complex formed on interaction of protonated cinchonidine (acetic acid as solvent) and methyl pyruvate, which upon hydrogenation would yield *(R)*-methyl lactate and *(S)*-methyl lactate, respectively. The pro *(R)*-methyl lactate complex was found to be energetically favoured[15]. It is suggested to be adsorbed in a planar π-bonding mode on the Pt surface via the aromatic quinoline ring, without hindering the interaction of the carbonyl groups of methyl pyruvate with the Pt surface. Indirect evidence for the flat adsorption of the quinoline ring emerged from hydrogen (H/D) isotope exchange studies[16]. Note that in the complex affording *(R)*-methyl lactate, both carbonyl moieties of the reactant methyl pyruvate lie in a plane parallel to the Pt surface providing optimal adsorptive interaction.

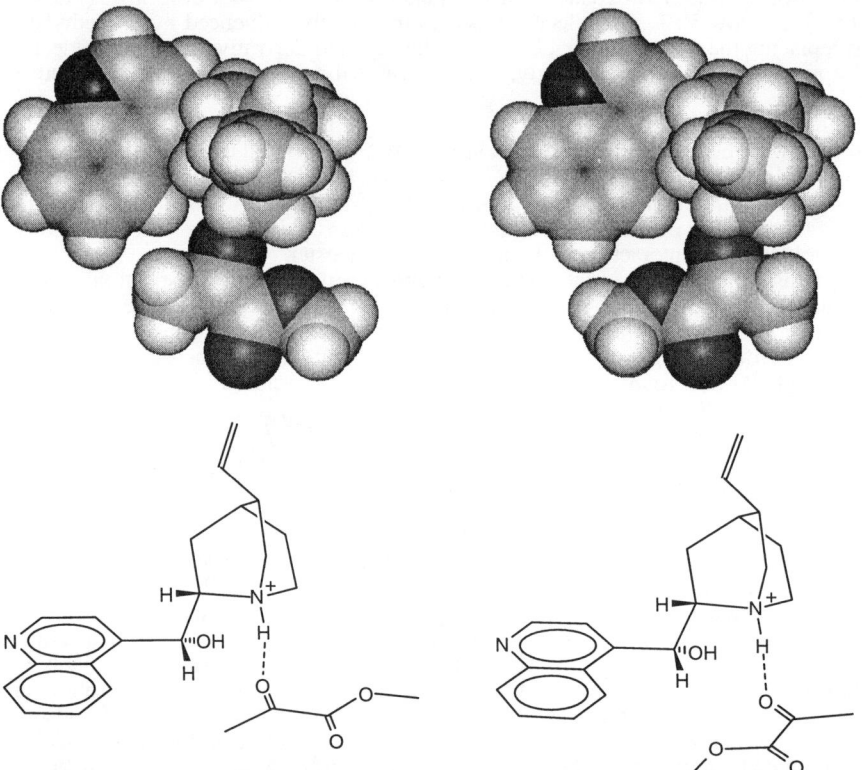

Figure 1 *Proposed structures of enantiodifferentiating complexes formed between protonated cinchonidine and methyl pyruvate which yield (R)-(+)-methyl lactate and (S)-(-) methyl lactate, respectively, upon hydrogenation. CD is adsorbed via π-bonding of the quinoline ring on the Pt surface. Top: Platinum surface lies parallel to drawing plane*

This adsorption mode is sterically hindered in the pro *(S)*-methyl lactate complex. The opposite behaviour is found when the complexes formed between protonated cinchonine (the near-enantiomer of cinchonidine) and methyl pyruvate are optimised . In this case, the complex leading to *(S)*-methyl lactate upon hydrogenation is energetically favoured and can be adsorbed without significant steric hindrance, while the one affording *(R)*-methyl lactate is sterically hindered. Thus calculations showed that - in agreement with the experimental observations - a change in the chirality of the stereogenic region (C-8, C-9) of the cinchona alkaloid results in a corresponding change in the chirality of the product lactate. Note that in the transition complexes the α-ketoester is stabilised in its half-hydrogenated form[10,15].

The origin of the hydrogen responsible for the stabilising interaction between the cinchona alkaloid and the ketoester may change depending on the solvent. The hydrogen can either come from the solvent (protonation of basic nitrogen, acetic acid) or from dissociatively adsorbed hydrogen. The enantiodiscrimination is considered to occur due to favourable stabilisation of one of the diastereomeric transition complexes.

The molecular modelling approach, taking into account the methyl pyruvate—cinchona alkaloid interaction and the steric constraints imposed by the adsorption on the platinum surface, leads to a reasonable explanation for the enantiodifferentiation. Although the predictions of the complexes formed between MP and the cinchona modifiers have been made for an ideal case (neglect of solvent interactions and quantum description of adsorptive interaction with platinum), this approach proved to be extremely useful in the search for new modifiers[7], which will be considered next.

3 DESIGN OF NEW ENANTIODIFFERENTIATING SYSTEMS

3.1 New Modifiers

The search strategy, which included a systematic reduction of the cinchona alkaloid structure to the essential functional parts and rationalisation of the structure of the corresponding complex formed between the new modifier and methyl pyruvate by means of molecular modelling, indicated that simple chiral aminoalcohols should be promising substitutes for cinchona alkaloid modifiers[7]. An example of an efficient modifier emerging from this search is 2-(1-pyrrolidinyl)-1-(1-naphthyl) ethanol[17] (PNE). The optimised structures of the enantiodifferentiating complexes yielding *(R)*- and *(S)*-methyl lactate, respectively, are illustrated in Fig. 2. The similarity of these complexes to the corresponding transition complexes formed between methyl pyruvate and protonated CD (Fig. 1) is striking, both complexes are stabilised by a N-H-O interaction and parallel adsorption of the two carbonyl moieties is sterically hindered in the pro *(S)*- complex. Besides PNE, a series of 2-hydroxy-2-aryl ethyl amines and other structurally related chiral compounds were evaluated by modelling and tested concerning their potential as chiral modifiers in the enantioselective hydrogenation of EP over Pt/alumina[7,18-21].

3.2 New Reactants

The search strategy based on the calculation of the structure of the transition complex was also applied to explore whether reactants other than α-ketoesters can be enantioselectively hydrogenated using the platinum-cinchona system. Recent examples include the hydrogenation of ketopantolactone[22], 2,2,2-trifluoroacetophenone[23] and various keto amides[24]. Figure 3 shows the optimised structures of the diastereomeric transition complexes suggested to be relevant in this hydrogenation. In all cases the pro *(R)*-complex is energetically favoured. Note that with the ketoamide (N-ethyl pyruvamide) the hydrogen bonding between the quinuclidine N and the oxygen of the amidecarbonyl is energetically favoured compared to the corresponding bonding involving the oxygen of the α-carbonyl group.

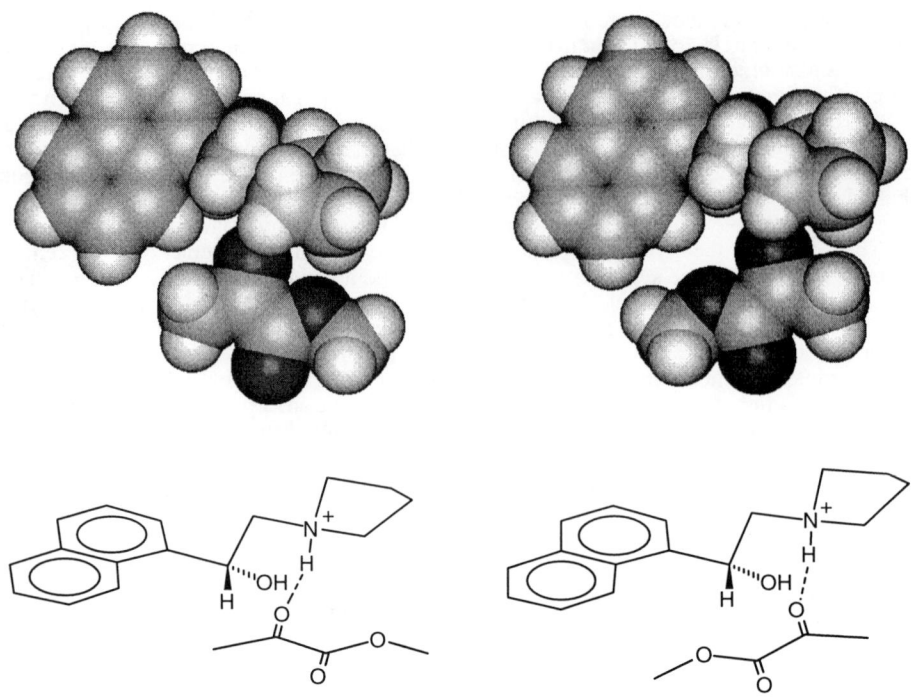

Figure 2 *Proposed structures of complexes formed between protonated (R)-2-(1-pyrrolidinyl)-1-(1-naphthyl)ethanol (PNE) and methyl pyruvate (MP) which yield upon hydrogenation (R)-(+)-methyl lactate and (S)-(-) methyl lactate, respectively. Top: Platinum surface lies parallel to drawing plane. NPE is adsorbed via π- bonding of the naphthyl ring on the Pt surface. Adsorption of the pro (R)-methyl lactate complex is sterically favoured*

4   CATALYTIC BEHAVIOUR OF NEW MODIFIERS AND REACTANTS

4.1 Modifiers

Some of the new modifiers and their enantiodifferentiating potential in EP hydrogenation are summarised in Fig. 4. The naphthalene derivative PNE [17] proved to be a remarkably efficient modifier inducing at low hydrogen pressures (1-10 bar) enantiomeric excesses as high as 75%, which is comparable to that obtained with cinchonidine under these conditions. This corroborates the predictive potential of the molecular modelling studies (Fig. 2) which indicated a diastereomeric transition complex resembling the one calculated for EP—CD. However, in contrast to CD, which is more effective at higher pressures (70-100 bar), the enantioselectivity of PNE decreases at higher hydrogen pressures due to partial hydrogenation of the naphthalene ring and concomitant loss of the π-bonding. Replacing the naphthyl by a quinolyl anchoring group has no marked effect on the e.e. under otherwise similar conditions, indicating that the quinolyl nitrogen does not play an important role, neither for adsorption nor for enantiodifferentiation.

   The importance of the extended aromatic ring system for suitable anchoring is illustrated by the fact that corresponding benzene and pyridine derivatives are ineffective (e.e. ≈ 0%). However, a striking increase in e.e. from 75 to 87% is achieved when the naphthyl anchoring moiety is substituted by an anthracenyl group resulting in 1-(9-anthracenyl)-2-(1-

pyrrolidinyl)ethanol [21] (APE, Fig. 4). This behaviour was not predictable by means of the model calculations, because the electronic interaction of the anchoring group with the platinum surface could not be taken into account. Based on the different enantiodiscrimination of the modifiers with benzyl-, naphthyl- and anthracenyl anchoring group, it seems that a suitable adsorption of the transition complex on the platinum surface is extremely important. The geometrical constraints imposed on the transition complex are obviously insufficient when a benzyl of pyridyl ring is used for anchoring. Substituting the 9-anthracenyl group by a 9-triptycenyl moiety (1-(9-triptycenyl)-2(1-pyrrolidinyl)ethanol, TPE, Fig. 4) also results in a complete loss of enantiodifferentiation[21], demonstrating further that the extended flat aromatic ring system is a crucial structural element of efficient modifiers for α-ketoester hydrogenation.

An attractive alternative to the novel aminoalcohol type modifiers is the use of 1-(1-naphthyl)ethylamine (NEA, Fig. 4) and derivatives thereof as chiral modifiers[19]. Trace quantities of *(R)*- or *(S)*-1-(1-naphthyl)ethylamine induce up to 82% e.e in the hydrogenation of ethyl pyruvate over Pt/alumina. Note that naphthylethylamine is only a precursor of the actual modifier, which is formed in situ by reductive alkylation of NEA with the reactant ethyl pyruvate. This transformation (Fig. 4), which proceeds via imine formation and subsequent reduction of the C=N bond, is highly diastereoselective (d.e. >95%).

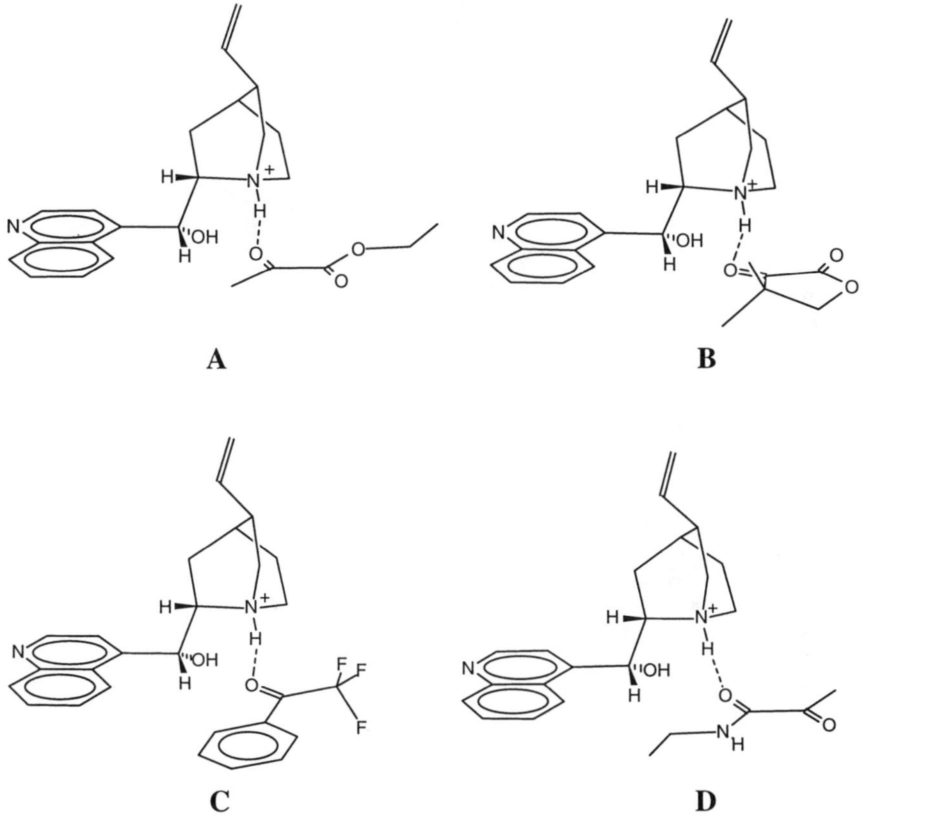

Figure 3 *Structures of enantiodifferentiating complexes between protonated CD and varoius reactants. (R)-products are formed upon hydrogenation. A: ethyl pyruvate—CD, B: trifluoroacetophenone—CD, C: ketopantolactone—CD, D: N-ethyl pyruvamide—CD*

Interestingly the absolute configuration at the stereogenic centre in α-position to the ester group has no influence on the enantioselection of modifiers **a** and **b** (Fig. 4, bottom). Reductive alkylation of NEA with different aldehydes or ketones provides easy access to a variety of related modifiers[19]. The enantioselection occurring with the modifiers derived from NEA could be rationalised with the same strategy of molecular modelling as demonstrated for the Pt—PNE and Pt-cinchona system.

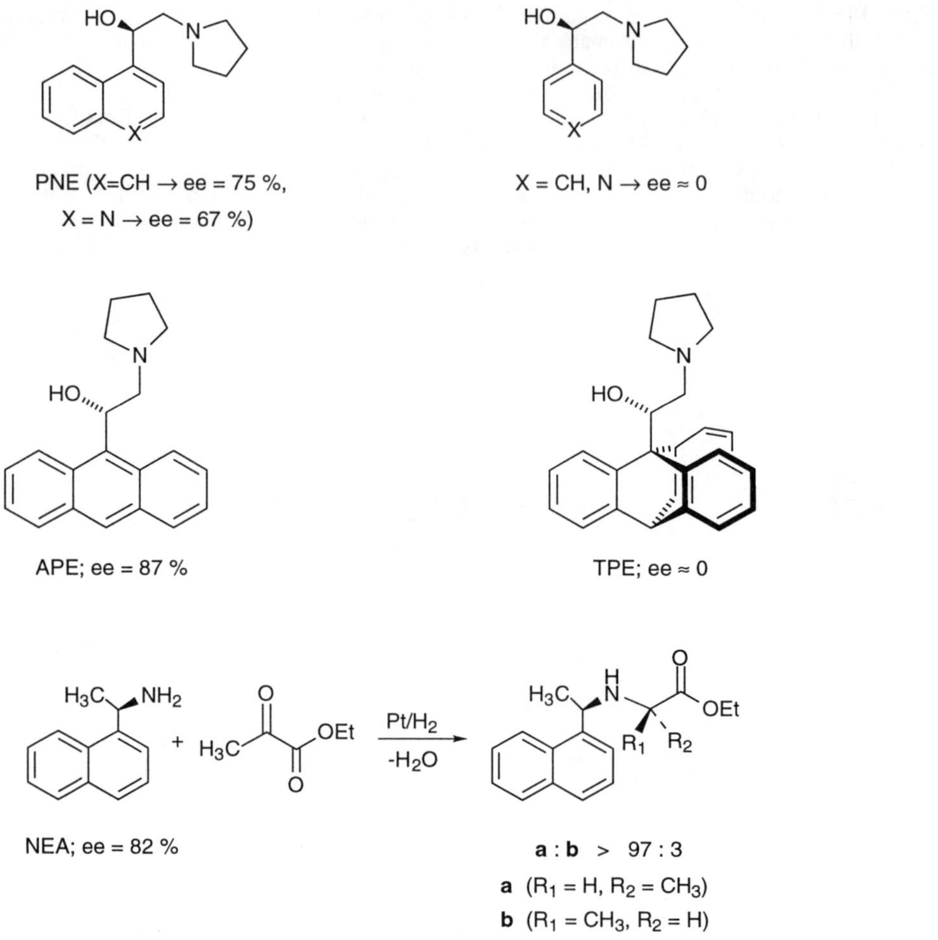

PNE (X=CH → ee = 75 %,

X = N → ee = 67 %)

X = CH, N → ee ≈ 0

APE; ee = 87 %

TPE; ee ≈ 0

NEA; ee = 82 %

a : b > 97 : 3

a (R_1 = H, R_2 = CH_3)

b (R_1 = CH_3, R_2 = H)

Figure 4 *Modifiers for the hydrogenation of α-ketoesters over Pt/alumina*

4.2 Reactants

Hydrogenation experiments with the reactants, ketopantolactone[21], trifluoroacetophenone[22] and N-ethyl pyruvamide[23], whose interaction complexes with CD resemble that of MP—CD, as proposed by molecular modeling (Fig. 3), were all successful. Enantiomeric excesses achieved were: ketopantolactone (e.e. = 79%), trifluoroacetophenone (61%), and N-ethyl pyruvamide (58%). This corroborated the theoretical studies which indicated that the pro (*R*)-complex between reactant and CD is energetically favoured.

5 CONCLUSIONS

Progress in the mechanistic understanding of the enantioselective hydrogenation of α–ketoesters has fostered the design of new modifiers and extended the scope of chirally modified platinum catalysts. Based on experimental and theoretical studies, structures of the enantiodiscriminating diastereomeric complexes formed between modifier and reactant have been rationalised. These findings have been successfully applied for designing new efficient modifiers and for extending the scope of reactants which can be hydrogenated enantio-selectively on chirally modified platinum.

Acknowledgment

Financial support by the Swiss National Science Foundation (Program Chiral 2) is gratefully acknowledged. The author thanks present and past co-workers whose names appear in the reference list for their valuable contributions.

References

1.   E. Erlenmeyer and H. Erlenmeyer, *Biochem. Zeitschr.* 1922, **233**, 52.
2.   G.M. Schwab and L. Rudolph, *Naturwiss.*, 1932, **20**, 363.
3.   A. Baiker and H. U. Blaser in 'Handbook of Catalysis', G. Ertl, H. Knözinger and J. Weitkamp, Eds., Vol. 5, Verlag Chemie, Weinheim, 1997, p. 2442.
4.   Y. Nakamura, *Bull. Chem. Soc. Jpn.*, 1941, **16**, 367. Y. Izumi, *Adv. Catal.*, 1983, **32**, 215.
5.   Y. Orito, S. Imai, S. Niwa and G-H. Nguyen , *J. Synth. Org. Chem. Jpn.*, 1979, **37**, 173. Y. Orito, S. Imai and S. Niwa, *J. Chem. Soc. Jpn.*, 1979, 1118.
6.   A. Tai and T. Harada, in 'Tailored Metal Catalysts', (Y. Iwasawa, Ed.) D. Reidel, Dordrecht, 1986, p. 265. G. Webb and P.B. Wells, *Catal.Today*, 1992, **12**, 319.
7.   A. Baiker, *J. Mol. Catal. A: Chemical*, 1997, **115**, 473.
8.   H.U. Blaser, H.P. Jalett, M. Müller, and M. Studer, *Catal. Today*, 1997, in press.
9.   H.U. Blaser, H.,P. Jalett, D.M. Monti, A. Baiker, and J.T. Wehrli, *Stud. Surf. Sci. Catal.*, 1991, **67**, 147.
10.   G. Bond, P.A. Meheux, A. Ibbotson, and P.B. Wells, *Catal. Today*, 1991, **10**, 371.
11.   J.T. Wehrli, A. Baiker, D.M. Monti and H.U. Blaser, *J. Mol. Catal.*, 1989, **49**, 195.
12.   J.T. Wehrli, A. Baiker, D.M. Monti, H.U. Blaser and H.P. Jalett, *J. Mol. Catal*, 1989, **57**, 245.
13.   H.U. Blaser, H.P. Jalett, J. Wiehl, *J. Mol. Catal.*, 1991, **68**, 215.
14.   J. Wang, C. LeBlond, C.J. Orella, Y. Sun, J. Bradley and D.G. Blackmond, *Stud. Surf. Sci. Catal.* 1997, **108**, 183.
15.   O. Schwalm, B. Minder, J. Weber, and A. Baiker, *Catal. Lett.*, 1994, **23**, 268.
16.   G. Bond and P.B. Wells, *J. Catal.*, 1994, **150**, 329.
17.   B. Minder, T. Mallat, A. Baiker, G. Wang, T. Heinz A. Pfaltz, *J. Catal.*, 1995, **154**, 371.
18.   K.E. Simons, G. Wang, T. Heinz, T. Mallat, A. Pfaltz and A. Baiker, *Tetrahedron Asym.*, 1995, **6** , 505.
19.   B. Minder, M. Schürch, T. Mallat, A. Baiker, G. Wang, T. Heinz, and A. Pfaltz, *J. Catal.*, 1996, **160**, 261.
20.   B. Minder, M. Schürch, T. Mallat, A. Baiker, *Catal. Lett.*, 1995, **31**, 143.
21.   M. Schürch, T. Heinz, R. Aeschimann, T. Mallat, A. Pfaltz, and A. Baiker, *J. Catal.*, submitted for publication
22.   M. Schürch, O. Schwalm, T. Mallat, J. Weber, and A. Baiker, *J. Catal.*, in press.
23.   M. Bodmer, T. Mallat, and A. Baiker, *Catal. Lett.*, 1997,.**44**, 95.
24.   G.Z. Wang, T. Mallat, and A. Baiker, *Tetrahedron Asym.*, 1997, **8**, 1.

AN INTEGRATED APPROACH TOWARDS THE SYNTHESIS OF NEW SILICA SUPPORTED CATALYSTS

A. Choplin, F. Quignard, P. Leyrit, S. dos Santos, C. McGill, O. Graziani, D. Sinou

Institut de Recherches sur la Catalyse, and Laboratoire de Synthèse Asymétrique, CNRS-Université Lyon-1, Villeurbanne, France

1 INTRODUCTION

A number of industrial reactions are catalyzed by organometallic complexes in homogeneous phase, despite the problems encountered for catalyst recovery. This problem being crucial when expansive ligands are involved as with enantioselective reactions, numerous studies have been performed in order to immobilize these catalysts. Different ways were envisaged, among which the most powerfull uses biphasic media, water/organic solvent being the most popular. Immobilization by heterogenization has so far lead to few convincing results. This must be attributed to the fact that the reaction between the organometallic catalyst and the surface of the solid often induces changes of the degree of oxidation and/or of the coordination sphere of the metal[1] and to the rupture of the bond between the complex and the support.

These latter observations prompted us to reconsider the strategy of synthesis of immobilized catalysts. The choice of the organometallic precursor complex, of the support and of the mode of immobilization must clearly be made in conjunction with all the critical parameters of the target reaction; these include the nature of the reactants, the products, the solvent, and the conditions of pressure, temperature... It must take into account the general rules of organometallic chemistry and the mechanisms of homogeneous (and in some cases heterogeneous) catalysis. We will illustrate this approach with two examples of silica supported catalysts based on zirconium and palladium respectively, differing by the mode of immobilization; the target reactions are as diverse as reduction/oxidation by hydrogen transfer, epoxidation of olefins, dehydropolymerisation of silanes and nucleophilic substitution of allylic substrates.

2 RESULTS

2.1 Silica Bonded Molecular Zirconium Catalysts

Among the reactions catalyzed by group 4 elements, we have selected on one hand the reduction/oxidation of ketones/alcohols by hydrogen transfer and the dehydrocoupling of silanes, on the other hand the epoxidation of olefins by H_2O_2. For these reactions, the efficiency of the catalysts is correlated to the capacity of the organometallic active complexes to remain mononuclear[2] or of the active heterogeneous sites to remain perfectly isolated.[3] Indeed, the dimerization of the complexes or the sintering of the sites into oxide particles are known deactivation processes. We have therefore chosen as organometallic precursor an homoleptic alkyl zirconium complex, i.e. tetra(neopentyl) zirconium, (ZrNp$_4$) and as support a partially dehydroxylated silica (silica evacuated at 500°C); their reaction leads to the formation of only one type of surface zirconium complex, **1** (eq.1) which is chemically attached to the solid via a strong covalent bond Zr-[O]$_s$.[4] This surface complex cannot as such self-associate; nevertheless, we have simultaneously kept its surface concentration low ($\leq$ 1% wt.).

$$\equiv\text{SiOH} + \text{ZrNp}_4 \rightarrow \equiv\text{SiO-ZrNp}_3 + \text{NpH} \qquad (1)$$
$$\mathbf{1} \qquad\qquad \text{Np = neopentyl}$$

The surface tris(neopentyl)zirconium complex (**1**) is not as such a catalyst; we have therefore modified it by introducing the ligands adapted to the target reaction. Species (**2**) - (**5**) (scheme 1) are obtained from (**1**) by mild chemical reactions (hydrolysis, alcoholysis and hydrogenolysis) and species(**6**) by a treatment at 673K under vacuum; these reactions all preserve the integrity of the anchoring bond. These surface complexes were identified by IR spectroscopy, elemental analysis and/or chemical reactivity.

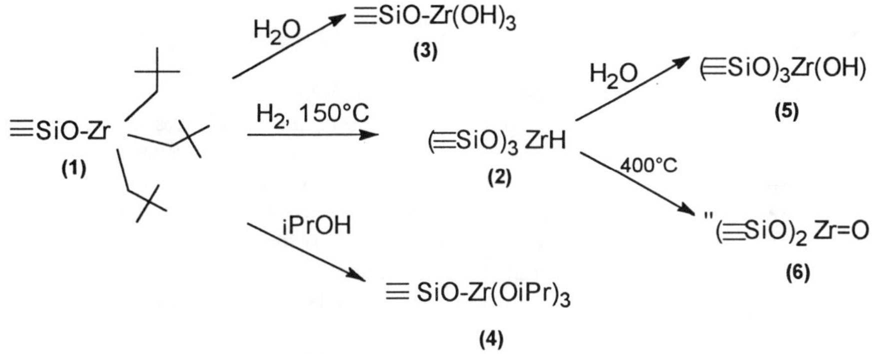

Scheme 1

2.1.1 Reduction and Oxidation by Hydrogen Transfer. These reactions, (eq. 2) are usually performed in presence of substoichiometric amounts of Al(OR)$_3$. More recently, it was shown that they can be catalyzed homogeneously by alkoxy complexes of lanthanides and elements of group 4 and heterogeneously by aqueous ZrO$_2$.[5]

In the surface complex ($\equiv$SiO)Zr(OiPr)$_3$, (**4**) Zr shows the proper coordination sphere, except for one alkoxy ligand which is substituted by a siloxy ligand. Table 1 shows some of the properties of (**4**).

Table 1 *Reduction /Oxidation by Hydrogen Transfer Catalyzed by ($\equiv$SiO)Zr(OiPr)$_3$*

substrate	reductant(oxidant)	[substrate]/[Zr]	T(K)	% conv.
	iPrOH[a]	72	353	75[c]
	iPrOH[a]	50	353	38[c]
	PhC(O)Me	40	383	49[b]
	PhC(O)Me	40	383	55[d]
	PhCHO	72	383	49[b]

a: solvent: iPrOH; b: after 6h, solvent: toluene;c: after 20h; d: solvent : octane

Several points deserve comment; (i): surface complex (4) is an active catalyst for both Meerwein-Ponndorf-Verley and Oppenauer reactions, the reactivities of the different substrates being explained by steric and electronic effects, and by the relative stability of the intermediate alkoxy Zr complexes towards alcohols;[5] (ii): Zr(OiPr)$_4$ under the same experimental conditions, is totally inactive, a fact which must be correlated to its self association; (iii):complex (4) is more active than silica supported Zr catalyst, obtained from Zr(OiPr)$_4$ (20% wt. Zr);[6] (iv): no significant Zr leaching was observed and catalyst (4) can be recycled without significant loss of activity by simple filtration and in some cases regeneration with boiling isopropanol.

These observations strongly suggest that (4) is a true mononuclear complex, self association being avoided by an initial good dispersion of Zr on the silica surface and by the stability of the Zr-[O]$_s$ bond under the reaction conditions.

2.1.2. Dehydrogenative Coupling of Silanes. The synthesis of polysilanes is industrially performed by Wurtz coupling of chlorosilanes.[7] It was shown recently that such polymers can be obtained via dehydrogenative coupling of silanes (eq. 3) catalyzed by zirconocene alkyl or hydride complexes.[8]

$$\equiv \text{Si-H} \ + \ \text{H-Si} \equiv \ \text{-----} \blacktriangleright \ \longrightarrow \ -\!\!\left(\!\underset{|}{\overset{|}{\text{Si}}}\!-\!\underset{|}{\overset{|}{\text{Si}}}\!\right)_n \ + \text{H}_2 \qquad (3)$$

Supported complex (2)[9] is fundamentally different from the hydride complexes, Cp$_2$ZrH$_2$: it is formally an 8e$^-$ species (compare to 16 e$^-$), therefore highly electrophilic and is unusually thermally stable (T$\leq$200°C). These two features explain the peculiar reactivity of (2), which activates the C-H bond of alkanes (even of methane)[10] and the Si-H bond of all types of silanes.[11] For this latter reaction, supported (2) presents a specific property: it can activate simultaneously the Si-H and the C-H bond of some silanes such as for example

Et$_3$SiH (demonstrated by H/D exchange followed by *in situ* IR spectroscopy). (Fig. 1) Silica supported (**2**) is more active and selective than the homogeneous analog complexes: no formation of cyclic oligomers and a narrow mass distribution. (Fig.2) These properties may be attributed to the high electrophilic character of Zr in (**2**).

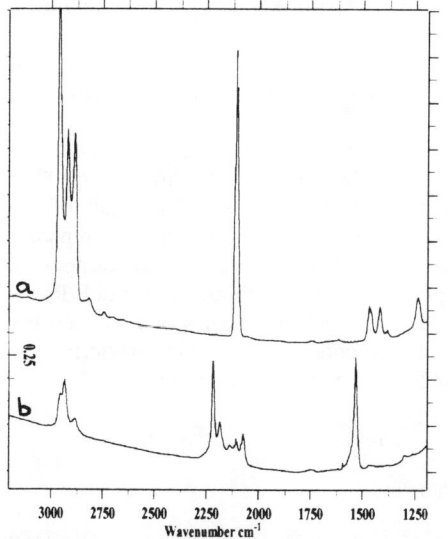 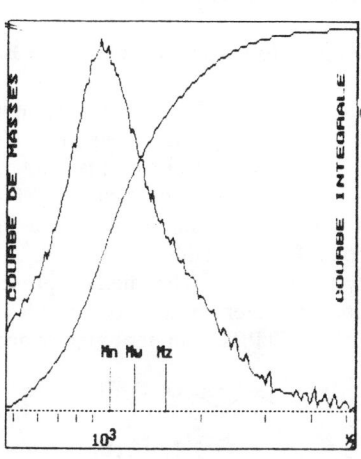

Figure 1: Gas phase IR spectrum of: (a): Et$_3$SiH + D$_2$; (b): (a) after contact with **1** (3h, 60°C)

Figure 2: Mass distribution of the polymers obtained with PhSiH$_3$ as determined by SEC

2.1.3 Epoxidation of Cyclohexene by H$_2$O$_2$. The best heterogeneous catalyst for the epoxidation of olefins, using H$_2$O$_2$ as oxidant, is presently the ENICHEM catalyst, a titanosilicalite, TS-1;[12] in this solid, Ti, substituted for Si in the framework of silicalite, is perfectly isolated (no Ti-O-Ti bonds). This structural property is considered as the key parameter of the high selectivity for epoxide formation. Although all reported data suggest that Zr based complexes or solids are rather poor catalysts for this same reaction,[13] we have tested the silica supported complexes (**3**), (**5**) and (**6**) for the epoxidation of cyclohexene. Hydroxy complexes are suggested intermediates in any among the proposed catalytic cycles; the latter surface entity, which is either a zirconyl complex ($\equiv$SiO)$_2$Zr=O or a Zr "incorporated" in the framework of silica ($\equiv$SiO)$_4$Zr[14] would be similar respectively to the presumed active sites of the Shell catalyst[15] or to a Zr analog of TS-1 respectively.[12]

Table 2: *Epoxidation of Cyclohexene by H$_2$O$_2$ on silica supported Zr Catalysts*

catalyst	Zr (%wt)	H$_2$O$_2$ conv.(%)	select. (%)		tot. select. (%)
			epoxide	diol	
3	0.93	76	63	8	71
5	0.67	68	59	6	65
6	0.37	72	70	5	75

Exp. cond.: m(cata)=2 g, cyclohexene: 0.5 mol; H$_2$O$_2$ (70%wt.): 25 mmol; solvent: diglyme (30mL), T=363K, t(reaction): 3h.

The three surface species are unexpectedly as active as conventional amorphous titanium based solids:(table 2) total epoxidation yields (diol included) are close to 55%. The best catalyst is solid **6** in terms of activity per Zr_s and selectivity. This seems to be correlated to the structure of the precursor surface entity, but may also be a function of Zr concentration, a factor which may govern the extent of site isolation. No Zr could be detected in the final solution.

2.2 Supported Aqueous Phase Pd Based Catalysts for the Trost-Tsuji Reaction

Pd(0) stabilized by phosphine ligands are well known catalysts for the reaction of allylic alkylation, the so-called Trost-Tsuji reaction.[15] This catalyst was "immobilized" in water (after substitution of the triphenylphosphine ligands by the trisodium salt of the tri(sulfonatedphenyl)phosphine (TPPTS)) and successfully used under biphasic water/nitrile conditions.[16] We have thus tested the procedure described as Supported Aqueous Phase Catalysis (SAPC),[17] an adaptation of the Supported Liquid Phase Catalysis (SLPC).[18] This method preserves the catalyst intact and allows one to keep the amount of water at a very low level. Table 3 presents the catalytic properties of (Pd(OAc)$_2$/5TPPTS) under biphasic and under SAP conditions for reaction (4).

$$Ph\diagdown\diagup\diagdown OCO_2Et + NuH \longrightarrow Ph\diagdown\diagup\diagdown Nu + EtOH + CO_2 \qquad (4)$$

NuH: ethyl acetoacetate, morpholine

Table 3 *Catalytic Properties of Pd(OAc)$_2$/5TPPTS in H$_2$O/nitrile (1/1) and under SAP Conditions*

nucleophile	solvent	catalyst type	T°C	conv. (%)	sel.(%)^d
ethylacetoacetate	CH$_3$CN	biphasic	80	50^a	50^e
"	CH$_3$CN	SAP (22% H$_2$O)	80	40^a	95
"	PhCN	biphasic	80	0^b	
"	PhCN	SAP (32% H$_2$O)	80	90^b	95^f
morpholine	CH$_3$CN	biphasic	50	85^c	100
"	CH$_3$CN	SAP(50% H$_2$O)	50	30^c	100
"	PhCN	biphasic	50	5^c	100
"	PhCN	SAP (50% H$_2$O)	50	95^c	100

^{a}:after 8h; ^{b} after 40mn; ^{c}: after 5mn; ^{d}: for the monoalkylated product; ^{e}: cinnamyl alcohol formed; ^{f}: dialkylated product formed. SAP catalyst is supported on a mesoporous (pore Ø: 24nm) silica from Grace.

With CH$_3$CN as a solvent, the SAP catalysts show an activity lower than the biphasic system, independly of the water content of the solid and of the nature of the nucleophile; this must be correlated to total miscibility of H$_2$O and CH$_3$CN. With ethylacetoacetate, the SAP catalyst is highly selective in accordance with the low water concentration (in H$_2$O/CH$_3$CN, cinnamyl alcohol forms).

With PhCN as a solvent, the SAP catalyst is superior to its truly biphasic analog for both nucleophiles and this whatever its water content; maximum activity of the SAP catalyst is observed with water contents close to ca. 50% wt. H$_2$O (which is close to that necessary to fill the pores of the silica support). Finally, the SAP catalyst is, in all cases, more stable towards degradation to metallic particles.

3 DISCUSSION. CONCLUSIONS

These two series of examples clearly show that an integrated approach of the design of immobilized molecular catalysts is necessary to get an efficient and stable catalyst. With chemically anchored complexes, no homogeneous counterpart is generally known; one introduces at least one "solid" siloxy ligand (on silica for example) whose chemical and thermal stability has so far not been extensively studied. With Zr, this bond is strong enough to remain intact under H_2O and ROH up to temperatures close to 100°C, so that we could perform various reactions with yields generally superior to those reported for organometallic Zr complexes. This superior activity must be correlated to the solid itself, which intervenes through electronic, steric and immobilizing effects. The Supported Aqueous Phase method is well adapted to catalysts which cannot suffer ligand substitution and to reactions which are sensitive to large amounts of water. In both cases, recyclability was checked: it is very good with silica supported Zr catalysts and is dependent on the water content with the Pd / SAP catalysts.

References

1. *see for example:* Yu. I. Yermakov, B.N. Kusnetsov and V.A. Zakharov, "Catalysis by Supported Complexes", Elsevier, Amsterdam, 1981; H. H. Lamb, B. C. Gates and H. Knözinger, *Angew. Chem. Intern. Ed. Engl.*, 1988, **27**, 1127; J. M. Basset, J. P. Candy, A. Choplin, C. Nédez, F. Quignard, C. C. Santini, *Mat. Chem. Phys.*, 1991, **29**, 5; A. Choplin and F. Quignard, *Trends in Inorg. Chem.*, 1993, **3**, 463.
2. B. Knauer and K. Krohn, *Liebigs Ann.*, 1995, 677.
3. D. R. C. Huybrechts, P. L. Buskens and P. A. Jacobs, *J. Mol. Catal.*, 1992, **71**, 129.
4. F. Quignard, C. Lécuyer, C. Bougault, F. Lefebvre, A. Choplin, D. Olivier and J. M. Basset, *Inorg. Chem.*, 1992, **31**, 928.
5. C. F. de Graauw, J. A. Peters, H. van Bekkum and J. Huskens, *Synthesis*, 1994, 1007.
6. K. Inada, M. Shibagaki, Y. Nakanishi and H. Matsushita, *Chem. Lett.*, 1993, 1795.
7. R. J. West, *J. Organomet. Chem.*, 1986, **300**, 327.
8. C. Aitken, J. F. Harrod and E. Samuel, *Can. J. Chem.*, 1987, 65, 1804.
9. F. Quignard, C. Lécuyer, A. Choplin and J. M. Basset, *J. Chem. Soc., Dalton Trans. 1*, 1994, 1153.
10. F. Quignard, A. Choplin and J. M. Basset, *J. Chem. Soc., Chem. Commun.*, 1991, 1589.
11 B. Coutant, F. Quignard and A. Choplin, *J. Chem. Soc., Chem. Commun.*, 1995,137.
12. M. Taramasso, G. Perego and B. Notari, US patent 4,410,501 (1983)
13. R.A. Sheldon and J. A. van Doorn, *J. Catal.*, 1973, **31**, 427.
14. F. Quignard, A. Choplin and R. Teissier, *J. Mol. Catal.*, 1997, **120**, 9.
15. J. Tsuji, H. Takashashi and M. Morikawa, *Tetrahedron Lett.*, 1965, 4387; B. M. Trost and T. R. Verhoeven, *J. Am. Chem. Soc.*, 1978, **100**, 3435.
16. E. Blart, J. P. Genêt, M.Safi, M. Savignac and D. Sinou, *Tetrahedron*, 1994, **50**, 505
17. K. T. Wan and M. E. Davis, *Nature*, 1994, **370**, 449.
18. G. J. K. Acres, G. C. Bond, B. J. Cooper and J. A. Dawson, *J. Catal.*, 1966, **6**, 139; P. R. Rony, *Chem. Eng. Sci.*, 1968, **23**, 1021; P. R. Rony, *J. Catal.*, 1969, **14**, 142.

REICHARDT'S DYE AS A PROBE FOR SURFACE POLARITY

**Stewart J. Tavener, James H. Clark*, Paul A. Heath,
Duncan J. Macquarrie and John Rafelt**
Department of Chemistry, University of York, Heslington, York, UK YO1 5DD

1   Introduction

Catalytic activity in heterogeneous systems may be strongly affected by the ability of the catalyst surface to adsorb compounds, which in turn is controlled by surface polarity. In heterogeneous catalysis, the reaction occurs either at the catalyst surface, inside pores, or in the slow moving liquid film at the surface and the adsorption of substrates or desorption of products is often rate limiting.[1] In certain cases, such as supported phase transfer catalysts, the compatability of the surface with the reactant phases is critical.[2] However, measurements of surface polarity are not easy and are rarely used in practice.

I

Reichardt's pyridinium-*N*-phenoxide betaine dye (**I**) holds the world record for solvatochromism and shows a range of energies for its $\pi \rightarrow \pi^*$ absorption band from 147 kJ mol^{-1} (810 nm) in diphenylether to 264 kJ mol^{-1} (453 nm) in water. This exceptional behaviour makes the dye a useful indicator of solvent polarity, with the measurement being made by human eye (the absorption range is almost entirely within the visible region), or, more quantitatively, by UV-vis spectrophotometry. The original solvent polarity scale based on this dye, proposed by Dimroth and Reichardt, was simply defined as the energy of the longest wavelength adsorption for the dye, measured in kcal mol^{-1}.[9] This scale may be normalised to avoid non-SI units (equation 1):

$$E_T^N = \frac{E_{T(solvent)} - E_{T(TMS)}}{E_{T(water)} - E_{T(TMS)}}$$ **Equation 1**

More polar solvents, such as water and *N,N*-dimethylformamide stabilise the charged zwitterionic ground state of the dye, but not the uncharged excited state, which leads to a larger energy change for the $\pi \rightarrow \pi^*$ transition than in less polar media. However, on this scale dimethyl sulfoxide has a much lower value than might otherwise be expected, suggesting that other specific interactions, as well as general solvent polarity, affect the transition energy. In this application, Reichardt's dye is acting as a reporter molecule, revealing information about its local environment via an easy to measure property. This environment-dependent transition energy should therefore also be useful for probing the surface of a material, giving an indication of the polarity of that surface.

Some earlier attempts have been made to measure solvatochromic parameters for the surfaces of inorganic oxides, but these investigations have usually involved the study of the solid in contact with a mobile liquid phase.[3] Silica samples studied as a suspension in 1,2-dichloroethane were found to have an E_T^N value of 0.798 to 0.842,[4] and were also found to have extremely high polarities as measured by the UV adsorption maximum of adsorbed 4-nitroanisole.[5]

2 Results and Discussion

We have recently reported that Reichardt's Dye is a convenient probe for the polarity of surfaces of chemically and thermally treated silica surfaces.[6] A solution of the dye (in dichloromethane) is simply added to the the material under study, the solvent removed on a rotary evaporator and the material dried on a vacuum line for ten minutes. Because dimerisation is known to occur in solution at higher concentrations the amount of dye added was restricted to 1 µmol g^{-1} (*circa* 2 nanomoles m^{-2}).[9] The choice of solvent used to download the dye is important because alcohols and ketones will react with or bind strongly to surface hydroxyls on silica surfaces[7], and the dye is insufficiently soluble in apolar solvents such as aliphatic hydrocarbons. Dichloromethane was found to be the most suitable solvent because of its high volatility and chemical stability. The $\pi \rightarrow \pi^*$ transition energy of the adsorbed dye is measured by diffuse reflectance UV-vis spectrophotometry, although the polarity may be estimated by eye: very polar surfaces such as untreated silicas are red, ranging through purple and blue for organosilicas, to green for non polar surfaces such as macroporous polystyrene. The spectrum of the undyed material is subtracted from that of the dyed material to remove any contributions from the solid itself. The values for E_T^N presented here are calculated using a modified version of equation 1, using $E_{T(surface)}$ instead of $E_{T(solvent)}$.

This method has the advantages of being very rapid, and no special equipment is needed (simply a UV-vis spectrometer fitted with an integrating sphere). Reichardt's dye is a particularly suitable probe molecule for this application as the entire range for the $\pi \rightarrow \pi^*$ transtion is within the visible region (many integrating spheres only have a limited range beyond the visible region). Because the dye is so strongly solvatochromic the spectrometer does not need to be particularly sensitive. The dye is expensive, but a 1g (~£60) bottle is sufficient for about 7000 measurements if used frugally.

The dye may not be used in strongly acidic environments as the oxygen centre becomes protonated, preventing the solvatochromic transition.[8] Residual adsorbed HCl prevents sensible values from being obtained for the surface polarities of silicas treated with chlorosilanes. This method has the limitation of being a single parameter measurement. Other solvatochromic scales use several dyes to distinguish between H-bond donor, acceptor, and other non specific interactions. Those probe molecules are weakly solvatochromic by comparison, and require sensitive measurements in the UV.[9]

A range of thermally treated silicas was prepared by heating commercial mesoporous silicas (Kieselgel 60 and Kieselgel 100, *ex Merck*) for 20 hours at the desired temperature under a flowing nitrogen atmosphere before being treated with the dye solution. Thermal dehydration of silica gel is known to have a large effect on surface chemistry of the silica, but comparatively little effect on its surface area.[10] Thermal treatment of both silicas was found to have a marked effect on their polarity, with the E_T^N value correlating well with thermogravimetric analysis data (figure 1). Initial heating causes a rapid loss of mass as loose, physisorbed water is removed from the surface, and a decrease in polarity as the adsorbed dye is no longer in a wet environment, but now sees a hydrogen bonded network of surface silanols. Further heating will remove some of these silanols, producing isolated polar groups whose polarity is not reduced by hydrogen bonding - this is reflected in the slight increase in polarity observed in the 300 to 400° C region for both silicas. As the temperature is raised to 700° C, more surface hydroxyls dehydrate, producing more and more siloxane bridges (figure 2). Both silicas show an increase in polarity on heating beyond 700° C. The cause of this is not clear but may be due to a low energy restructuring of the gel matrix, although no energy changes could be detected by differential scanning calorimetry.

Organosilane-treated silicas show a reverse linear correlation between degree of functionalisation and polarity as measured by the E_T^N values. Figure 3 shows the effect of

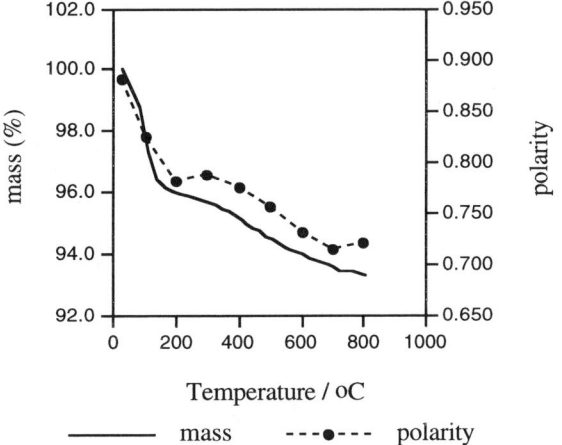

Figure 1 Thermogravimetric analysis and the effect of thermal treatment on the polarity (E_T^N) of Kieselgel 60 silica

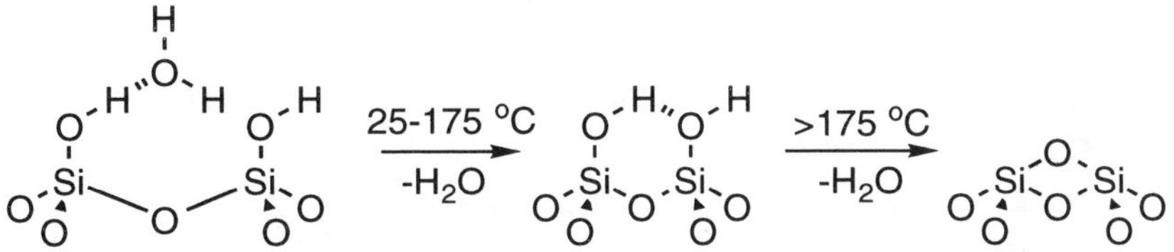

Figure 2 Thermal dehydration of a silica surface

the degree of functionalisation on the polarity of silicas treated with phenyltrimethoxysilane. For these materials the dye is exposed to non-polar organic functions as well as fewer polar surface hydroxyls, compared with the unmodified silica. The measured polarities of a range of organically modified silicas at specific degrees of functionalisation are given in table 1.

Although all of the organic groups studied reduce the polarity of silica, they do so by different degrees. The decrease in polarity per millimole of organic groups per gram of silica (ΔE_T^N) increases in the order: cyanoethyl < aminopropyl < imidazole < phenyl < chloropropyl < trimethylsiloxyl, which is broadly in agreement with the E_T^N values found for comparable compounds in the liquid phase: TMS = 0.000; benzene = 0.111; 1-chlorobutane = 0.191; n-butylamine = 0.213; propionitrile = 0.401; water = 1.00.[11]

This method was also used to measure the polarities of a range of organic and inorganic support materials (table 2). Hydroxylic silica is the most polar of these, and polystyrene, unsurprisingly, shows very little polarity. Calcination reduces the polarity of

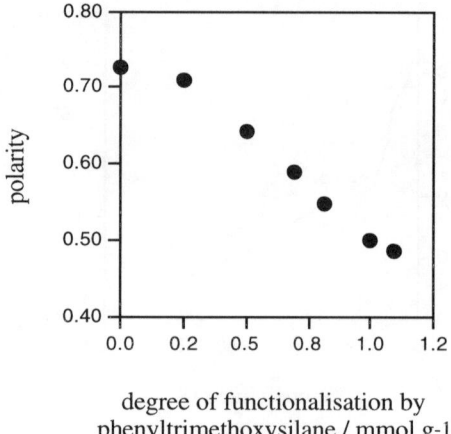

Figure 3 Effect of the degree of functionalisation on the polarity (E_T^N)
of silicas treated with phenyltrimethoxysilane

alumina, and the HMS material shows a value similar to that of silica treated at the same
temperature. A value could not be obtained for zirconia, because it is sufficiently acidic to
coordinate the oxygen centre on the dye and prevent the electronic transition.

Table 1 Polarities of a range of organically modified silicas

material [a]	degree of functionalisation[b] / mmol g^{-1}	λ_{max} / nm	E_T^N	ΔE_T^N / mmol^{-1} g
silica-OSi(CH$_3$)$_3$	0.53	560	0.630	-0.634
silica-Ph	1.00	610	0.500	-0.467
silica-(CH$_2$)$_3$Cl	0.64	554	0.645	-0.503
silica-(CH$_2$)$_3$NH$_2$	0.99	590	0.546	-0.425
silica-(CH$_2$)$_2$CN	0.99	557	0.636	-0.334
silica-(CH$_2$)$_3$N⟨N⟩ [d]	0.97	611	0.497	-0.485

a prepared by reaction with the appropriate organosilane, unless otherwise stated; *b* measured by C, N
elemental microanalysis; *c* Kieselgel 100, untreated; *d* for preparation method see reference 12

Table 2 Polarity measurements for a range of organic and inorganic support materials

material	E_T^N
silica: Kieselgel 100	0.967
silica: Kieselgel 60	0.880
alumina: basic	0.895
alumina: neutral	0.526
alumina: neutral, calcined	0.452
HMS (21Å)	0.764
titania	0.571
polystyrene: Amberlite XE305	0.278
polystyrene: Amberlite XAD4	0.225

3   Conclusions

Adsorption of Reichardt's Dye on solid surfaces is an easy and rapid method for reliably estimating their polarity. This method should prove to be a valuble technique in the study of adsoption and desorption in liquid - solid catalytic systems.

We thank the EPSRC clean technology unit for a ROPA award (to SJT), the Royal Academy of Engineering/EPSRC for a Fellowship (to JHC) and the Royal Society for a fellowship (to DJM).

4   References

1 M. Tomoi and W. T. Ford, J. Am. Chem. Soc., 1981, **103**, 3821.
2 S. L. Regen, D. Bolikal and C. Barcelon, *J. Org. Chem.*, 1981, **103**, 3828;
   P. Tundo and P. Venturello, *J. Am. Chem. Soc.*, 1979, **101**, 6606.
3 J. L. Jones and S. C. Rutan, *Anal. Chem.*, 1991, **63**, 1318; Z. Li and S. C. Rutan, *Analytica Chimica Acta*, 1995, **312**, 127; M. Kosmulski, *J.Colloid Interface Sci.*, 1996, **179**, 128.
4 S. Spange, A. Reuter and E. Vilsmeier, *Colloid Polym. Sci.*, 1996, **274**, 59.
5 C. G. Flowers, S. Lindley and J. E. Leffler, *Tetrahedron Lett.*, 1984, **25**, 4997.
6 S. J. Tavener, J. H. Clark, G. W. Gray, P. A. Heath and D. J. Macquarrie,
   J. Chem.Soc .,Chem.Commun., 1997, 1147.
7 R. K. Iler, *The Chemistry of Silica*, John Wiley and Sons, 1979.
8 C. Reichardt, *Chem. Soc. Rev.*, 1992, 147.
9 C. Reichardt, *Solvents and Solvent Effects in Organic Chemistry, second edition*, VCH Publishers, 1988.
10 W. K Lowen and E. C. Broge, *J. Phys. Chem.*, 1961, **65**, 16.
11 Y. Marcus, *Chem. Soc. Rev.*, 1993, 409.
12 O. Leal, D. L. Anderson, R. G. Bowman, F. Basolo and R. L. Burwell Jr., *J. Am.Chem. Soc.*, 1975, **97**, 5125.

(1R, 2S)-EPHEDRINE LINKED TO MTS MESOPOROUS SILICAS AS NEW SUPPORTED CHIRAL CATALYSTS FOR ENANTIOSELECTIVE ADDITION OF DIETHYLZINC TO BENZALDEHYDE

N. Bellocq, D. Brunel, M. Laspéras and P. Moreau

Laboratoire de Matériaux Catalytiques et Catalyse en Chimie Organique - UMR 5618
Ecole Nationale Supérieure de Chimie de Montpellier - CNRS
8, rue de l'Ecole Normale 34296 MONTPELLIER Cedex 5 FRANCE
Tel : +33 (0)4 67 14 43 23 ; Fax : +33 (0)4 67 14 43 49

1 INTRODUCTION

Several methods have been described for the asymmetric alkylation of aldehydes with dialkylzincs using chiral aminoalcohols as catalysts, either in homogeneous[1,2] or heterogeneous conditions.[3,4] Among the various types of β-aminoalcohols used, N-alkylephedrines or N,N-dialkylnorephedrines have been extensively studied as typical chiral auxiliaries.

Compared with their homogeneous counterparts, heterogeneous catalysts present many advantages, such as easier separation and recovery from the reaction mixture and thus enhanced recycling possibilities, which are now well established in fine organic synthesis. In order to benefit from such advantages, the immobilization of the above mentioned chiral aminoalcohols on solid supports such as polymers[3] and alumina or silica gel[4] has been reported for the enantioselective addition of dialkylzinc to aldehydes. In the former case, high enantiomeric excesses are obtained, while in the latter case, these ee s are only moderate.

The increasing interest focused on supported catalysts led us to study the immobilization of (1R, 2S)-ephedrine on the new generation of MTS (Micelle Templated Silicas) type mesoporous silicas. These solids are distinguished from amorphous silicas by their regular porosity, consisting of uniformly sized channels with pore diameters within a mesoporous range of 20-100 Å.[5] Taking into account their characteristic structure, the insertion and grafting of functional molecules is possible as it has been shown recently in our laboratory.[6] It thus appears that the use of such solids as supports for chiral auxiliaries should afford an improvment of the asymmetric induction in a confined environment. In order to understand the role of the nature of the surface and the effect of site proximity, the addition of diethylzinc to benzaldehyde was studied as a model reaction, using (-)-ephedrine immobilized on mesoporous MTS type silicas. We report herein the results concerning the activity of these new supported catalysts.

2 RESULTS AND DISCUSSION

2.1 Chemical Modification of MTS

2.1.1 Synthesis of the materials. The immobilization of ephedrine has been carried out over two MTS type silicas, MTS (1) and MTS (1'). MTS (1) is characterized by a 31 Å pore diameter, expressed by 4V/S, and by a 10 Å wall thickness. MTS (1') was obtained by a post-synthesis treatment of an original MTS (S = 749 m²/g, V_{mes} = 0.56 mL/g, pore

Table 1 *Characterization of Supported MTS Silicas*

		Grafted CPS			Grafted chiral ephedrine				
	Solids	N_X x 10^4	S (m²/g)	V_{mp} (mL/g)	Solids	N_X x 10^4	N_E x 10^4	S (m²/g)	V_{mp} (mL/g)
MTS (1)	(1)		962	0.74					
	(2a)	12.8	739	0.42	(3a)	2.9	8.8	631	0.29
	(2b)	14.3	701	0.41	(3b)	6.1	8.3	620	0.28
	(2c)	16.1	608	0.27	(3c)	4.8	11.3	311	0.13
MTS (1')	(1')		684	0.40					
	(2d)	8.8	571	0.23	(3d)	0.6	9.4	373	0.16
	(2e)	4.6	598	0.30	(3e)	0.4	5.8	576	0.27
	(2f)	4.3	563	0.24	(3f)	0.9	3.4	554	0.26

diameter = 30 Å, wall thickness = 12 Å) in a saturated water atmosphere for ten days, leading to a partial loss of the regular mesoporosity, as shown in Table 1.

Solids (2) were obtained by conventional silanization of (1) or (1') with excess of 3-chloro or 3-iodopropyltrialkoxysilane (CPS). Nucleophilic substitution of the halogen by the basic amino group of (-)-ephedrine then led to the catalysts (3). The reaction was carried out with excess ephedrine; the elimination of ephedrine chlorhydrate was performed by careful washing with methanol (Scheme 1).[7]

In order to study the effect of either internal diffusion on the initial rate or site proximity[8] on the ee s, the dispersion of active sites on the silica surface was investigated. With this aim, 3-iodopropyltrialkoxysilane was grafted on solid (1') in competition with ethyltrimethoxysilane leading to the solid (2e), which was then reacted with ephedrine to give (3e) (Scheme 1). Moreover, in order to passivate the surface and to avoid the formation of a less active ephedrine, the support was treated with hexamethyldisilazane[9] (HMDS) before halogen substitution by ephedrine to yield the functionalized solid (3f) (Scheme 1).

2.1.2 Characterization of the Materials. Amounts (mol/g) of grafted moieties for solids bearing either an halogeno function (2) or an ephedrine function (3), respectively N_X and N_E, were calculated from elemental and thermogravimetric analyses. These two methods were in good agreement; the average results are listed in Table 1. It is worth noting that, for solids (2), the coupling reagent $(MeO)_3Si(CH_2)_3Cl$ gives rise to a more efficient grafting than its ethoxy analog. This result can be explained by a greater reactivity of the methoxy groups.

Table 1 shows that, whatever the solid, the substitution of the halogen atom by ephedrine is not quantitative. Some halogen moieties indeed remain on the final chiral supported catalysts. In most of the cases, conservation of the total number of grafted species is observed. The sum of the number of grafted ephedrine and of remaining halogeno moieties is nearly equal to the initial number of grafted coupling functions. This result confirms that, in those cases, ephedrine essentially susbtitutes the halogen atom. Nevertheless, in the case of (3e) for which this number is higher, the excess of total grafted species is explained by the presence of ephedrine directly bound to the silica surface through hydroxyl group attack, as shown in Scheme 1. This grafting mode may result from the presence of surface defects much more favorable for the alkoxylation reaction than the regular mesoporous surface.

Infrared spectra of solids (3), where aromatic C-H stretching vibrations are observed, confirm the immobilization of ephedrine.

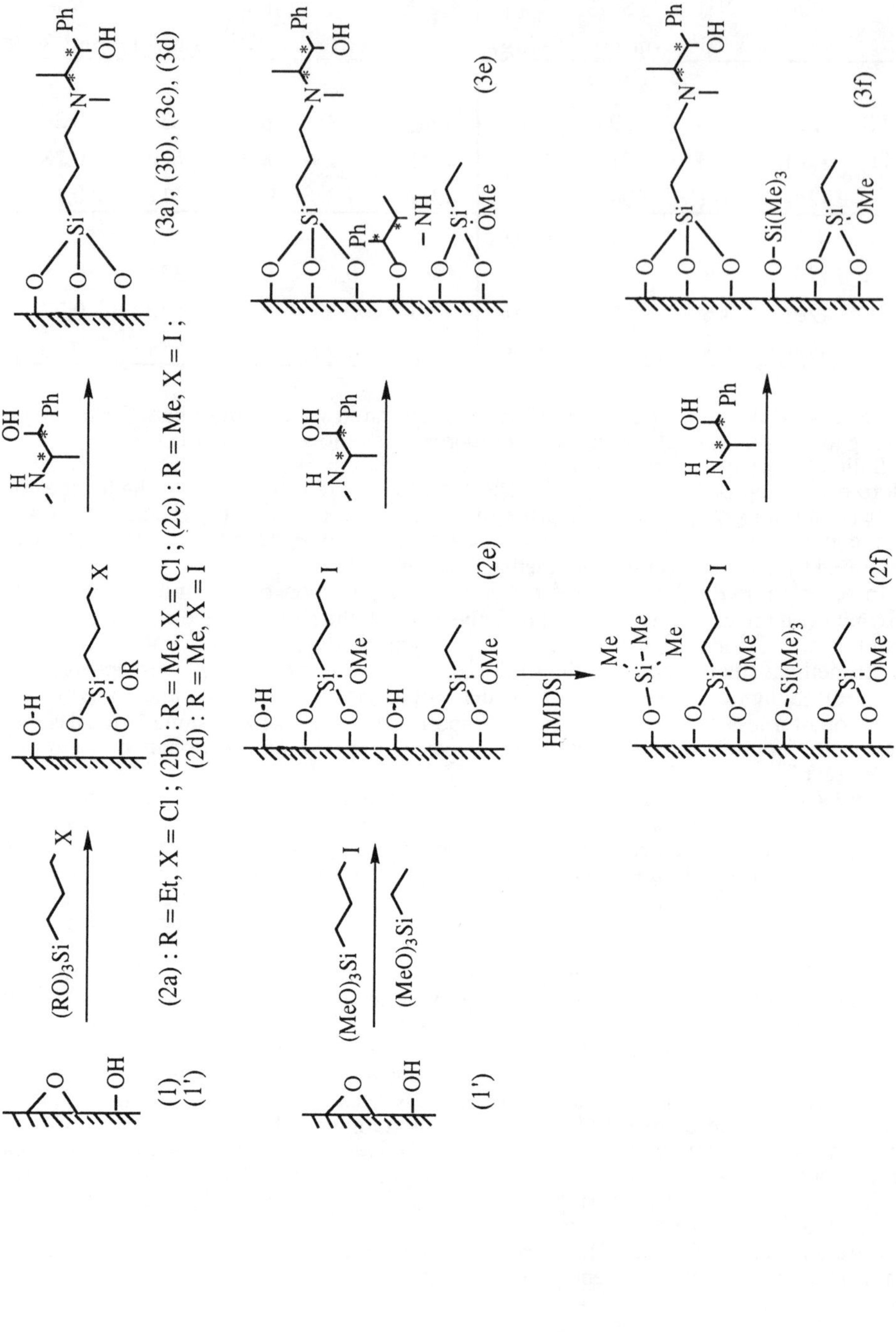
(1)
(1')
(RO)$_3$Si
X
(2a) : R = Et, X = Cl ; (2b) : R = Me, X = Cl ; (2c) : R = Me, X = I ;
(2d) : R = Me, X = I
(3a), (3b), (3c), (3d)
(MeO)$_3$Si
(2e)
(3e)
HMDS
O–Si(Me)$_3$
OMe
(2f)
(3f)
Ph
OH

Scheme 1

Surface characterization and mesoporous volume were determined by nitrogen volumetry (Table 1). As already reported[7], surface and mesoporous volume decrease with grafting of the halogeno function. It is worth noting that an increase of the number of grafted functions is followed by a regular decrease of the surface area and of the mesoporous volume for (2a)-(2c) solids for which no spacers are used. The relative effect is less pronounced for (2e) for which spacers are grafted and for (2f) for which catalytic site spacing and surface passivation are carried out. The substitution of the halogen by ephedrine generates another diminution of surface areas and of available mesoporous volumes, specially for (3a)-(3c) solids. Table 1 shows that the relative loss of mesoporous volume is lower when spacers are used (3e), (3f) than for solids (3a)-(3c).

2.2 Enantioselective Addition of Diethylzinc to Benzaldehyde

Supported (-)-ephedrine MTS (3) have been tested in the enantioselective addition of diethylzinc to benzaldehyde. The catalytic activity and selectivities have been compared with those obtained for the reaction in homogeneous conditions, using the same amount of catalyst (8.5×10^{-2} mmol) according to a previously described procedure[7] (Scheme 2).

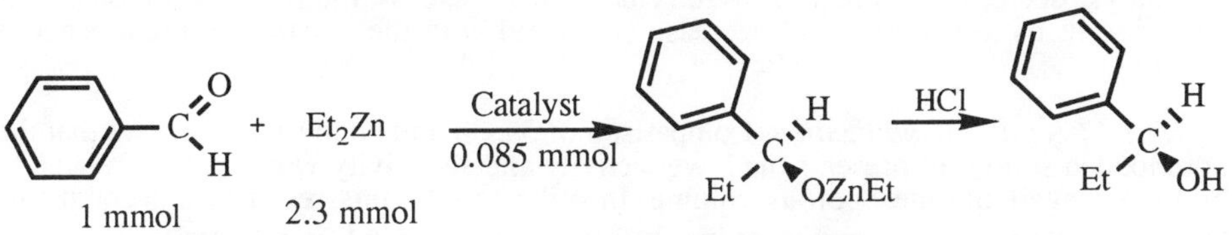

Scheme 2 *Enantioselective addition of diethylzinc to benzaldehyde*

The (R)-1-phenyl-propan-1-ol is obtained with moderate to good ee s. Moreover, benzyl alcohol is formed as a by-product in a competitive way.[1] Table 2 shows the results obtained for both enantioselectivity and selectivity together with the initial rate of the reaction. Toluene (entries 0-2, 4, 5) or hexane (entries 3, 6-9) have been used as solvents. As previously reported[10], higher rates are obtained with hexane (for exemple entries 2 and 3). Results clearly indicate that binding (-)-ephedrine to the MTS support leads to a significant reduction of its effectiveness in terms of initial rate of (R) + (S)-1-phenyl-propan-1-ol formation (entries 1 and 2). Moreover, an increase of the rate is observed when spacers are present (entry 8) and when the passivation of the surface is carried out (entry 9). On the other hand, the yield in benzyl alcohol increases when the activity of

Table 2 *Results of Asymmetric Ethylation of Benzaldehyde*

Entry	Catalyst	ee (%)	Selectivity (%)	r_0 (mol.L^{-1}.h^{-1}) $\times 10^3$
0	without*	0	45	0.6
1	(-)-ephedrine*	67	97	97.7
2	3a$^+$	37	87	5.0
3	3a$^\#$	35	90	11.3
4	3b$^+$	37	84	5.5
5	3c$^+$	23	58	2.9
6	3c$^\#$	23	61	3.9
7	3d$^\#$	33	84	6.4
8	3e$^\#$	26	85	13.6
9	3f$^\#$	33	89	19.0

* Homogeneous conditions in toluene; + toluene as solvent; # hexane as solvent.

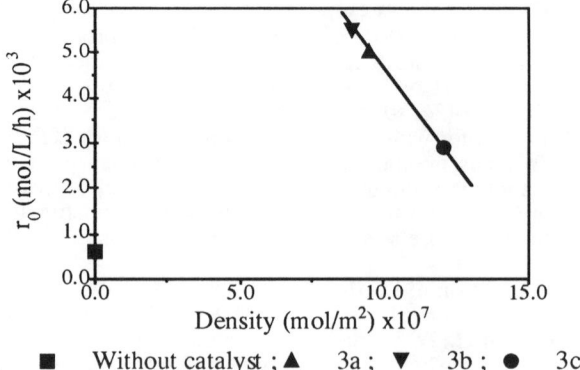

Figure 1 *Variation of initial rate with density of ephedrine moieties*

the catalyst decreases. As regards to the enantiomeric excess, the immobilization of (-)-ephedrine leads to a substantial decrease compared with the reaction in homogeneous conditions.

2.2.1 Effect of Immobilization. Comparison of 3a, 3b and 3c in toluene shows that the more loaded solid (3c) presents the lower activity and selectivity which can be related to the lowest available mesoporous volume. In order to take this result into account, the density of catalytic sites, expressed by N_E/S (mol/m^2, where S is relative to the parent MTS) has been considered. The variation of the initial rate with such a density is shown in Figure 1. The observed regular decrease of r_0 can be explained by a reduced accessibility to catalytic sites when residual porous volumes greatly decrease after functionalization. Activity of solids 3a (entry 2) and 3b (entry 4) prepared from ethoxy or methoxy chloropropylsilane appears to depend only on this density. As already mentioned, selectivity can be directly related to the initial rate. In the same manner, a decrease of the enantiomeric excess for (3c) (entry 5) is explained by the low activity of the catalyst and the higher participation of the uncatalyzed reaction (entry 0).

2.2.2 Effect of Dispersion and of Surface Passivation. Such a study has been carried out with MTS (1') supported catalysts (Table 2). Solids (3e) and (3f) have been used in the alkylation reaction in order to control the role of the density of the catalytic sites and to study the effect of potential residual silanols, respectively.

A ratio of around one to three ephedrine and two to three spacers are present at the

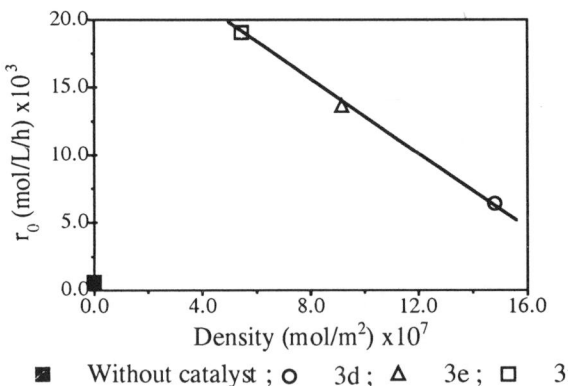

Figure 2 *Variation of initial rate with density of ephedrine moieties*

solid surface as determined by elemental analysis. In both cases the rate of the reaction is increased (entries 6 and 7). However, the same regular decrease of r_0 with the density of grafted (-)-ephedrine as above (Figure 1) is obtained whatever the solid treatment, as shown on Figure 2. Thus, the effect on initial rate can be explained by the site accessibility for all the supported solids. However, the rates obtained with the solids (3e) and (3f) remain relatively low compared with the reaction rate with unsupported ephedrine (5 to 6 times lower). This result indicates that the modification with spacers of these MTS silicas of around 30 Å diameter does not prevent diffusional limitations. On the other hand, the enhancement of the initial rate for the catalyst (3f) is not accompanied by a corresponding effect on the enantiomeric excess. Hence, the effect of the neutralization of residual silanols of the surface on the enantioselectivity of the reaction is not obvious. On the contrary, grafting of (-)-ephedrine through the hydroxyl group for the catalyst (3e) induces a decrease of the enantiomeric excess when reaction is performed at a constant number of catalytic sites.

3 CONCLUSION

The immobilization of (-)-ephedrine on mesoporous MTS type silicas leads to new chiral supported catalysts which are active in the enantioselective alkylation of benzaldehyde with diethylzinc. The initial rate of the reaction is directly related with the density of catalytic sites: r_0 increases when density decreases. The same effect is observed concerning the selectivity. However, the results obtained show that diffusional limitations cannot be excluded even in the case of competitive grafting of spacers. The enhancement of initial rates with decreasing densities is not correlated with a corresponding effect on enantiomeric excesses. Work by using MTS silicas with higher pore diameters is in progress to provide a better understanding of these results.

References

1   R. Noyori, S. Suga, K. Kawai, S. Okada, N. Kitamura, N. Oguni, M. Hayashi, T. Kaneko and M. Matsuda, *J. Organomet. Chem.,* 1990, **382**, 19
2   K. Soai and S. Niwa, *Chem. Rev.*, 1992, **92**, 833
3   S. Itsuno, Y. Sakurai, K. Ito, T. Maruyama, S. Nakahama and J. M. J. Fréchet, *J. Org. Chem.*, 1990, **55**, 304 ; M. Watanabe, K. Soai, *J. Chem. Soc., Perkin Trans 1*, 1994, 837 ; N. El Moualij and C. Caze, *Eur. Polym. J.*, 1995, **31**, 193.
4   K. Soai, M. Watanabe and A. Yamamoto, *J. Org. Chem.*, 1990, **55**, 4832
5   J. S. Beck, J. C. Vartuli, W. J. Roth, M. E Leonowicz, C. T. Kresge, K. D. Schmitt, C. T-W. Chu, D. H. Olson, E. W. Sheppard, S. B. Mc Cullen, J. B. Higgins and J. L. Schlenker, *J. Amer. Chem. Soc.*, 1992, **114**, 10834.
6   D. Brunel, A. Cauvel, F. Fajula and F. Di Renzo, *Stud. Surf. Sci. Catal.*, 1995, **97**, 173.
7   N. Bellocq, D. Brunel, M. Laspéras and P. Moreau, *Stud. Surf. Sci. Catal.*, 1997, in press
8   B. Pugin and M. Müller, *Stud. Surf. Sci. Catal.*, 1993, **78**, 107
9   A. Cauvel, G. Renard and D. Brunel, *J. Org. Chem.*, 1997, **62**, 749
10 K. Soai, S. Niwa and M. Watanabe, *J. Chem. Soc., Perkin Trans 1*, 1989, 109

THE FORMATION OF GEL PRECURSORS TO MESOPOROUS SILICON OXIDES: FAST SYNTHETIC ROUTES AND DOPING IN THESE MATERIALS

S. O'Brien, T. R. Spalding, S. E. Lawrence and M. A. Morris(*)

Materials and Inorganic Sections
Department of Chemistry
University College Cork
Cork, Ireland

* author to whom correspondence should be addressed

1 ABSTRACT

Mesoporous silicate materials could be prepared using a micelle-like species of a quaternary ammonium salt ($N(CH_2)_3R^+X^-$) as a template for oxide development so that the solid was formed with a well-defined pore structure. R, here was a 16 carbon atom linear chain and X a bromide species. This preparation was modified by substituting the quaternary salt for a long chain amine type headed chain. The use of the amine compound reduced gelation times from several days to a few minutes and extremely high quality, transparent, air-stable gels were synthesised. The gel formation was followed by MAS-NMR and DSC techniques. The use of the amine compound also facilitated the inclusion of transition metal atoms into the gel matrix and then into the calcined silicate material.

The gels were aged at increasing temperatures to yield high surface area materials and evidence for the mesoporous solids is seen in high resolution TEM micrographs. The resultant solids exhibited high temperature stability and materials calcined at 823K have surface areas of the order of 200 m^2 g^{-1}. The materials were amorphous in nature until calcination temperatures of between 773 and 973K when crystallisation was observed. DSC could be used to monitor the change in long-range order and it was found that the incorporation of boron into the solid increases the crystallisation temperature.

2 INTRODUCTION

MCM-41 compounds are mesoporous alumina silicates first developed by the Mobil Oil Corporation[1]. These materials facilitated the adsorption and reaction of large molecules within a pore structure whose size was greater than the 1nm or less exhibited by zeolites. The MCM compounds have an inorganic matrix formed over aligned cylindrical surfactant micelles arranged into hexagonal arrays. The pore size is of the order of 1.6 to 12 nm. These materials should prove useful catalysts and supports for industrial use since they exhibit high surface areas and high thermal stability; surface areas as high as 900 m^2 g^{-1} have been recorded after ageing at 813K[2].

The product of the reaction of alkoxysilanes is a solid gel which appears to be formed by hydrolysis and condensation of the silane that is the source of silicon and oxygen. Typical silanes include tetramethoxysilane (TMOS) and hydrolysis is performed by water in basic conditions where ethanol is used as a solvent[3,4]. However, reaction times of the order of two weeks at 368K are reported[3,4]. Such reaction conditions would make the preparation of large quantities of commercial catalysts impractical and prohibitively expensive. In this paper, methodology is developed for the preparation of mesoporous silicates which is quick and can also be used for the 'impregnation' of these materials with active transition metal centres.

3 EXPERIMENTAL

Initially, gels were prepared in the same way as described previously[3]. An aqueous solution of template (hexadecyltrimethylammoniumbromide, HDTMAB) and ethylene-diamine mixture was prepared to which was added a TMOS and boric acid mixture (concentrations as required). The mixture was heated in polypropylene bottles for 14 days before yielding a gel in which was suspended a white solid precipitate which was harvested after washing. Transition metal ions could be added to the solid by use of a salt, e.g. $CuCl_2$, soluble in water.

The reaction was modified by use of an amine type template instead of the salt above. Here, the template (hexadecylamine, HAD) was dissolved in ethanol to which was added an ethanolic solution of TMOS, these were mixed and water added. The pH of the solution was between 9 and 10 as required for base hydrolysis. Ethanol was used as a solvent to allow optimum mixing. Within minutes (dependent on the exact conditions) a gel-like substance was formed. On standing, this became glass-like in appearance. Nickel and copper were incorporated into these materials by the addition of soluble compounds (here chlorides) to the template solution. Thus, amine groups were directly co-ordinated to the metal cations and so are directly added to the final calcined oxide materials.

Materials were analysed using PXRD, SEM, TEM and MAS-NMR coupled with surface area and density measurements. θ-2θ PXRD profiles were collected on a Philips PW 3710 MPD apparatus using $CuK\alpha_1$ radiation. MAS-NMR data were collected on a Chemagnetics CMX-Lite apparatus. Samples were mounted in a zirconia ceramic 6mm tube within a ceramic holder/spinner; spin rates of 5kHz were normally used. Samples could be treated by heating between room temperature and 550K within the tube. Flip angles of 45 degrees and pulse widths of 4μs were routinely used. Recycle times of between 2.5 and 20s were used for ^{29}Si CP-MASNMR spectra, Chemical shifts are referenced to HMB and TMOS standards as appropriate. SEM micrographs were collected on a Jeöl (JSM-35) instrument using acceleration voltages of 20 kV. TEM micrographs (Jeöl, 120 kV acceleration voltages) were collected by 'painting' gel materials onto standard gold gauzes and allowing to dry.

4 RESULTS AND DISCUSSION

Products formed by the reaction of HDTMAB and TMOS are similar to those reported in the literature[3]. SEM and PXRD suggests that hexagonal crystals of a mesoporous silicate are formed on calcination to 823K. Surface areas, densities and TEM indicate a pore size of around 2-4nm. Results are described in detail elsewhere[4]. ^{29}Si CP-MASNMR data are shown in fig.1 from a sample of borosilicate (approximately 50Si:1B ratio in the finished material). Peaks are indicated at -102.8 and -110.8 ppm typical of Q^3 (silicon atoms bound to three SiO_4 tetrahedra and one hydroxyl group) and Q^4 (four SiO_4 links) coordinations respectively. There is no evidence for Q^2 links and the measured ratio of Q^4 to Q^3 is 1.79. These results are typical of MCM-41 materials[5].

The amine template route provided almost instant (2-3 mins, room temperature) gelation which yielded a homogeneous, rigid glass-like material on standing for 5-10 minutes. The following amounts were used:- 0.02 mole TMOS, 10^{-4} moles of template, 0.08 moles of water and 0.09 moles of EtOH. In conditions where the solvent concentration was raised viz:- 5.6 moles EtOH and 0.55 moles of H_2O, the product was an inhomogeneous, non-rigid gel with a white precipitate. This second material is not discussed further here. Instead, only samples produced by the former route are discussed. This is because it would provide an inexpensive (100% yield) and convenient

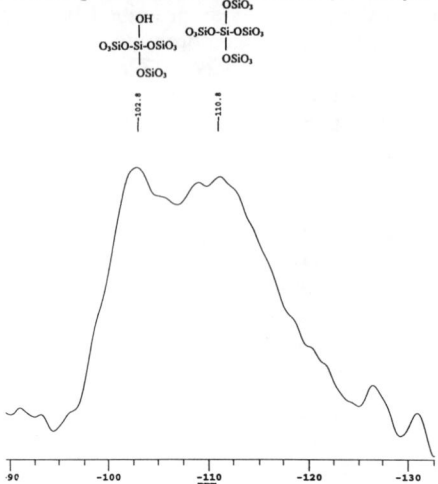

Figure 1. *^{29}Si CP MASNMR data of a low boron concentration borosilicate (precursor calcined at 823K for 5 hours). Peak at -102.8 ppm is Q^3 and peak at 110.8 ppm is Q^4*

catalyst preparation. The rigid gel precursor was calcined at 823K for 5 hours to yield a white oxide product. The surface area of this solid was measured at $220m^2g^{-1}$.

The decrease in initial reaction times appears to be due to the presence of the amine group. Maintaining constant pH and water concentrations, gel times increased by a factor of 6.5 on decreasing the initial template concentration by 4. Further, molar replacement of the HAD by ammonium hydroxide solution has little effect on measured gelation times. The gelation process was followed using solid state NMR spectroscopy in static conditions. During these experiments the sample was not spun to prevent loss of material. Reagents were mixed (as described below) and then added to the rotor tubes; ^1H data were collected as a function of time.

Figure 2(a) shows spectra collected from a mixture of the template in EtOH. Peaks centered at about 1ppm are due to the H in $-CH_3$ groups in ethanol whilst between 3.7 and 3.3 ppm the $-CH_2$ type hydrogens can be seen. At 5.1-5.3ppm the $-NH_2$ hydrogens are resolved. On addition of TMOS (figure 2(b)) There is rapid development of intensity at 3.5 ppm due to the addition of silane methoxy hydrogens. Note also the complete loss of $-NH_2$ signal. From these data it is suggested that the amine group of the template is reacting to give an initial product with Si-N bonds. Further evidence is seen from a peak envelope shape in the region of 5.0ppm typical of methanol (data not reported here) which is formed in this reaction. On water addition there are dramatic changes accompanying gel formation (data are shown in figure 2(c), 7 mins after mixing). The peak at about 3.5ppm (silane methoxy species) is lost and a strong peak develops at 3.3ppm assigned to hydrogens in the CH_3 group in methanol. Also seen is the development of intensity at 4.7 to 4.9 ppm which is due to -Si-OH hydrogens. These data can only be due to hydrolysis. The methanol $-CH_3$ type signals dominate because of the amount of methanol released in the hydrolysis reaction. If the sample is left for several hours, the spectra shown in fig 2(d) result. This is because the gel has become rigid and the decrease in species mobility has led to highly broadened features.

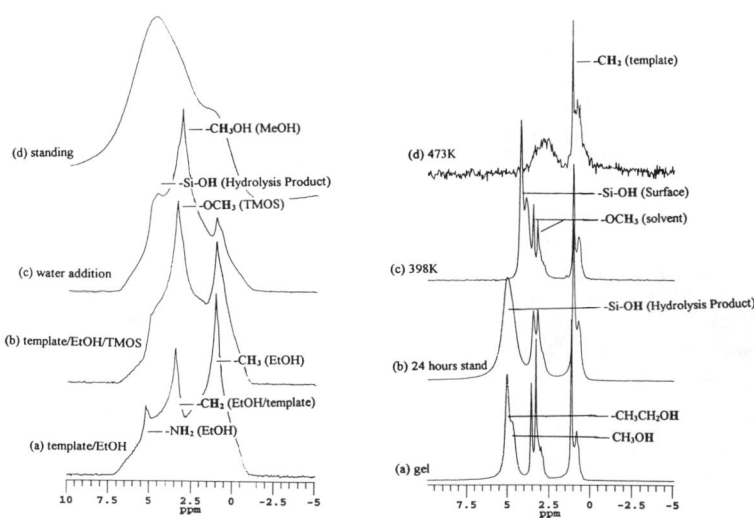

Figure 2 (left). *^1H NMR spectra taken at various times prior to and during the gelation process*
Figure 3 (right) *^1H MAS-NMR spectra collected at 5kHz spin rates of gel materials at various times and temperatures*

The structure of the rigid gels can be probed by spinning the sample. Data are reported in fig. 3. Spinning the sample from fig. 2(d). reveals well resolved features due to methanol and ethanol -OH groups at 4.8 and 5.1 from trapped solvent. Also seen are signals due to -CH_2 (EtOH derived, 3.6 ppm) and -CH_3 (MeOH derived, 3.3ppm). Peaks at 0.9, 3.0-2.9 are due to -CH_3 and -CH_2 species respectively from the template molecule. 24 hours of further standing only results in further solvent loss as seen by the decrease in signal around 3ppm. The large feature at 4.7-5.1ppm is due to Si-OH hydrogens. Ageing within the NMR probe at 398K radically changes the spectra. It is believed that this is due to the initial formation of silicate type bonding in the solid. The largest peak in the spectrum is seen at 4.3ppm, typical of -OH adsorbed at silica surfaces[6]. The peaks from 3.3 to 3.9 ppm suggest some methoxy and ethoxy groups are still held within the molecular framework but these features could also be due to -OH at different surface sites. On further heating to 473K the only strong ^1H signals seen are at 1.1 to 0.7 ppm and between 2.7 and 3.4ppm. These data strongly suggest that the template peak is still within the framework as silicate formation occurs. Further evidence for this assertion is given below.

It has been suggested that the use of amine compounds as templates is problematic and that the removal of amine results in the collapse of the pore structure[7]. There is no evidence of such reactions here. Figure 4 shows a TEM micrograph of the dried gel.

This was obtained by coating gold gauze substrates with the gel and allowing to dry. The sample was then subject to the conditions within the microscope which result in oxide formation. Pores of average size 9.5-10.5nm can be easily seen. The pore diameter and density within the solid this is completely consistent with a surface area of $220m^2g^{-1}$ measured by nitrogen adsorption (BET) at samples calcined at 823K for 12 hours and this is clear indication that the pore structure is extremely stable. The micrograph is more complex than might be predicted because the samples investigated could not be orientated in the microscope and 3D structure is being viewed. Similar micrographs were collected from all of the materials discussed here again suggesting that no gross structural changes occur on removal of the template. The pore size is larger than that exhibited by samples prepared by more usual routes but this may be related to the use of gel precursors; it is likely that high shrinkage that occurs on solvent (reaction product solvent) removal which will lead to larger pore sizes. The reported pore size is consistent with that of mesoporous material (pore size range of between 2 and 100nm).

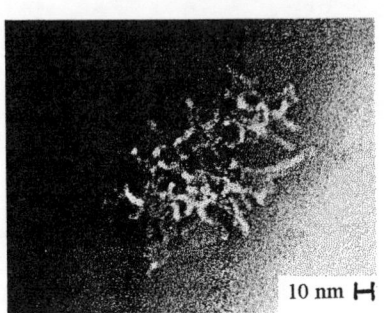

Figure 4. *TEM micrograph of the solid material following gelation*

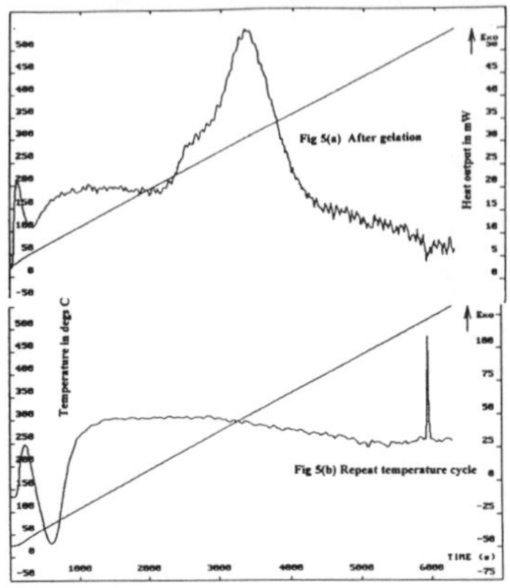

Figure 5. *DSC traces of material heated in nitrogen, a) after gelation b) repeat cycle*

Further evidence for the temperature stability of the materials can be derived from DSC measurements as no changes in structure can be detected on heating gel precursors to temperatures around 873K. Figure 5 shows DSC measurements recorded in flowing N_2. Fig.5(a) shows a typical trace of an initial amine derived gel material heated to 829K. There is an endotherm at about 330K (due to water desorption) followed by a large exotherm at 523K with a distinct shoulder at 493K. The results are consistent with previous TGA analysis reported in the literature[3,8]. The peak at 493K can be associated with oxide formation by condensation etc. The feature at 523K can be ascribed to loss of the template and further reaction. Note that no carbon or nitrogen could be detected by bulk chemical analysis following this treatment. DSC provides no evidence for other peaks associated with pore structure changes. If the sample is subjected to another treatment in the same conditions (following cooling) there is no evidence for peaks at low temperature (besides the desorption of water) indicative that the first heating resulted in the complete removal of template and solvent, fig 5(b). Instead there is a well-resolved feature at 793K. By comparison with PXRD profiles (not shown) the appearance of this peak is consistent with the formation of well-defined crystalline phases which suggests this is the lowest temperature at which the pore structure could be lost. Addition of boron to the gelling materials (as boric acid) raises the temperature of the crystalline feature above 873K which suggests that the addition of boron would strongly affect sintering rates for catalysis applications and could be used to effect stabilization.

The use of an amine is also advantageous in the preparation of metal ion loaded materials. The amine template was added to an ethanolic solution of either copper or nickel chloride. The metal ions are strongly coordinated to amine nitrogen in these conditions. The gel could be prepared in the usual manner with loadings of 5% by weight ($CuO:SiO_2$ or $NiO:SiO_2$) forming good transparent gels. Additions of 10% by weight resulted in the precipitation of copper or nickel rich particles and degradation in the transparency of the gel. The 5% loaded materials were dried at 333K for 12 hours (yielding a white powder) and washed with a water/ethanol mixture. From quantifiable analysis of the filtrate and solid, no metal ions were removed by washing suggesting that the metal ions are intimate with the gel product. Further, calcination to 823K did not result in a colour change associated with the oxidation and sintering of transition metal oxide phases to large 3D particles of oxide. It is suggested that this route may prove useful in the preparation of active transition metal loaded materials.

5 SUMMARY

Mesoporous silicate materials with a well-defined pore size of about 10nm can be prepared via a convenient method. This results in the formation of a glass-like product. An amine template was used to confer a mesoporous structure. The product, on calcination to 823K, is essentially an amorphous oxide which showed some small degree of crystallisation when heated to around 793K. This temperature could be increased by the addition of boron to the gel formation step.

The amine route was a simple one-step reaction which led to the immediate formation of a gel. It is suggested that this route would be extremely advantageous in the preparation of large quantities for commercial use. Further, the gel like properties could be used in the preparation of materials for coating substrates. An additional benefit is that transition metal ions could be directly complexed to the amine for the preparation of active materials. It was seen that these transition metal impregnated materials did not produce large 3D islands of transition metal oxide when calcined. These materials are currently being tested as catalysts for NO_x reduction.

6. REFERENCES

1. C. T. Kresge, M. E. Leonowicz, W. J. Roth, J. C. Vartuli and J. S. Beck, *Nature*, 1992, **359**, 710
2. D. Trong On, P. N. Joshi and S. Kaliaguine, *J. Phys. Chem.*, 1996, **100**, 6743
3. U. Oberhagemann, I. Kinski, I. Dierdorf, B. Marler and H. Gies, *J. Non-Crys. Solids*, 1996, **197**, 145
4. S. O'Brien, T. R. Spalding, S. E. Lawrence and M. A. Morris, to be presented at the *3rd Inter. Conf. Materials Chemistry*, Exeter, 1997
5. W. Kolodziejski, A. Corma, M-T. Navarro and J. Perez-Pariente, *Sol. St., Mag. Res.*, 1993, **2**, 253
6. T-C. Sheng, S. Lang, B. A. Morrow and I. D. Gray, *J. Cat.*, 1994
7. N. Ulagappan and C. N. R. Rao, *Chem. Commun.*, 1996, 1685
8. C A Koh, R Nooney and S Tahir, reported at *11th International Congress on Catalysis*, Baltimore, 1997.

NEW SOLID BASES DERIVED FROM ORGANICALLY-MODIFIED SILICAS AND THEIR USE IN THE KNOEVENAGEL REACTION

Duncan J Macquarrie*, James H Clark, Dominic B Jackson, Arnold Lambert, James E. G. Mdoe and Andrew Priest
Department of Chemistry, University of York, Heslington, York YO1 5DD, UK

1 INTRODUCTION

The search for novel heterogeneous catalysts capable of carrying out highly selective transformations has centred mainly on novel solid acids and oxidation catalysts. Much less progress has been made in the field of solid bases, particularly in liquid phase chemistry.

As part of an initiative centred on the use of Chemically Modified Surfaces as novel heterogeneous catalysts, we have investigated the use of amines covalently bound to silica e.g. γ-aminopropylsilica (AMPS)[1] and Controlled Pore Organosilicates (CPOS)[2] as basic catalysts for the Knoevenagel reaction. Venturello et al.[3] have already shown that AMPS is a useful catalyst for the Knoevenagel reaction, although their system was based on a column reactor with large quantities of catalyst being used. Results obtained did, however, indicate good activity with a range of substrates. The aim of our investigations was initially to define more thoroughly the scope and limitations of this catalyst, and then to extend our work to the novel CPOS family of materials, which show promise as highly stable materials with easily controllable loading and porosity. This paper discusses our preliminary findings.

2 EXPERIMENTAL

Catalysts were characterised by Diffuse Reflectance FTIR (Perkin-Elmer 1720 Infra-red Spectrometer) as a mixture with KBr. Spectra were recorded under vacuum (650 N m^{-2}) and at 130°C in order to remove water from the sample. ^{13}C-CP-MAS NMR spectra were recorded on a Bruker MSL300MHz CP-MAS NMR. The loading of aminopropyl groups was determined by Elemental Analysis. Porosity measurements were carried out using a Coulter SA3100 Porosimeter and Specific Surface Area was calculated using the BET method. Surface polarity was measured by the Reichardt's dye method[4]. Catalysts **1** and **2** were prepared according to literature methods[1,2]. The preparation of **2** was modified to provide 2.5mmol g^{-1}, 3.7nm AMP-CPOS (**3**) (using a 4 : 1 molar ratio of TEOS to γ-aminopropyl(trimethoxy)silane), and 1.2mmol g^{-1}, 1.8nm AMP-CPOS (**4**) (9 : 1 ratio of silanes with *n*-octylamine template).

2.1 Characterisation of catalysts

All catalysts display similar IR spectra to those already published[1,2]. **1** has a typical ^{13}C spectrum[5], and the CPOS materials have additional peaks due to OEt groups. Other parameters are collated in Table 1. Representative Pore Size Distribution and Adsorption Isotherm data are shown for **2** in Figure 1.

Table 1. *Selected Analytical Data for AMPS and CPOS catalysts*

Catalyst	Loading (mmol g^{-1})	Specific Surface Area (m^2 g^{-1})	Pore Size (nm)	Surface Polarity E_T^N
1	0.95	254	1-12	0.56
2	1.2	756	3.6	0.90
3	2.5	745	3.7	0.82
4	1.2	715	1.8	0.86

Figure 1 *Porosity measurements for* **2**

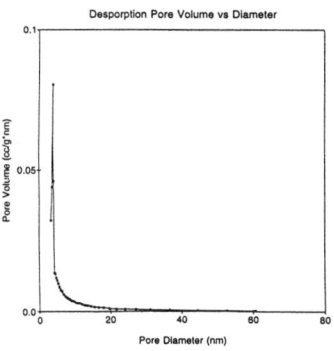
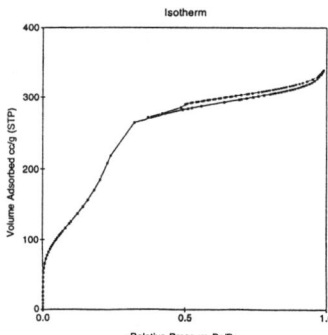

2.2 Reaction studies

A typical procedure is as follows. The catalyst (0.25g) and *n*-dodecane (0.25g) are added to the appropriate solvent (25ml) and heated to reflux in a two-necked flask fitted with a Dean and Stark trap. At reflux 20mmol of the active methylene compound and 20mmol of the carbonyl compound are added simultaneously and rapidly. Samples are taken and filtered before analysis by GC. Work-up consists of filtration of catalyst followed by evaporation of solvent. In most cases this yields essentially pure product. Products were further purified by column chromatography where necessary. GC yields are quoted on the basis of internal standard calibrated with pure samples of the appropriate product.

3 RESULTS AND DISCUSSION

3.1 AMPS (1)

1 was prepared by refluxing γ-aminopropyl(trimethoxy)silane (30 mmol) with Merck Kieselgel 100 (activated overnight at 300°C, 10g) in toluene, followed by thorough washing[1].The loading achieved was 0.95mmol g^{-1}, and the majority of the Si-OMe groups have hydrolysed during the preparation. It has not proved possible to increase the loading beyond this level. The surface polarity[4] (E_T^N)is 0.56, a figure which lies approximately midway between silica (0.967 - 0.78, depending on extent of drying) and simple primary amines (0.18 - 0.25). This is presumably due to the surface being a mixture of exposed silica and amine groups. The much higher values obtained for the CPOS materials, and the fact that other differently functionalised CPOS (chloropropyl-, cyanoethyl-CPOS) have almost identical values, indicates that the higher surface areas of these materials means that the average polarity seen is much closer to that of bare silica. Thus these materials have a much lower proportion of their surface covered than **1**. The increased polarity of the surface may have important implications for their reactivity (see below).

The piperidine catalysed reaction between 3-pentanone and ethyl cyanoacetate was chosen as a reasonable model for the homogeneous system. In refluxing cyclohexane the reaction proceeded to give the desired product **5**, together with significant amounts of piperidinyl cyanoacetate **6** and the double condensation product **7** arising from deprotonation of the vinylogous CH_2 group in **5** (Scheme 1). The best system was found to be 90 mol% piperidine as catalyst, under which conditions a conversion to **5** of 18% was achieved after 6 hours. This represented an optimum conversion within the range of 20mol% to 400mol%.

Scheme 1 *Condensation of 3-pentanone with ethyl cyanoacetate under homogeneous conditions (piperidine catalysis)*

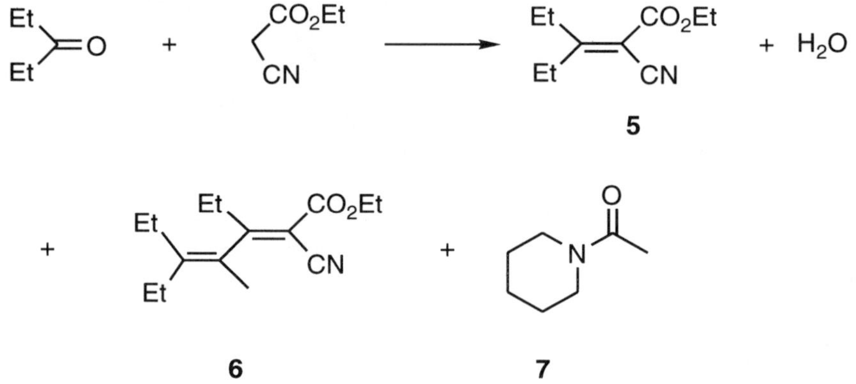

The reactivity of AMPS, on the other hand, was extremely selective with a range of substrates. The reaction is reversible and is slowed significantly by the presence of water. Efficient removal of water during the reaction by means of a Dean and Stark water separator allowed reactions to proceed smoothly to completion. This requirement restricts the choice of solvents, but, as will be seen later, this does not represent a major problem. Initial reactions were carried out in cyclohexane and results are given in Table 2.

As can be seen from Table 2, the reaction is quite general, with a wide range of carbonyl compounds reacting. The reactions are exceptionally selective, with only the desired product being formed. Relative reactivities of the carbonyl compounds are typical, with aldehydes being very reactive, and ketones much less so. Of the ketones, cyclohexanone is the most active, and bulky ketones and aromatic ketones are much less active (Entries 8-10). Again this is as expected, although the activity seen with ketones is particularly good, with even acetophenone giving good yields. Turnover numbers are generally good. No reaction was seen using silica itself as the catalyst, even after reaction times 10 times longer than those shown, nor was reaction observed after removal of AMPS from the reaction system. These observations demonstrate that the catalysis is due to surface bound groups, and that no leaching occurs during the reaction.

Table 2 *Initial Screening of Knoevenagel Condensations using AMPS. Effect of Carbonyl Component*

Entry	R	R'	Temperature (C)	Time (h)	Yield(%)	TON[a]
1	Ph	H	25	4	99[b]	
2.	nC_5H_{11}	H	25	7	97[b]	
3.	nC_7H_{15}	H	25	8	98[b]	
4	cC_5H_{10}		82	1	98	650
5	Et	Et	82	2	97	265
6	Et	Et	82	4	65[b]	
7	n-C_4H_{10}	Me	82	4	98	
8	t-C_4H_{10}	Me	82	24	22	
9	Me	Ph	82	24	68	250
10	Ph	Ph	82	72	8	

All yields are GC yields using internal standards and are corrected for response factors. In all cases, isolated yields are 3-7% lower.
(a) Turnover Number; measured as moles product per mole amino group during reactions with multiple addition of substrates.
(b) Reaction carried out without removal of water.

The reaction was studied further by investigating the activity of AMPS towards a variety of CH acids using benzaldehyde and 3-pentanone. Results are shown in Table 3 and Table 4.

Table 3 *Initial Screening of Knoevenagel Condensations using AMPS. Effect of carbon acid component. Reactions with benzaldehyde*

Entry	R	R'	Temperature (C)	Time (h)	Yield(%)
1.	CN	CN	25	0.5	99
2.	CN	CO_2Et	82	0.05	95
3.	CO_2Et	CO_2Et	82	24	30
4.	CO_2Et	$COCF_3$	82	2	57
5.	MeCO	MeCO	82	17	25
6.	MeCO	CO_2Et	82	19	35

All yields are GC yields using internal standards and are corrected for response factors. In all cases, isolated yields are 3-7% lower.

Table 4 *Initial Screening of Knoevenagel Condensations using AMPS. Effect of carbon acid component. Reactions with 3-pentanone*

Entry	R	R'	Temperature (C)	Time (h)	Yield(%)
1	CN	CN	82	0.25	60
2	CN	CO$_2$Et	82	2	95
3	CO$_2$Et	CO$_2$Et	82	24	trace
4	CO$_2$Et	COCF$_3$	82	12	0

All yields are GC yields using internal standards and are corrected for response factors. In all cases, isolated yields are 3-7% lower.

Thus it can be seen that, for benzaldehyde, a variety of carbon acids can be converted in modest to excellent yields. It should be noted that reaction conditions have not been optimised for these compounds, and work is ongoing to improve yields. In all cases no other products are detectable by GC. Reactivity in the case of 3-pentanone is significantly more restricted, and this reflects the substantially less reactive nature of the carbonyl component. In the case of malonodinitrile (Table 4, entry 1), the low yield may be due to the low thermal stability of the nitrile.

The role of solvent is often of great importance in reactions occurring under heterogeneous catalysis[6], and we have thus also studied the influence of a range of different solvents in the reaction of ethyl cyanoacetate with cyclohexanone or 3-pentanone. The choice of solvent was limited to those which can azeotropically remove water from the reaction mixture. Reactions were thus carried out at the respective boiling points of the solvents in question. Reaction profiles are shown in Figure 2.

Thus it can be seen that the best solvent for AMPS is cyclohexane. It appears from the data that the polarity of the solvent is of vital importance, since toluene sustains a significantly lower rate than cyclohexane even though the reaction is running 30°C hotter. Similarly, more polar solvents such as chlorobenzene and 1,2-dichloroethane are also remarkably poor. These results can be explained by the partitioning of reactants between the (polar) catalyst surface and the solvent. With increasing polarity, reactants favour the liquid phase and effective concentrations on the catalyst surface decrease, leading to lower rates. With this logic in mind we investigated the use of higher boiling alkanes (heptane and octane). While these gave, initially, higher rates than cyclohexane, they were more rapidly deactivated.

The deactivation of AMPS has also been studied. Spent catalyst was removed from the reaction mixture after it had ceased to be active. After thorough washing with methanol, acetone and ether, the catalyst was analysed by DRIFT and CP-MAS NMR. Whereas the NMR spectra indicated the presence of AMPS units and additional material, no conclusive evidence was gained regarding poisoning. On the other hand, the DRIFT IR spectra provided strong evidence for a deactivation mechanism consisting of slow amide formation between the amino group and the ester on ethyl cyanoacetate[1].

Figure 2 *Solvent effects in the reaction of 3-pentanone with ethyl cyanoacetate with AMPS*

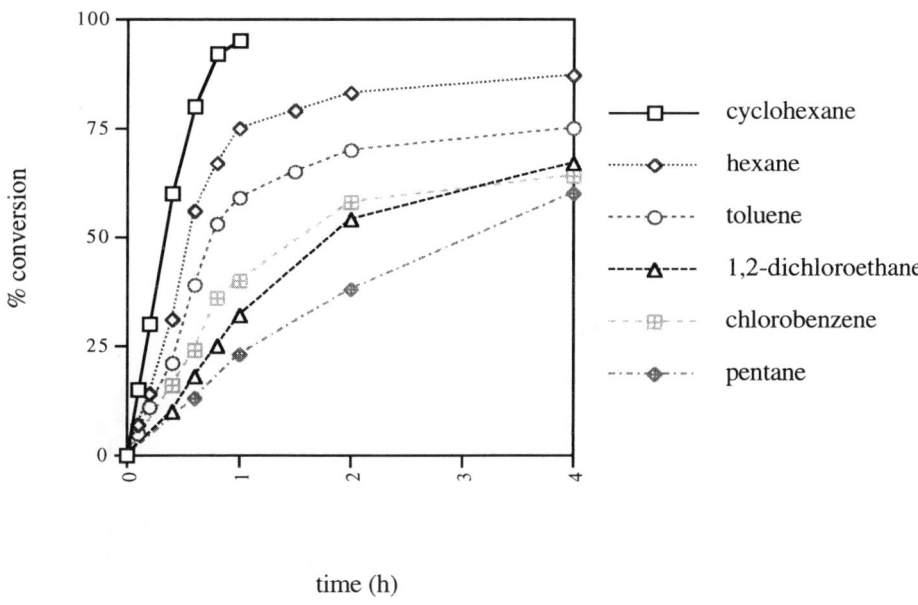

time (h)

3.2 Controlled Pore Organosilicates (CPOS)

The preparation of CPOS has already been described[2]. The principle has been extended successfully to include smaller pore versions (1.8nm) as well as higher loading materials (2.5mmol g^{-1}). The latter is ca. 2.5 times higher than is achievable with conventional materials, and represents a potentially important advantage of these materials. Surface areas of amino-containing CPOS are typically 700-750m^{2}g^{-1}, higher by a factor of 2-3 than the amorphous silicas, but less than a typical MCM. Non-amino-containing CPOS tend to have substantially higher surface areas, in the range 1350 - 1600m^{2}g^{-1}. While pore size distributions and adsorption isotherms are typical of the MCMs, powder diffraction studies indicate less long range order than for genuine MCMs. While the long range aim of this aspect of the work is to provide shape selective base catalysts, we have initially concentrated on defining the basic activity of these materials. Initial results from CPOS testing are given in Table 5.

Several points can be made regarding these results. All reactions are clean and produce only the desired product. Secondly, the general trend in activity with respect to carbonyl component is essentially as expected, with one dramatic exception. While most substrates react slightly more slowly with catalyst **2** than with AMPS, benzaldehyde takes approximately 100 times as long to reach completion. A further, less dramatic difference is that acetophenone has a lower turnover number, whereas for all other reactions studied TONs were significantly higher. The reasons for this behaviour are not yet clear, but may be related to the electronic requirements of aryl carbonyl groups.

Table 5 *Reaction studies using aminopropyl-CPOS as catalysts*

$$\underset{\text{'R}}{\overset{\text{R}}{>}}=O \quad + \quad \overset{CO_2Et}{\underset{CN}{<}} \quad \longrightarrow \quad \underset{\text{'R}}{\overset{\text{R}}{>}}=\overset{CO_2Et}{\underset{CN}{<}} \quad + \quad H_2O$$

	R	R'	Catalyst	Solvent	Temp(C)	Time(h)	Yield(%)*	TON**
1	Ph	H	2	cyclohexane	82	36	94	
2	n-C$_7$H$_{15}$	H	2	cyclohexane	82	0.5	93	
3	cC$_5$H$_{10}$		2	cyclohexane	82	4	96	2450 (650)
4	cC$_5$H$_{10}$		2	cyclohexane	82	12	70^a	
5	cC$_5$H$_{10}$		2	toluene	110	2	92	
6	cC$_5$H$_{10}$		2	EtOAc	78	4	11	
7	cC$_5$H$_{10}$		2	CHCl$_3$	61	4	18	
8	cC$_5$H$_{10}$		2	DCE^b	81	4	16	
9	cC$_5$H$_{10}$		4	cyclohexane	82	4	49	
10	Et	Et	2	cyclohexane	82	30	89	1127 (265)
11	Et	Et	2	toluene	110	18	95	
12	Et	Et	3	toluene	110	4	97	
13	Me	n-Bu	3	toluene	110	4	95	1244
14	Me	Ph	2	toluene	110	72	49	55 (250)
15	Me	Ph	3	toluene	110	36	48	47

* GC yields with n-dodecane as internal standard. Isolated yields are 3-7% lower.
** Number of moles product per mole of NH$_2$ groups. Figures in brackets are for reactions using 1.0mmol g^{-1} aminopropyl-silica as catalyst.
a) Reaction carried out without removal of water.
b) 1,2-dichloroethane

Solvent effects are broadly similar, although one important change is that toluene is significantly better than cyclohexane. Thus, while solvent polarity is still a controlling factor, the much greater polarity of the catalyst surface in the case of the CPOS catalysts means that effective partitioning of substrate onto the catalyst surface can still occur with solvents as polar as toluene. Additionally, results obtained with the higher loading catalyst **3** indicate that increasing the loading of the catalyst can increase reaction rates quite significantly. The small pore catalyst **4** performs less well, although this appears to be more a case of rapid deactivation than diffusion limitations, since the early stages of its reactions proceed at similar rates to those seen for **2**. Finally, turnover numbers are typically higher than for **1** by a factor of 4-5. Interestingly, the poisoning mechanism is different, with evidence for organic contamination of the surface, but no nitrile peaks are apparent in the IR spectra of used catalysts.

The combination of all these factors means that these catalysts differ in several respects from the apparently similar **1**. Optimum conditions for the CPOS catalysts gives results which are superior in some cases to those obtained for **1** (similar reactivity and increased catalyst lifetime).

4 CONCLUSIONS AND FUTURE WORK

The activity of four aminopropyl-functionalised catalysts has been studied in depth. The aminopropyl silica (AMPS, **1**) is a versatile, highly selective and active catalyst for the Knoevenagel reaction. It functions best in non-polar solvents, with cyclohexane being optimum. Deactivation takes place by amide formation on the surface. Aminopropyl-functionalised controlled pore organosilicates (CPOS) are slightly less active in general under identical conditions, but can be used in a wider range of solvents and at higher temperatures. Higher loadings are also achievable. The deactivation mechanism is currently unknown, but is not the same as for AMPS. Turnover numbers are higher.

Future work will consist of a more detailed kinetic study of both classes of catalyst, as well as increasing the range and loading of basic groups.

REFERENCES

1. D. J. Macquarrie, J. H. Clark, , A Lambert, J. E. G. Mdoe and A. Priest, *React. Funct. Polymers*, in Press
2. D. J. Macquarrie, *Chem Commun*, **1996** 1961
3. E. Angeletti, C Canepa, G. Martinetti and P. Venturello, *J Chem Soc., Perkin 1*, **1989**, 105
4. S. J. Tavener, J. H. Clark. G. W. Gray, P. A. Heath and D. J. Macquarrie, *Chem. Commun.*, **1997** 1147
5. K. C. Vrancken, P. Van Der Voort, K. Possemiers, P. Grobet and E. F. Vansant in "Chemically Modified Surfaces" Ed. J. J. Pesek and I. E Leigh, Royal Society of Chemistry, 1994, p46.
6. see e.g. D. J. Macquarrie, *Chem. Commun.*, **1997** 601 and R. Sreekumar, R. Padmakumar and P. Rugmini, *Chem. Commun.*, **1997** 1133.

METAL CATALYSIS INSIDE MICROPOROUS SYNTHETIC RESINS: SOME RECENT RESULTS

B. Coraina[a,b], A. A. D'Archivio[a], L. Galantini[a], K. Jeřàbek[c], M. Králik[d], S. Lora[e], G. Palma[f], M. Zecca[g]

[a] Universita' di L'Aquila, Dipartimento di Chimica, Ingegneria Chimica e Materiali, via Vetoio, 67010 L'Aquila, Italy; [b] Centro per lo Studio della Stabilità e Reattività dei Composti di Coordinazione, C.N.R., via Marzolo 1, 35131 Padova, Italy; [c] Institute of Chemical Processes Fundamentals, Academy of Sciences of the Czech Republic, Rozvojova 135, 165 02 Suchdol , Praha 6, Czech Republic; [d] Department of Organic Technology, S.T.U., Radlinskeho 9, 812 37 Bratislava, Slovak Republic; [e] Istituto F.R.A.E., C.N.R.,via Romea 4, 35020 Legnaro, Italy; [f] Dipartimento di Chimica Fisica, via Loredan 2, 35131 Padova, Italy; [g]Dipartimento di Chimica Inorganica, Metallorganica e Analitica, via Marzolo 1, 35131 Padova, Italy.

1 INTRODUCTION

Macroporous (MaR) and microporous (MiR) synthetic resins became commercially available in the late fifties and met a paramount popularity largely as synthetic ion-exchangers[1]. MaR   and MiR materials (mostly polystyrene cross-linked with divinylbenzene) differ significantly in their chemical composition and largely in their micro- and nanostructure[2]. MiR are usually 2-8 % cross-linked, while MaR are 8-20 % cross-linked materials and are normally synthesized in the presence of porogenic components[1]. In the dry state MiR materials do not possess any porosity in the solid state, but they develop a extensive nanoporosity in the swollen state[2]. On the contrary, MaR do possess a permanent microporosity even in the dry   state[1], which supplements in the swollen state the nanoporosity developed by the swelling process[2].

Macro- and microporous resins are currently utilized in important industrial processes as either solid acid catalysts or accessible supports for enzymes and cells in industrial biocatalysis[3]. In chemical applications they are used as beads (0.2 to 1.25 mm) or powders, in fixed-bed reactors or suspension reactors (often operated batchwise) or more frequently in flow-through reactors. Working temperatures range from room temperature up to about 120°C. The mechanical strength of these solids is relatively poor, but this drawback can be managed through various technical solutions[1b].

In spite of these successful applications, not much attention in the open literature has apparently been paid to the exploitation of *metal catalysts* supported *on synthetic resins* and, to the best of our knowledge, only two industrial processes based on such catalysts are currently practised, i.e. the removal of dissolved dioxygen from industrial waters (heating circuits, ultrapure water for microelectronics industry, process water in the chemical industry, etc.)[1] and the synthesis of MIBK from acetone[1].

We have been investigating in recent years the preparation of *metal palladium* catalysts (metal crystallites size ca. 2-4 nm [4b,4c]) supported on *microporous ion-exchange resins* [4] and we have also set up a method for the assessment of quantitative relationships between their nanoscopic morphology (nanostructure) and molecular accessibility in the swollen state, based on Inverse Steric Exclusion Chromatography (ISEC) [5], ESR [6] and field-gradient spin-echo NMR [7] spectroscopies (Scheme 1).

Scheme 1

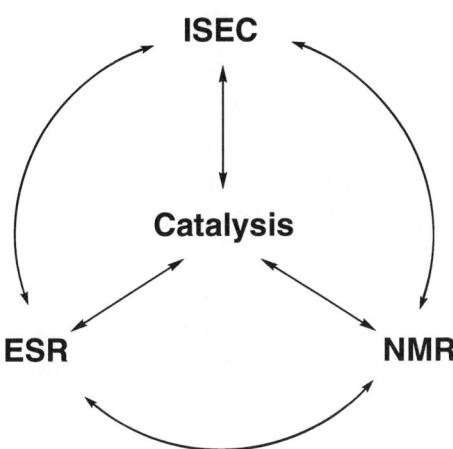

2 RESULTS

We have recently tested some of these catalysts featured by different metal loading and cross-linking degrees, in the hydrogenation of cyclohexene (1M) under mild conditions (T = 25 °C; P = 0.5 MPa)[8] in methanol , at a Pd analytical concentration equal to 2. 5 x 10^{-4} M. The metal crystallites were dispersed inside of amphiphilic[9] microporous resins based on styrene and 2-methacryloxyethylsulfonic acid, cross-linked with methylenebisacrylamide (1-6 %, mol)[10] and fully characterised, *inter alia* , by means of the above mentioned techniques. Atomic dispersion (*vide infra*) of Pd^{II} was achieved by reacting the acidic resins with $PdAc_2$[4] and the subsequent reduction to metal palladium was achieved on the basis of an established protocol set up in our laboratories[4], which enables an even dispersion of metal crystallites through the polymer network. In the activated catalysts all protons are replaced by sodium ions (see Figure 1).

We focused our attention on two pairs of catalysts based on resins with 1 % and 3 % cross-linking degree (**P1** and **P3**), containing 2.2 % Pd (**P1Pd2, P3Pd2**) or 0.22 % Pd

(**P1Pd0.2**, **P3Pd0.2**). Some relevant physical data on these materials are collected in Table 1.

Table 1. Some physical data referring to catalysts **P1Pd2**, **P3Pd2**.

catalyst	av. pcc[a]	τ[b] (ps)	D^c x 10^{+5}	τ / τ_0[d]	D_0 / D[d]
P1Pd2	0. 16	60±17	1. 8±0. 20	1. 6	1. 4
P3Pd2	0. 24	115±18	1. 1±0.15	3. 1	2. 3

[a] Weighed average polymer chain concentration in water (nm/nm^3) (ref. 5). [b] Rotational correlation time for the paramagnetic probe TEMPONE rotating in methanol constrained inside the catalyst network (ref. 6). [c] Fick's diffusion coefficient (cm^{2}s^{-1}) of methanol constrained inside the catalyst network. [d] τ_0 and D_0 refer to rotation and diffusion in unconstrained solvent.

It is seen that the polymer networks **P1** and **P3** are featured by substantially different polymer chain concentrations and that the increase of polymer chain density parallels the increase of rotational correlation time (i. e. viscosity increases with chain concentration[6]) of the spin label utilized as a molecular probe for exploring the microporous network, as well as the expected decrease of the solvent molecule diffusion coefficient. As a relevant comment we point out that the nanoscopic parameters collected for **P1Pd2** and **P3Pd2** are not expected to appreciably change for **P1Pd0.2**, and **P3Pd0.2** in view of the expected very scarce relevance of the metal percentage to the macromolecular nanostructure. Moreover, XRMA reveals that palladium distribution in **P1Pd2** and **P3Pd2** is quite homogeneous through the particle body of both catalysts. Direct XRMA measurements on **P1Pd0.2** and **P3Pd0.2** were not possible (owing to the lower Pd percentage) but, in view of the reduction protocol employed[4], we assume that metal distribution is also homogeneous.

Our catalytic tests were performed upon using particles featured by well defined sizes, i. e. less than 0. 1 mm, from 0. 1 to 0. 3 mm and from 0. 3 to 0. 7 mm. The catalytic materials, easily filtrable when particle size is larger than 0. 1 mm, are reusable at least three times without appreciable loss of activity, if recovered under hydrogen.

We find that for the quantitative hydrogenation of the substrate, the kinetic behaviour of the examined four catalysts depends on support nanostructure, metal loading and catalyst particle size. As it is shown in Fig. 1, for particle size < 0. 1 mm, the hydrogen consumption profile for catalysts **P1Pd0.2** and **P3Pd0.2** appears to be kinetically controlled and independent of cross-linking degree, i. e. of nanostructure of the polymer matrix (see Tab. 1). On the contrary, and remarkably, for catalysts **P1Pd2** and **P3Pd2** and particle size ranging from 0. 3 to 0. 7 mm, the hydrogen consumptiopn profile corresponds to a *diffusion controlled course of the reaction* and it is consequently nanostructure-dependent.

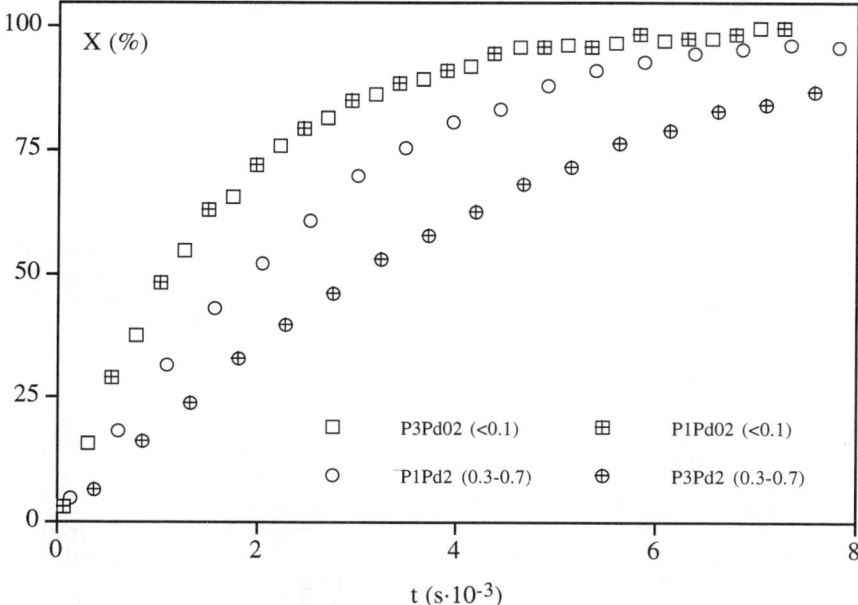

Figure 1. Conversion profiles for the quantitative hydrogenation of cyclohexene (1M) in methanol at T = 25 °C; P = 0.5 MPa. Shaken (6 Hz) constant pressure reactor, ref. (4,) at a Pd analytical concentration equal to 2. 5 x 10^{-4} M. The particle sizes are indicated in parentheses inside the frame of the figure.

The results presented in this report show very clearly that the somewhat pervading belief that metal crystallites dispersed inside microporous resins are to be considered "buried" in an essentially impenetrable maze of polymer chains, *is just a bias* . A proper choice of polymer chain concentration in the support ("nanoreactor" size) and of catalyst particles dimensions ("microreactor"particle size) can enable the attainment of clean kinetically-controlled conditions and consequently to the exploitation of the whole of the dispersed nanocrystallites, without a perceivable kinetic trouble produced by the support. On the other hand, it appears to be possible to attain also diffusion-controlled conditions, which may be useful in some specific cases.Our results clearly illustrate, albeit at a preliminary level, these important properties of resin-supported metal catalysts and our results militate in favour of a more extended consideration of these catalytic materials. They offer the advantage of the production of properly sized metal nanophases and of promising designable nanoscopic features of the polymer network, expectedly leading to i) reagent size effects (*vide infra*), ii) to the possible exploitation of hydrophobic-hydrophilic interactions and iii) to the facile achievement of polyfunctional catalysis[1b].

References

1.   a) A. Guyot in "Synthesis and Separations using Functional Polymers" , D.C. Sherrington, P. Hodge eds.; John Wiley, New York; 1988; p. 1; b) A. Guyot, *Pure & Appl. Chem.*, 1988, **60**, 365.b) H. Widdecke in "Synthesis and separations using   functional polymers", H.W. Sherrington, P. Hodge, eds.; John Wiley: New York;   1988; pp 149-179; c) R. Wagner, P.M. Lange, *Erdöl, Erdgas, Kohle,* 1989, **105**,   414.
2.   B. Corain and K. Jeřàbek, *Chim. Ind. Milan* , **1996**, *78*, 563 and references therein.
3.   a) "Industrial Application of Immobilized Biocatalysts"A. Tanaka, T. Tosa, T. Kobayashi, eds, , Marcel Dekker, 1993 ; b) E. Katchalski-Katzir,*TIBTECH*, 1993, **11**, 471; c) A. Wiseman, *J. Chem. Ed.* ,1996, **73** , 55.
4.   (a) M. Kralik, M. Hronec, S. Lora, G. Palma, M. Zecca, A. Biffis, B. Corain, *J. Mol. Catal. A: Chemical* , 1995, **97**, 145; (b) M. Kralik, M. Hronec, V. Jorik, S. Lora, G. Palma, M. Zecca, A. Biffis, B. Corain, *J. Mol. Catal. A: Chemical.*, 1995, **101**, 143; c) A. Biffis, B. Corain, Z. Cvengrosova, M. Hronec, K Jer àbek,   M. Králik, *Appl. Catal.*, 1995, **124**, 355; d) A. Biffis, B. Corain, Z. Cvengrosova, M. Hronec, K. K Jeràbek, M. Králik, *Appl. Catal.*, 1996, **142**, 327.
5.   K. Jeràbek, *Anal. Chem.*, 1985, **57**, 1598.
6.   A. Biffis, B. Corain, M. Zecca, C. Corvaja, K. Jeràbek *J. Am. Chem. Soc.*, 1995, **117**, 1603 and references therein.
7.   S. K. Ghosh, E. Tettamanti, A. Ricchiuto, *Chem. Phys. Letters* , 1983, **101**, 499 and references therein.
8.   M. Králik, M. Zecca. A.A. D'Archivio, L, Galantini, B. Corain, submitted to *J. Mol. Catal.*
9.   R. Arshady, *Adv. Mater.*, 1991, **3**, 182.
10.   M. Zecca, M. Králik, M. Boaro, G. Palma, S. Lora, M. Zancato, B. Corain, submitted to *J. Mol. Catal. A.*

OXIDATION OF ETHYLBENZENE WITH DIOXYGEN USING HOMOGENEOUS AND SUPPORTED IRON(III) PORPHYRINS AS CATALYSTS

J.H. Clark, S. Evans and J.R. Lindsay Smith*

Department of Chemistry
The University of York
York, YO1 5DD
UK

1 INTRODUCTION

There have been several recent investigations of the clean and potentially useful iron(III) porphyrin-catalysed oxidations of alkanes with dioxygen.[1,2] The general consensus is that these reactions occur by free-radical processes although whether they involve an iron(II)/iron(III) redox catalyst system[2] and/or oxoiron(IV) species remains to be resolved.

The present study was initiated with the aim of optimising the catalytic oxidation system through an improved mechanistic understanding of the reactions involved. A subsequent aim was to support the catalyst and to optimise the system further. Some preliminary results from our studies, using ethylbenzene as substrate, are described below.

2 RESULTS AND DISCUSSION

2.1 The Oxidation System and Analytical Methods

Iron(III) tetrapentafluorophenylporphyrin [$Fe^{III}(TF_5PP)Cl$] (Figure 1) was selected as the catalyst since it is easily prepared and readily modified by nucleophilic replacement of the *para*-fluorines and/or substitution of the ß-pyrrole hydrogens.[3,4,5] Thereby allowing both structure/activity studies on the catalyst and a means of anchoring the catalyst to the support (Figure 1). The reactions in this investigation have been carried out in neat ethylbenzene, under 1 atmosphere of dioxygen, in the temperature range 30-110 °C.

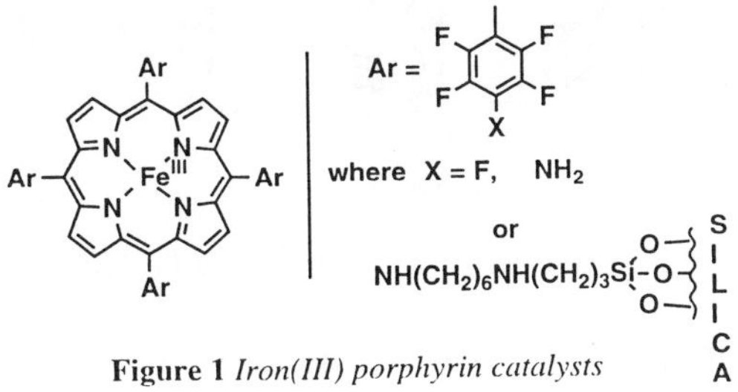

Figure 1 *Iron(III) porphyrin catalysts*

The reactions were followed using g.c. analysis, to measure the yields of the oxidation products, and UV-Vis spectroscopy to monitor the fate of the catalyst (Figure 2). Three major products were detected in all the reactions, 1-phenylethanol, acetophenone and 1-phenylethyl hydroperoxide (Scheme 1), although their relative yields depend on the reaction conditions and the catalyst employed. Other products, including benzaldehyde, benzoic acid and di(1-phenylethyl) ether were only present in trace quantities.

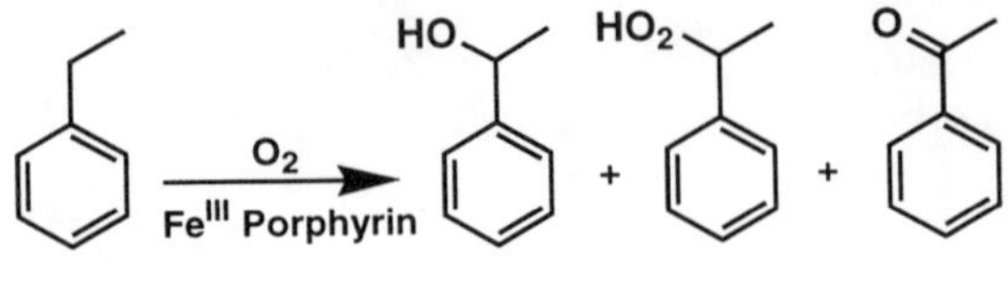

$$\text{(ethylbenzene)} \xrightarrow[\text{Fe}^{III} \text{ Porphyrin}]{\text{O}_2} \text{(1-phenylethanol, HO)} + \text{(HO}_2\text{, hydroperoxide)} + \text{(acetophenone, O)}$$

Scheme 1

1-Phenylethyl hydroperoxide is readily dehydrated to acetophenone during g.c. analysis and consequently the yields of neither of these products could be determined directly. Hydroperoxides, however, are readily reduced to alcohols by triphenylphosphine[6] and, in this way, by carrying out g.c. analyses of the reaction mixture before and after treatment with PPh$_3$, we were able to monitor the growth of the three major products during the oxidations.

A FAB-mass spectrometric analysis of a homogeneous reaction mixture, in which the catalyst [Fe^{III}(TF$_5$PP)CI] was partially degraded, revealed ions with m/z 1148 and 1045, equivalent to 1-phenylethoxyl and hydroxyl derivatives of the iron porphyrin. We believe these species arise from reaction of the iron porphyrin with 1-phenylethoxyl and hydroxyl radicals present in the oxidation system and are intermediates leading to the total destruction of the catalyst.

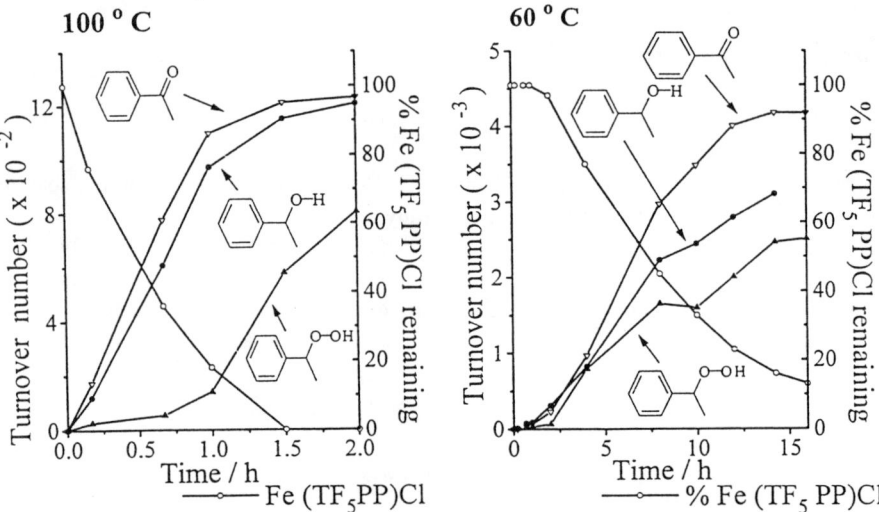

Figures 2 and 3   *Time course of oxidations with Fe^{III} (TF$_5$PP)Cl at 100 and 60 °C*

2.2 The Influence of Temperature on the Homogeneous Oxidation

Lowering the temperature of the homogeneous reactions catalysed by Fe^{III}(TF$_5$PP)Cl resulted in the following changes:

(i)   a decrease in the rate of reaction, and an increase in the catalyst lifetime and induction period (Figure 3),

(ii)   an increase in the overall yield of products and a change in product distribution (Figure 4),

(iii)   the appearance of 1-phenylethyl hydroperoxide from the start rather than towards the end of the reaction (Figures 2 and 3).

We conclude, in agreement with previous studies,[1,2,7] that the oxidations involve a free radical chain reaction: the main chain carriers are 1-phenylethyl and 1-phenylethylperoxyl radicals and the initial product is 1-phenylethyl hydroperoxide (Scheme 2). Catalysis arises through the initiation of new radicals by redox reactions between the iron porphyrin and hydroperoxide (Scheme 2). At high temperature the latter reactions are fast, consequently hydroperoxide does not build up in the reaction (while the catalyst is active), and the radical flux is high, radical chains are short and catalyst destruction is fast. By contrast, at low temperatures the iron porphyrin/hydroperoxide reactions are relatively slow, the radical concentrations are very low and as a result the chains are long, catalyst destruction is slow and catalyst turnovers are high.

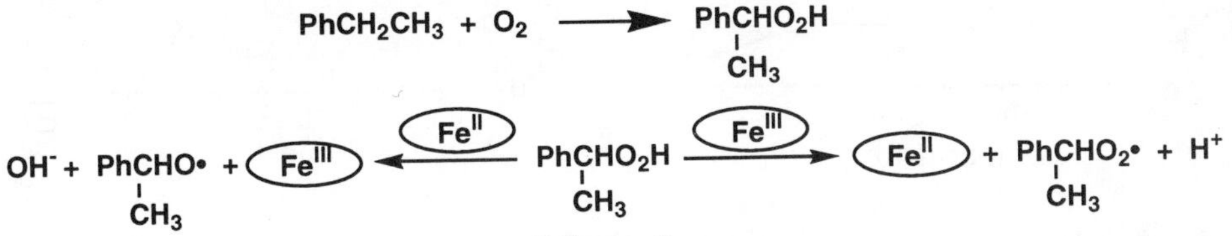

Scheme 2

When $Fe^{III}(TF_5PP)Cl$ was replaced by $Fe^{III}(TF_4APP)Cl$, as a model for the supported iron porphyrins, the oxidations were slower, the catalyst was more stable and the yields were approximately an order of magnitude higher (Figure 5). The increased catalytic efficiency was unexpected since improved performance is generally associated with electron withdrawing groups stabilising the catalyst to self oxidation;[5,8] electron donating substituents, such as NMe_2, have been reported to result in inferior catalysts for oxidations using iodosylbenzene as the oxidant.[9] The origin of this effect is currently under investigation.

2.3 Oxidation Using Supported Catalysts

The supported catalysts (Scheme 3) were obtained by reaction of $Fe^{III}(TF_5PP)Cl$ with 3-(N-1,6-diaminohexyl)propylated silica (prepared from 1,6-diaminohexane and 3-chloropropylated silica). These were further modified either with Me_3SiCl to convert surface Si-OH to Si-OSiMe$_3$ or with acid washing to protonate free amino groups. Catalyst loadings were 10, 85 and 172 mg g^{-1} of silica support.

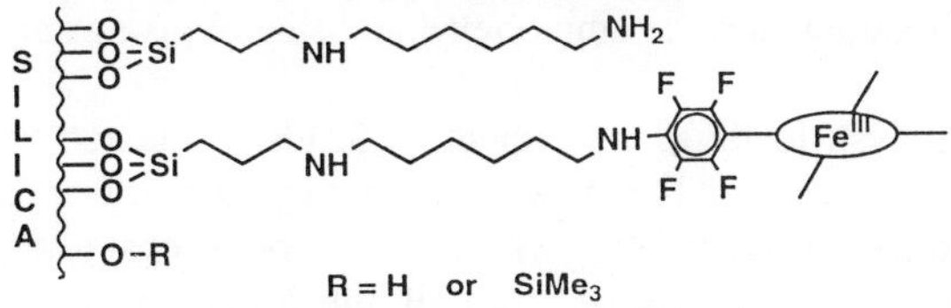

Scheme 3

Comparison of homogeneous oxidations with an analogous reaction with supported catalyst reveals significant differences (e.g. Figure 6; loading 10 mg g^{-1} at 100 °C). For the latter the rate is slower, the induction period longer, the catalyst is more stable and 1-phenylethyl hydroperoxide is a major product; in many respects the reaction profile resembles more closely a low temperature homogeneous oxidation.

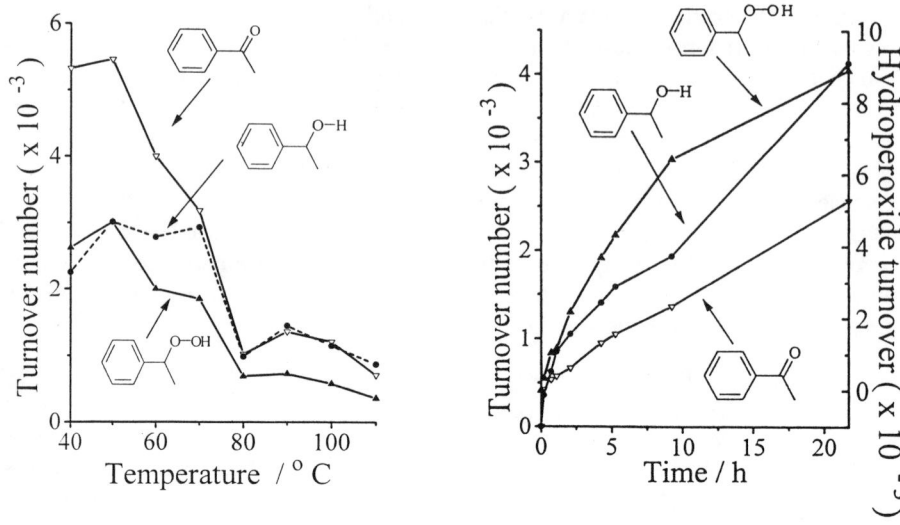

Figure 4 *Dependence of product yields on oxidation temperature*

Figure 5 *Time course of oxidation with Fe^{III}(TF$_4$APP)Cl at 100 °C*

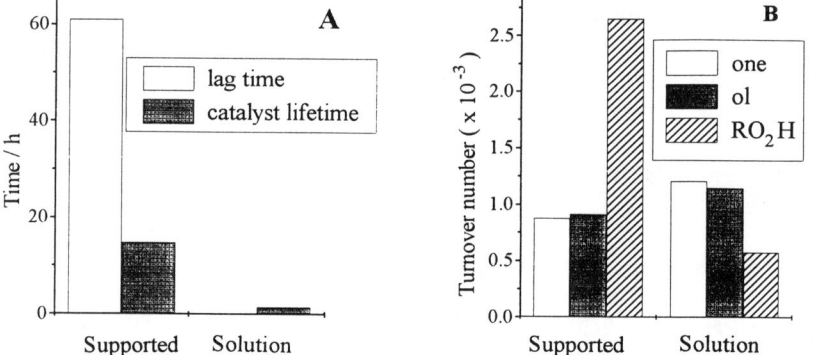

Figure 6 *Comparison of homogeneous and heterogeneous reactions at 100 °C*

To obtain further information on the important factors that affect the catalytic activity, the influence of loading, surface aminohexyl and silanol groups and support polarity were investigated.

Increased catalyst loading, as expected, leads to faster reactions and a shorter induction period; however, the catalyst turnovers are lower and catalyst lifetime is dramatically reduced (Figure 7). We believe these changes, which resemble those observed on increasing the temperature of the homogeneous reactions, arise from a higher concentration of free radicals in the reactions with a higher loading of catalyst.

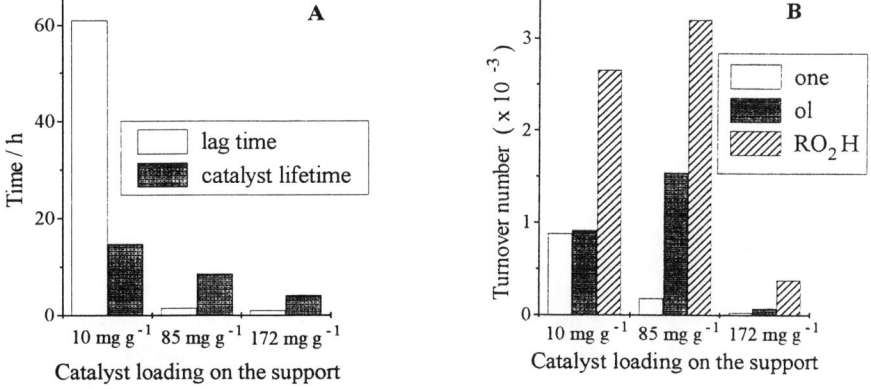

Figure 7 *Effect of catalyst loading on induction period and catalyst lifetime (A) and product yields (B)*

Amines are used as antioxidants and are also known to reduce $Fe^{III}(TF_5PP)Cl$ to bis-ligated $Fe^{II}(TF_5PP)$.[10] The latter complex might be a poor catalyst for the oxidations. For these reasons we argued that the lower reactivity of the supported catalyst might arise from the amino groups of the linker units bound to the silica but not involved in covalent binding to the porphyrin. This was confirmed by the following experiments:

(i)   repeating the oxidation at 100 °C with supported catalyst which had been washed with acid, to protonate any free amino groups, resulted in a reduced induction period and an increased yield.

(ii)   UV-Vis analyses on the supported iron porphyrin showed it to be in the iron(II) state before acid washing; however, after acid treatment which would protonate and remove the amine ligands it reverted to the iron(III) species.

(iii)   homogeneous reactions were carried out with $Fe^{III}(TF_5PP)Cl$ at 70 °C in the presence of 1,3-diaminopropane, diaminohexane-modified silica and acid washed diaminohexane-modified silica. The first oxidation was totally inhibited and gave *bis*-ligated $Fe^{II}(TF_5PP)$, the second was only partially effective (2,600 turnovers) and led to rapid catalyst destruction (< 2 h) and the third was comparable to the standard homogeneous process.

Trimethylsilylating the residual silanol groups on the supported catalyst has a beneficial effect on catalyst performance. It leads to increased yields and a decreased induction period although the catalyst's lifetime is unaffected. A possible explanation for this effect, based on support polarity is discussed below.

The polarities of the supports used in this study were measured using a method, based on Reichardt's E_T^N scale, very recently reported by Tavener *et al*,[11] and are compared with those of ethylbenzene and two of the products, acetophenone and 1-phenylethanol, in Table 1. This reveals that all the supports are polar with polarities closer to those of the products than that of the substrate. We believe that this may be a possible cause of the inhibiting effect of the support on the catalysed oxidations. The chain reactions are maintained by redox reactions between 1-phenylethyl hydroperoxide and iron porphyrin which with supported catalysts will generate alkoxyl and peroxyl radicals close to the surface of the support. For oxidation to continue those relatively polar radicals have to diffuse away from the polar surface into the non-polar bulk medium to initiate new radical chains. We propose that, instead, a significant proportion of these radicals are

Table 1   *Polarities (E_T^N values) of supports, ethylbenzene and products*

Silica Modification	E_T^N	Liquid reagents	E_T^N
None	0.91	1-Phenylethanol	0.53
Diaminohexyl	0.69	Acetophenone	0.40
TMS/diaminohexyl	0.63	Ethylbenzene	0.11
TMS	0.51		

competitively trapped by the polar support.   On this basis, less polar supports (e.g. the trimethylsilylated diaminohexylsilica) should lead to more effective catalysis.

3 CONCLUSIONS

(i)   Iron(III) porphyrin-catalysed oxidations of ethylbenzene by dioxygen are free radical autoxidations and are most efficient at low temperatures.

(ii)   Catalyst destruction arises mainly from reaction with alkoxyl and hydroxyl radicals and is more apparent at relatively high radical concentrations.

(iii)   Iron(III) porphyrins covalently bound to silica *via* amino linker units catalyse ethylbenzene autoxidation .

(iv)   Removal of surface amino or silanol groups improves the efficiency of the supported iron(III) porphyrin catalyst.

References

1.   J.E. Lyons, P.E. Ellis and H.K. Myers, *J. Catal.*, 1995, **155**, 59.

2.   M.W. Grimstaff, M.G. Hill, E.R. Birnbaum, W.P. Schaefer, J.A. Labinger and H.B. Gray, *Inorg. Chem.*, 1995, **34**, 4896.

3.   P. Battioni, J.F. Bartolli, D. Mansuy, Y.S. Byun and T.G. Traylor, *J. Chem. Soc. Chem. Commun.*, 1991, 1051.

4.   Q. La, R. Richards and G. M. Miskelly, *Inorg. Chem.*, 1994, **33**, 3159.

5.   D. Dolphin, T.G. Traylor and L.Y. Xie, *Acc. Chem. Res.*, 1997, **30**, 251.

6.   A.G. Rowley, in 'Organophosphorus Reagents in Organic Chemistry', ed. J.I.G. Cadogan, Acad. Press, London, 1979, Ch. 7.

7.   J.A. Labinger, *Catal. Lett.*, 1994, **26**, 95.

8.   B. Meunier, *Chem. Rev.*, 1992, **92**, 1411.

9.   D. Bouy-Debec, O. Brigaud, P. Leduc, P. Battioni and D. Mansuy, *Gaz. Chim. Ital.*, 1996, **126**, 233.

10.   P.R. Cooke, C. Gilmartin, G.W. Gray and J.R. Lindsay Smith, *J. Chem. Soc. Perkin Trans. 2*, 1995, 1573.

11.   S.J. Tavener, J.H. Clark, G.W. Gray, P.A. Heath and D.J. Macquarrie, *J. Chem. Soc. Chem. Commun.*, 1997, 1147.

Acknowledgements: S.E. thanks the EPSRC (Clean Technology) for a studentship.

ReO$_x$/γ-Al$_2$O$_3$-CATALYSED METATHESIS: SIMPLE, CROSS AND CYCLIZING

B. Kondratowicz, J. Sithamparapillai, A. Subramaniam and P.A. Sermon[*]

Fractal Solids and Surfaces Research Group, Department of Chemistry, Brunel University, Uxbridge, Middlesex, UB8 3PH, UK

ABSTRACT

Mesoporous ReO$_x$/γ-Al$_2$O$_3$ (promoted by halogenated and activated in N$_2$) is shown to be particularly effective in catalysis of α,ω–diene cyclization-metathesis (at least for some chain lengths), propene metathesis and ethene/butene cross-metathesis; mechanisms and modes of optimisation are considered.

1 INTRODUCTION

Heterogeneous catalysis[1] of alkene metathesis can be catalysed by ReO$_x$/alumina (with and without halogen[2] or borate moderation-promotion[3]). Such catalysts are active in ethene-cycloocta-1,5-diene cross-metathesis[4], as are complexes of transition metals. Banks and Bailey[5] were first to see the heterogeneously-catalysed reaction, which possibly occurs via *quasicyclobutanic* or *carbene*[6] mechanisms. Alkene metathesis is close to being thermally neutral, of low activation energy and is often first-order with respect to the alkene(s). At a molecular level the reaction has to compete with the more favourable isomerisation and dimerisation reactions[7] at say 300-500K, despite uncertainties in the free energy calculations for isomers[7].

Metathesis of course may be more complex (e.g. cycloalkenes can so form larger cyclic dienes[8] and dienes can form[10] cycloalkenes). Interestingly, a diene[9] biradical[10] has been associated with the Diels-Alder reaction of butadiene with a dienophile[11]. However, the enthalpies of transformation of gaseous hexadiene to cyclobutene and ethene is significantly positive at 298K (i.e. 20.5kJ/mol hexadiene converted) for diene ring-closing metathesis and so this is unfavourable as simple propene metathesis at modest temperatures.

This interesting area has been explored here for promoted and activated mesoporous alumina-supported rhenia.

2 EXPERIMENTAL

2.1 Materials

The precursor salts (from Aldrich) were ammonium perrhenate (NH$_4$ReO$_4$; AP; 99.9%

purity) or perrhenic acid ($HReO_4$; PA; 99.9% purity) in aqueous solution. γ-Alumina (99.8%, 9.9nm average pore size and $157m^2/g$) was obtained from UOP. For simple metathesis and cross-metathesis 3% propene/nitrogen (99.9% purity; BOC), 5% ethene/nitrogen (99.9% purity; BOC), 5% cis-but-2-ene/nitrogen (99.9% purity; BOC) and 5% trans-but-2-ene/nitrogen (99.9% purity; BOC) were used. The following acyclic dienes were also used: 1,5-hexadiene (97% purity; Fluka), 1,6-heptadiene (99% purity; Fluka), 1,7-octadiene (97% purity; Fluka) and 1,9-decadiene (96% purity; Aldrich).

2.2 Catalyst Preparation

The alumina-supported rhenia catalysts were prepared[2] by impregnation of the γ-Al_2O_3 support by an aqueous solution of the precursor which was evaporated to dryness (at room temperature) with constant stirring, dried for 12h at 393K and calcined at 673K for 2h in flowing air (100 cm^3/min). Some catalyst supports were chlorinated (by dilute HCl) or fluorinated (by dilute HF) before impregnation. The catalysts (shown in Table 1) were finally pretreated in-situ prior to catalysis (see Section 2.3). Here low ReO_x loadings were selected for use in propene simple metathesis (SM) and ethene/butene cross metathesis (CM), and 3.5 or 8.0%ReO_x (supports were modified by both Cl and F) for diene ring closure metathesis (RCM). Such catalysts were characterised by N_2 adsorption at 77K and showed the average pore sizes and total surface areas shown in Table 1.

Table 1 Catalyst compositions, average pore sizes (d_{pore}), total surface areas (S_{BET}), precursors (AP= ammonium perrhenate and PA=perrhenic acid) and propene metathesis activity at 423K

%Re as ReO_x	d_{pore}/nm	$S_{BET}/m^2\,g^{-1}$	Precursor	Modifier	r/% C_3H_6 conv
1.2	7	180	AP	-	6.0
1.2	7	184	PA	-	6.9
1.2	7	174	PA	Cl	15.9
1.2	6	155	PA	F	24.0
3.5	7	164	PA	Cl	17.7
8.0	8	167	PA	Cl	21.8

2.3 Catalysis

N_2, 6%O_2/N_2 and air were used for pretreatment after drying over 4A molecular sieve and N_2 was also passed over MnO_x/celite[10] to remove oxygen down to 0.001ppm. The alkenes, (propene, ethene and butene) were dried over 4A molecular sieve (chosen because propene and butene are not retained by the 4A bed, although some ethene *is* retained[11]). For SM and CM the sample (0.3g) of the catalyst was pretreated in air at 973K for 2h and cooled in N_2 to 298K (see also Figure 2). The alkene stream (3kPa propene for SM) was then introduced at constant molecular residence time (τ=3.6s) with the catalyst at 423K and the products were analysed by GC and GC-MS as described in Section 2.4.

In the diene RCM, the catalyst sample (0.5 g) was calcined in-situ for 2 h at 973K in a stream of air (100 cm^3/min) and was then cooled to room temperature. The reactor was transferred into a water bath (313-373K) and the diene (10 cm^3) was then added into the pre-heated catalyst and was stirred (30 rpm) for up to 4 h. The products were then analysed by GC-MS.

2.4 Product Analysis

Samples (0.1cm^3) of the gaseous product stream in SM and CM were injected into a gc (Pye Unicam GC fitted with a 3% squalane on alumina column at 353K) using a Negretti sampling valve. Retention times were 2.0, 5.3, 14.1, 15.0 and 18.0 min for ethene, propene, but-1-ene, trans-but-2-ene, cis-but-2-ene. A GC-MS analysis (Hewlett Packard 5890; Plot alumina column at 393-483K) was also used. Analyses of the products of diene reaction were carried out by GC-MS after filtering off the catalysts and product dilution in n-hexane.

3 RESULTS ON SIMPLE AND CROSS METATHESIS

3.1 Propene SM

Table 1 and Figure 1a show that over three different alumina-supported rhenia catalysts that activity in propene metathesis was greatest at 423K over the fluorinated catalyst and with this decreased most noticeably with reaction time. It was over the fluorinated catalyst that the ethene-to-butene product ratio was most depressed below the 1:1 value expected for SM. In these experiments the average molecular residence time τ (i.e. 3.6s) was far from those where mass transport control was expected to be important. Nevertheless, the metathesis activation energy was found to be very low (13-15kJ/mol); presumably it was diffusion limited.

3.2 But-2-ene and Ethene CM

In the cross metathesis of cis-but-2-ene and ethene at 423K over one non-halogenated catalyst derived from the AP or PA precursors at different reactant ratios, but constant τ (3.6s), the ethene was far less reactive than the butene at all reactant ratios and isomerisation of cis-but-2-ene was faster than its cross metathesis with ethene. Dimerisation has been considered paralleling SM and now one has to consider isomerisation paralleling CM. Deactivation of this catalyst was very slow compared to the more active fluorinated catalyst.

It was possible that the extent of the parallel isomerisation in CM was dictated by the instability of cis-but-2-ene in relation to trans-but-2-ene and to test this, CM of trans-but-ene with ethene was also investigated at 423K at constant τ. Even for this over a PA-derived catalyst the ethene was still rather unreactive, although isomerisation of trans-but-ene was more modest than that of cis-but-2-ene. The PA- and AP-derived alumina-supported rhenia catalysts were equally dominated by but-2-ene isomerisation rather than ethene/but-2-ene CM.

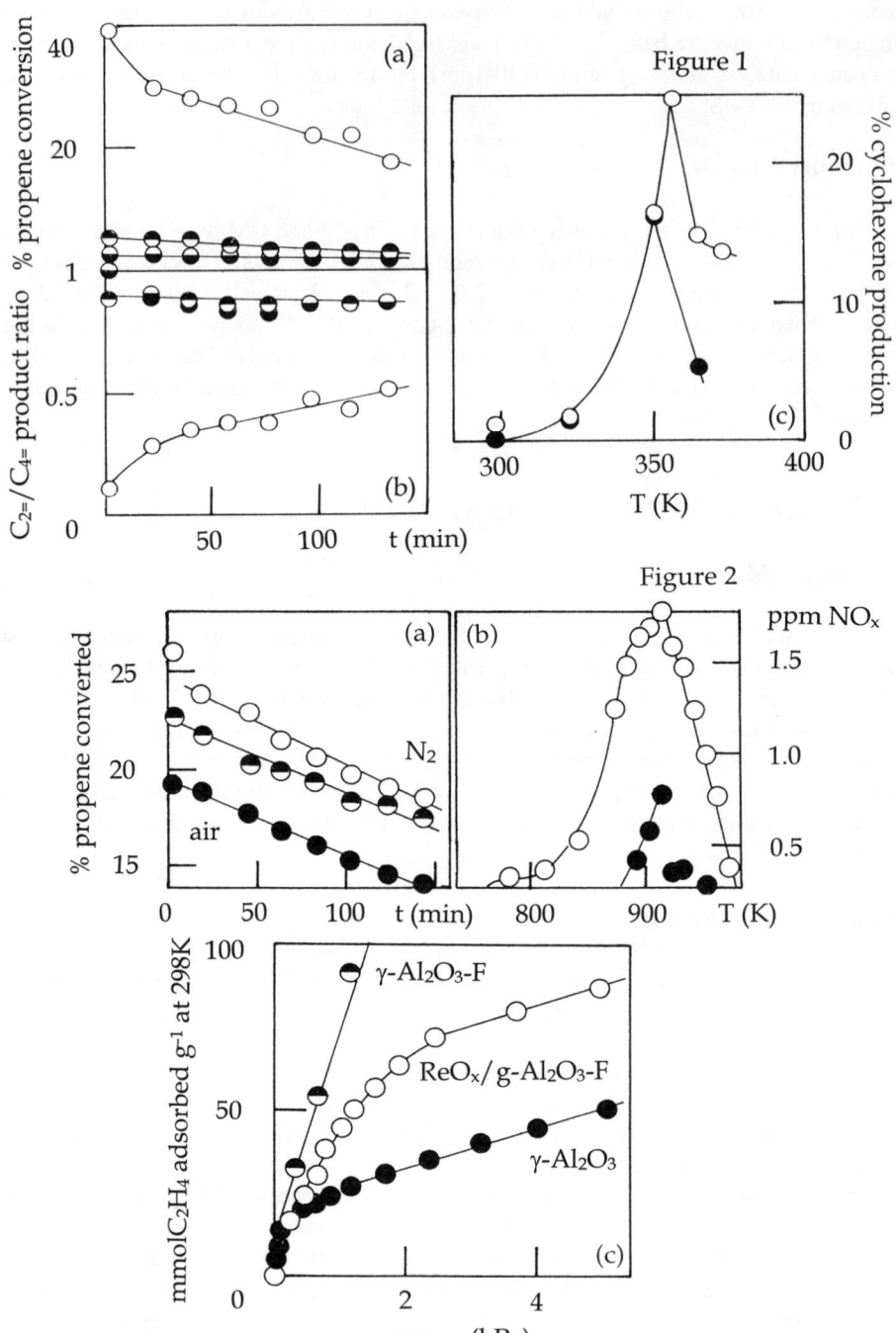

Figure 1. (a,b) *% propene conversion (a) for 1.2%ReO$_x$/γ-Al$_2$O$_3$ unhalogenated (◑), chlorinated (◐) and fluorinated (○) and C$_2$/C$_4$ product ratio (b) for 1.2%ReO$_x$/γ-Al$_2$O$_3$ unhalogenated (◑), chlorinated (◐) and fluorinated (○) at 423K. (c) Concentration of cyclohexene produced during 1,7-octadiene conversion for 4h over 3.5%ReO$_x$/γ-Al$_2$O$_3$-F (●) and 8.0%ReO$_x$/γ-Al$_2$O$_3$-F (○) at various temperatures.*

Figure 2. (a) *Propene metathesis activities over 1.2%ReO$_x$/γ-Al$_2$O$_3$-Cl after N$_2$ (○), 6%O$_2$/N$_2$ (◐) or air (●) treatment at 700K for 2h. (b) Release of NO$_x$ into a N$_2$ stream when this was passed over 1.2%ReO$_x$/γ-Al$_2$O$_3$-Cl (○) or γ-Al$_2$O$_3$ alone (●) at 80cm^3/min during temperature programming. (c) Adsorption isotherms for ethene for 1.2%ReO$_x$/γ-Al$_2$O$_3$ (○), γ-Al$_2$O$_3$-F (◐) and γ-Al$_2$O$_3$ (●).*

3.3 Diene RCM

Cross metathesis of 1,5-cyclo-octadiene with isobutene *occurs* in measurements over a 22h period over the present catalysts (despite their pore sizes being 6-7nm) and ring opening metathesis is also reported over Mo complexes[12]. Here diene metathesis was considered for the fluorinated and chlorinated alumina-supported rhenia type of catalyst, the former was found to be most active in propene SM (see Figure 1a and Table 1).

Analysis of the gas and liquid phase (after catalyst separation) components in metathesis of 1,5-hexadiene showed that these contained ethene as a product and that the percentage conversion of the diene at 318K for 4 h was 8%. A similar analysis revealed that with 1,6-heptadiene the percentage conversion of the diene at 323K after 4h over the same catalyst was higher (i.e. 10%). For RCM of 1,9-decadiene ethene, propene and a higher molecular mass product appeared, but the percentage conversion of the diene at 373K in the presence of the same catalyst for 4 h was higher still (i.e. 12%). Lowering the rhenia loading decreased the extent of diene cyclization by a third. For 1,7-octadiene analysis of the gas and liquid phase components showed that these contained ethene and cyclohexene and that the percentage conversions of the diene after 4 h at different temperatures were raised still further (see Figure 1c, where the percentage conversion reaches about 25% at about 355K) over two different catalysts. Hence the rate or extent of RCM increased with the diene reactant (or the cycloalkene product) carbon number.

3.4 In-situ Characterisation

Figure 2a shows that *N$_2$ pretreatment* was most effective in maximising subsequent propene metathesis, while Figure 2b shows that a significant concentration (relative to that from the support alone) of NO$_x$ may be produced in the N$_2$ pretreatment process (i.e. O is lost from the surface rhenia species). It was surprising that ethene adsorption at 298K (see Figure 2c) was greater on fluorinated/alumina than the alumina alone or the ReO$_x$-loaded alumina, since this means that its interaction with the surface is anything but specific to the ReO$_x$.

4 DISCUSSION

On the surface of the defect spinel γ-alumina must be found Re_2O_7 (yellow; melting point 573K) containing ReO_4 and ReO_6 building blocks. Its structure must be stabilised via the formation of Al-O-Re linkages. Samples of the catalysts when pretreated at 973K over 3h in N_2 lost some ligand O as NO_x (as shown in Figure 2b) and so active sites are *generated* during the reaction. AP and PA-derived catalysts show similar catalytic properties and this is not surprising since the active sites after pretreatment in-situ will be likely to be the same. The most active catalysts were fluorinated (presumably because Re_2O_7 interacts with Lewis acid sites, modified by F) but these also deactivate most rapidly.

It seems that since monocycloalkenes, rather than cyclodienes were seen, that *intra*molecular rather than *inter*molecular metathesis was taking place here.

The C_2:C_4 ratio in propene SM was not 1:1 as expected, but was depressed with regard to ethene. Dimerisation of ethene is unlikely, since ethene is *less* reactive in cross metathesis than but-2-enes (which undergo parallel isomerisation rather readily, with cis-to-trans conversion easier than trans-to-cis conversion). Alternatively, the low residence time of ethene on the surface could result from rapid $CH_2=Re_s$ adsorbate-adsorbate interactions *producing* ethene more rapidly than higher alkenes. One could also envisage isomerisation and diene cyclisation via a dicarbene mechanism:

$(CH_2)_{n-2}$- $=Re_s$

$(CH_2)_{m-2}$- $=Re_s$

One now needs to know how the rate of ethene production compares with that of cycloalkene production. It would be possible to consider the optimum diene reactant in terms of the strain in the product cycloalkene. It is not surprising that diene cyclisation is most effective when C_6-cycloalkenes are produced since (although as m and n increased so would the effective molecular size of the reactants and products and their boiling points, making mass transport into the porous catalyst more difficult) the formation of these cycloalkenes is always more favourable. If propene metathesis over the present catalysts involves a low activation energy (13-15kJ/mol) then how much lower will be the value for 1,7-octadiene (see Figure 1c) and how much more will the reaction be diffusion controlled? The free energy of cyclisation may at least be weakly positive[13]. This mass transport control (which presumably contributes to the observation of T_{opt} for 1,7-octadiene conversion at 358K) clearly needs optimisation. Here one needs to recall molecular sizes deduced from van der Waal's radii are all small compared to the pore sizes in Table 1: 0.518nm for ethene, 0.853nm for dec-1-ene and 0.650-0.689 for cyclohexene.

It is recognised that side reactions (including double bond shift) will occur and that active sites[14] may be modified by these reaction products with significant deactivation. For example dimerisation may have occurred and left products on the surface blocking active sites (rather than impurity O or H_2O removing active sites).

Although ethene adsorption is not specific to ReO_x, but occurs on the halogenated support even more effectively, it would be wrong to call the present solids simple reagents rather than catalysts; the truth is that they are a mixture of the two.

References

1.   P. B. Wells, Chem.Britain 989,Nov.1992

2.   A. Subramaniam, PhD thesis Brunel Univ. 1995
3.   M. Sibeijn, J. A. R. van Veen, A. Blick and J. A. Moulijn, J.Catal. 1994,**145**,416.
4.   J. Hietala, A. Root and P. Knuuttila, J.Catal. 1994,**150**,46.
5.   R. L. Banks and G. C. Bailey, Ind.Eng.Chem. Prod.Res.Devel. 1964, **3**,170.
6.   W. R. Kroll and G. Doyle, J.Chem.Soc.Chem.Commun. 1971,839.
7.   R. A. Alberty, Ind.Eng.Chem.Fundam. 1983,**22**,318.
8.   E. Wasserman, D. A. Ben-Efraim and R. Wolovsky, J.Amer.Chem.Soc. 1968,**90**,3286.
9.   S. F. Martin and A. S. Wagman, Tetra.Lett. 1995,**36**,1169; S. Stinson Chem.Eng.New April 10 1995,38; S. F. Martin, Y. Liao, Y. Wong and T. Rein, Tetra.Lett. 1994,**35**,691.
10.   S. W. Benson, J.Chem.Phys. 1967,**46**,4920.
11.   B. Eisler and A. Wasserman, Disc.Far.Soc. 1951,**10**,235.
12.   W. B. Hughes, J.Amer.Chem.Soc. 1970,**92**,532; J. Lai and R. R. Smith, J.Org.Chem.1975,**40**,775.
13.   R. H. Grubbs and G. C. Fu, J.Amer.Chem.Soc. 1992,**114**,5426
14.   R. H. Grubbs, S. J. Miller and S. H. Kim, J.Amer.Chem.Soc. 1995,**117**,2108.

OXOAMMONIUM SALT OXIDATIONS OF ALCOHOLS IN THE PRESENCE OF, AND ON THE SURFACES OF, AN ELECTRODE, AN ION EXCHANGE RESIN, AND SILICA GEL

J. M. Bobbitt, Z. Ma and D. Bolz

Department of Chemistry
University of Connecticut
Storrs, CT. 06269-3060, USA

T. Osa, Y. Kashiwagi, Y. Yanagisawa, F. Kurashima and J. Anzai

Pharmaceutical Institute
Tohoku University
Ayobayama, Sendai 980-77, Japan

J. E. Tacorante-Morales

Fac. Quimica
Universidad de la Habana
C. Habana, Cuba 10400

1 INTRODUCTION

The selective oxidations of primary alcohols to aldehydes and of secondary alcohols to ketones are two of the most used reactions in organic chemistry, and many reagents have been developed for such oxidations. One of the less developed systems involves the use of oxoammonium salts such as (2).[1,2] Oxoammonium salts are prepared by the further oxidations of free oxygen radicals such as 2,2,6,6-tetramethylpiperidin-1-oxyl (1, NHR = H; this material is commonly known as TEMPO). They can also be prepared by acid disproportionation as shown in Scheme 1, can be used for the oxidations of alcohols as shown in Scheme 2.

Scheme 1 *Oxoammonium salt formation and reactions*

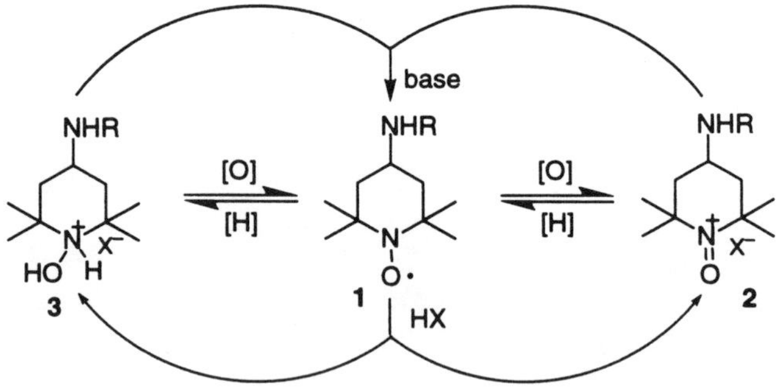

Scheme 2 *Oxidation of alcohols with oxoammonium salts*

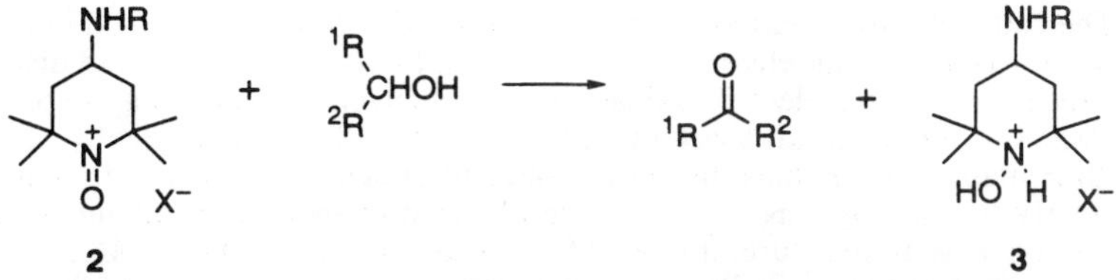

The reaction can be used in two ways. First, oxidations can be carried out using a secondary oxidant such as hypochlorite or an electrode surface in the presence of a catalytic amount of nitroxide radical. In a related method, the oxoammonium salts are prepared in situ by a disproportionation reaction.[3] Second, the reactions may be carried out using a stoichiometric amount of the oxoammonium salt. In either case, the yields are quantitative, and the products resulting from the reduction of the oxidants are hydroxyamine salts (**3**). In the catalytic reactions, this hydroxyamine is reoxidized by the secondary oxidant to the salt. In the stoichiometric reactions, the hydroxyamine salt is isolated (and can be reoxidized back to the salt). The major advantages of the method are good yields, convenient reaction conditions and isolations, and the fact that no heavy metals (such as the normal chromium or manganese reagents) are used.

We have prepared electrode surfaces coated with a nitroxide catalyst, polymeric nitroxides, and chiral polymeric nitroxides. For the stoichiometric reactions, we have prepared a new, stable, inexpensive oxoammonium salt, and found that the rates of the oxidation reactions are tremendously enhanced *in the presence of silica gel*. We have also found that the oxoammonum salts can be deposited on silica gel and used for oxidations.

2 ELECTRODE STUDIES

Graphite felt electrodes were coated with polyacrylic acid and crosslinked with hexamethylenediamine in the presence of dicyclohexylcarbodiimide (DCC). The electrode coating was then allowed to react with 4-amino-2,2,6,6-tetramethylpiperidinyl-1-oxyl (**1**, R = H) in the presence of DCC. The remaining carboxyl groups were butylated with

Scheme 3 *Preparation of coated electrodes*

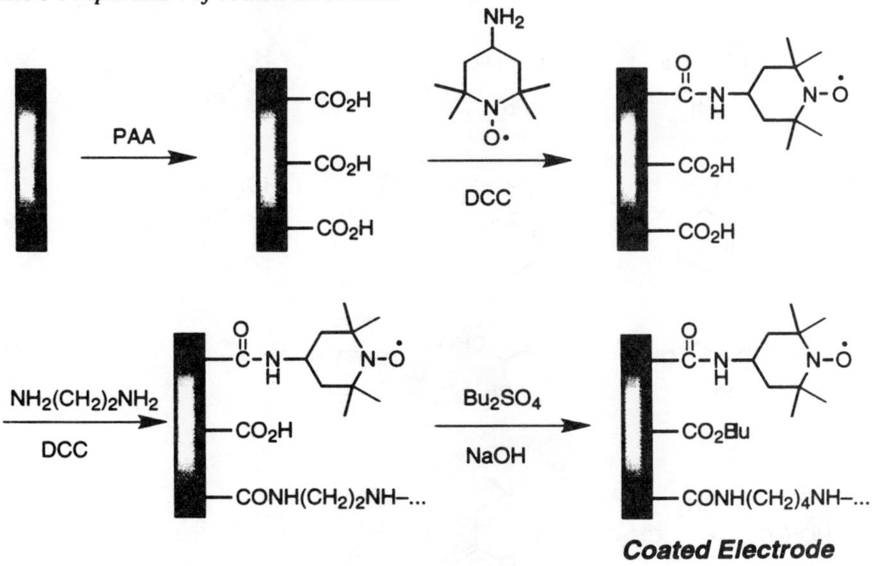

Coated Electrode

dibutyl sulfate and base (Scheme 3).[4,5] These coated electrodes have been used as anodes in electrochemical cells with acetonitrile as solvent, $NaClO_4$ as electrolyte, and 2,6-lutidine as base. The cathodes were platinum, and the cell was divided by a Nafion 117 membrane. Alcohols were oxidized to aldehydes or ketones;[6] and thiols were oxidized to disulfides.[7]

More recently, enantioselective oxidations have been carried out using these electrodes in the presence of the alkaloid, sparteine, as a chiral auxiliary. Specifically, 2-naphthols and 2-methoxynaphthalenes have been dimerized to optically active 1,1-dinaphthyls with enantioselectivities as high as 98%.[5] Several racemic secondary alcohols have been oxidized to give essentially pure, unreacted, optically active alcohol.[8] *Meso* substituted 1,4-diols were oxidized to optically active δ-lactones.[9] Some examples of these reactions are given in Table 1.

Thus, we appear to have achieved a chemically selective electrode surface. In related work, 4-acryloxy-TEMPO (1, RNH = $OCH_2CH=CH_2$) was polymerized, deposited on a carbon cloth electrode, and used for the oxidation of amines.[10]

Table 1 *Chiral Oxidations on a Coated Electrode in the Presence of (—)-Sparteine*

Substrate	Product	Product Remaining	% Ee
	(99 %)		98 [5]
	(92 %)		93 [5]
		(52 %)	99.6 [8]
		(50 %)	99.8 [8]
	(52 %)		0.2 [8]
	(94 %)		98 [9]

3 POLYMERIC NITROXIDES

As noted above, polymeric nitroxides have been prepared by polymerization. We have prepared polymeric nitroxides by condensing the carboxylic acid ion-exchange resin IRC-50 with 4-amino-TEMPO in the presence of DCC. The resulting polymeric nitroxide could be methylated with diazomethane to give the hydrophobic material (**5**) or left unchanged to give a more hydrophilic one. The polymeric materials were used with *m*-chloroperbenzoic acid as a secondary oxidant for the oxidation of several alcohols.[10,11,12] Chiral polymeric nitroxides (**6**) were also prepared, although the observed enantioselectivities did not exceed about 30%. Chiral nitroxides based on a binaphthyl system, such as (**7**) gave good enantioselectivities, although the nitroxides were not on a polymer.[13]

4 STOICHIOMETRIC OXIDATIONS

5

6

7

Polymeric Nitroxides

We have prepared 4-acetylamino-2,2,6,6-tetramethylpiperidine-1-oxonium perchlorate (**2**, R = Ac and X$^-$ = ClO$_4^-$) for the first time. Although there are many known oxoammonium salts,[1] they have a reputation for being unstable and hygroscopic. Our salt has been stable for over four years, open to the atmosphere, and can be easily prepared by a disproportionation reaction of (**1**, R = Ac)[3] with perchloric acid and water. The reagent is convenient to use in that it is bright yellow and *almost* insoluble in CH$_2$Cl$_2$, but soluble enough to react. The product from the oxidant, (**3**, R = Ac, X$^-$ = ClO$_4^-$) is white and completely insoluble in CH$_2$Cl$_2$. Thus, a substrate that is soluble or partially soluble in CH$_2$Cl$_2$ is stirred with a slurry of oxidant. If the reaction mixture turns white, reaction has taken place, and the product resulting from the oxidant can be completely removed by filtration. Evaporation of solvent gives pure aldehyde or ketone in quantitative yields. No acid has ever been noted, and anhydrous conditions are not necessary. A few examples are given in Table 2.

Table 2 *Stoichiometric Oxidations Using 2 (R = NHAc, X– = ClO$_4^-$)*

Substrate	Product	Yield
		79 %
		98 %
CH$_3$–C≡C–CH$_2$OH	CH$_3$–C≡C–CHO	100 %

We found that the system worked very well for allyl alcohols, benzyl alcohols and secondary alcohols, but aliphatic primary alcohols were very slow, often taking several days. *However, we found that the addition of small amounts of chromatographic grade silica gel greatly enhanced all of the oxidations and made the oxidations of primary alcohols possible in a few hours.* When the amount of silica gel was increased, the rate of reaction increased proportionately as shown in the Figure 1. We found that alumina and Florisil were almost as good as silica gel, but that the oxidations were not so clean. Powdered charcoal and powdered sugar showed little, if any effect. When the amount of silica gel was increased, the rate of reaction increased proportionately as shown in the Figure 2.

Figures 1 & 2

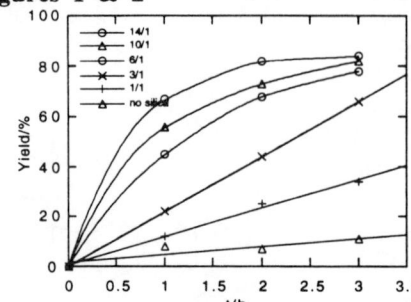 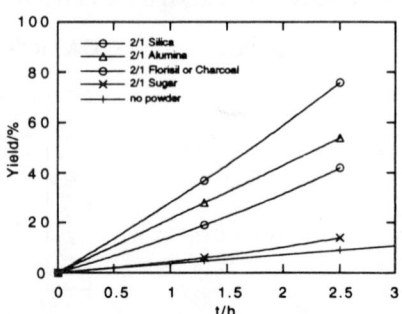

Silica gel has been found to enhance many reactions in organic chemistry, but the exact reasons for this do not seem to be well understood. One recent review has appeared,[14] and the work of Kropp is most important.[15] Although we have little evidence, we believe that the reactants are concentrated on the polar silica gel surface, and that the enhanced rates result from this concentration. It is also possible that the acidity of the gel has a role, but we note similar results on the basic alumina surface.

We had trouble trying to oxidize steroids such as cholesterol because they were almost insoluble in CH_2Cl_2. Dr. Tacorante-Morales in Cuba has deposited our oxoammonium salt on silica gel and used the resulting powder for the oxidation of steroids in acetonitrile (Scheme 4).

Scheme 4 *Oxidations of sapogenins with oxoammonium salt on silica gel*

References

1. J. M. Bobbitt and M. C. L. Flores, *Heterocycles,* 1988, **27,** 509.
2. A. E. J. de Nooy, A. C. Besemer, and H. van Bekkum, *Synthesis,* 1996, 1153.
3. Z. Ma and J. M. Bobbitt, *J. Org. Chem.* 1991, **56**, 6110.
4. T. Osa, Y. Kashiwagi, J. M. Bobbitt, and Z. Ma, 'Electroorganic Synthesis, Festschrift for M. M. Baizer', R. D. Little and N. L. Wineberg. Eds., Marcel Dekker, New York, 1991, p. 343. This is a review of all known types of organic electrode catalysts.

5. T. Osa, Y. Kashiwagi, Y. Yanagisawa, and J. M. Bobbitt, *J. Chem. Soc., Chem. Commun.*, 1994, 2535.
6. T. Osa, Y. Kashiwagi, K. Mukai, A. Ohsawa and J. M. Bobbitt, *Chem. Letters,* 1990, 75.
7. Y. Kashiwagi, A. Ohsawa, T. Osa, Z. Ma, and J. M. Bobbitt, *Chem. Letters,* 1991, 581.
8. Y. Kashiwagi, Y. Yanagisawa, F. Kurashima, J.-i. Anzai, T. Osa and J. M. Bobbitt, *J. Chem. Soc. ,Chem. Commun ,* 1996, 2745.
9. Y. Yanagisawa, Y. Kashiwagi, F. Kurashima, J.-i. Anzai, T. Osa and J. M. Bobbitt, *Chem. Letters,* 1996. 1043.
10. F. MacCorquodale, J. A. Crayston, J. C. Walton, and D. J. Worsfold, *Tetrahedron Lett.,* 1990, **31**,771.
11. Z. Ma, Ph. D. Dissertation, University of Connecticut, 1991.
12. Z. Ma, Q. Huang, and J. M. Bobbitt, *J. Org. Chem.* 1993, **58**, 4837.
13. S. D. Rychnovsky, T. L. McLernon and H. Rajapakse, *J. Org. Chem.* 1996, **61**, 1194.
14. B. C. Ranu, S. Bhar, R. Chakraborty, A. R. Das, M. Saha, A. Sarkar, R. Chakraborti and D. C. Sarkar, *J. Indian Inst. Sci.* 1994, **74**, 15.
15. P. J. Kropp, G. W. Breton, S. L. Craig, S. D. Crawford, W. F. Durland, Jr., J. E. Jones, III and J. S. Raleigh, *J. Org. Chem.,* 1995, **60**, 4146.

THIAZOLIUM SALTS ON CALCIUM SILICATE: SUPPORTED ORGANIC COVALENT CATALYSTS

Timothy P. Smyth* and Orla Kennedy

Department of Chemical and Environmental Sciences, University of Limerick, National Technological Park, County Limerick, Ireland

1 INTRODUCTION

Small organic molecules frequently occur as enzyme cofactors. Typically, they allow a reactivity impasse to be overcome which is not readily surmountable by the normal functionalities present in enzymes. The quintessential thiazolium compound, thiamine diphosphate (**1a**), is one such structure. It is a cofactor for the enzyme pyruvate decarboxylase. The problem to be solved in the decarboxylation of pyruvate is one of stabilisation of an acyl anion. The necessary stabilisation is provided by formation of a covalent adduct between the thiazolium-derived ylide and pyruvate, wherein the thiazolium ring acts as an electron sink (acyl anion equivalent in Scheme 1). The decarboxylated adduct readily cleaves thus allowing the catalytic cycle to continue: the nature of the covalent catalysis in this system is thus apparent as is the pivotal role of the thiazolium-derived ylide.[1]

Simple thiazolium salts behave as effective catalysts in non-enzymatic reactions. The prototypic example is the benzoin condensation[2] (Scheme 2). This reaction is one of carbon-carbon bond formation and formally involves the addition of an acyl anion derived from benzaldehyde to a second molecule of benzaldehyde. The performance of thiazolium systems can be conveniently evaluated by determining their efficiency as catalysts in this condensation reaction. We were interested in examining the interaction of thiazolium salts with basic solid supports for two reasons. Firstly, the evaluation of how an organic covalent catalyst would perform on a solid support[3] was of fundamental interest. Secondly, the possibility of forming a thiazolium-derived ylide or ylide complex, in a sequestered manner on a solid surface, and hence of direct observation of such a species, was also of intrinsic value in terms of increasing our knowledge of this key catalytic entity and its reactivity.[4] We chose to work with calcium silicate ($CaSiO_3$) as the solid support because of its documented use in base catalysed reactions[5] and also because of its free-flowing physical property even when a large amount of liquid has been added to it.[6]

1.1 Catalytic Activity of Supported Thiazolium Salts

The thiazolium salt **1b** is readily available and was used as the initial test compound. Preliminary evaluation was carried out in light petroleum (b.p. 100-120 °C) to ensure that the thiazolium salt did not dissolve off the solid surface and so a heterogeneous reaction system was maintained; the general experimental detail has been given

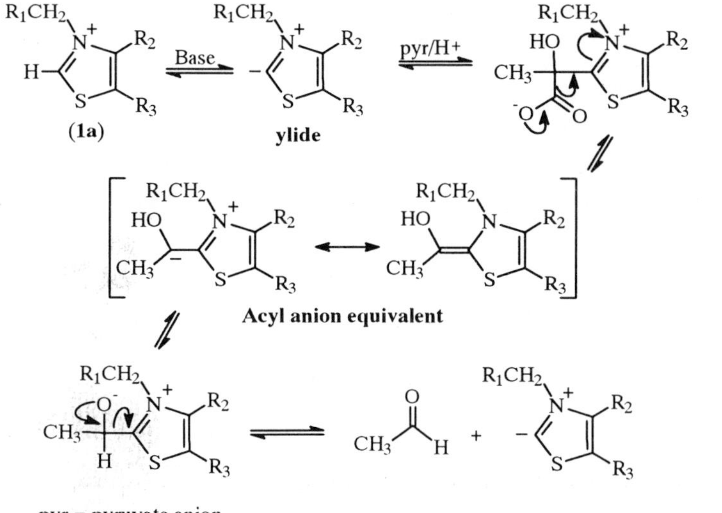

pyr = pyruvate anion

1a: R_1 = 5-(2-methyl-4-aminopyrimidinyl); R_2 = CH_3; R_3 = $CH_2CH_2OP_2O_5{}^3$

Scheme 1

elsewhere.[7] When supported on calcium silicate at a loading of 10% (w/w) of thiazolium salt, **1b** was found to be quite effective. The yield of benzoin (63%) was comparable to that obtained (65%) in the traditional solution phase process (ethanol and triethylamine). More interesting, however, was the finding that a higher yield (87%) was obtained in the total absence of solvent. The conversion of benzaldehyde to benzoin is an equilibrium process. In the absence of solvent benzoin is formed as a solid, and so the equilibrium is pushed further to the product side than in solution. A variety of thiazolium salts were prepared and evaluated as catalysts when supported on calcium silicate (10% w/w) in the total absence of solvent and these findings are given in Table 1.

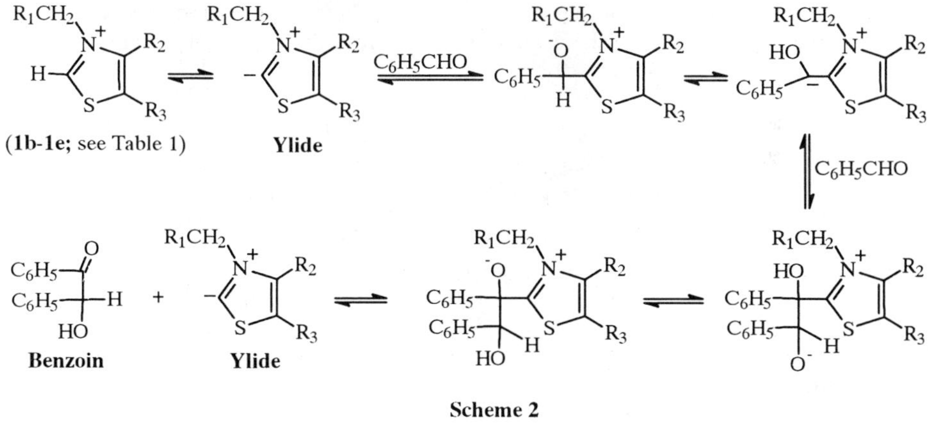

Scheme 2

Table 1 *Variation in yield of benzoin for thiazolium salts **1b** - **1e**.*[a]

	X^{-b}	R_1	R_2	R_3	t/h	Yield (%)
1b	Cl	C_6H_5	CH_3	CH_2CH_2OH	24	87
1b					8	63
1b'	Br	C_6H_5	CH_3	CH_2CH_2OH	24	55
1b''	I	C_6H_5	CH_3	CH_2CH_2OH	24	0
1c	Cl	C_6H_5	CH_3	$CH_2CH_2OC(O)C_2H_5$	8	82
1d	Cl	H	CH_3	CH_2CH_2OH	24	44
1e	Cl	H	CH_3	$CH_2CH_2OC(O)C_2H_5$	24	58

[a]All reactions were carried out at 50 °C without solvent. [b]Thiazolium counter ion.

A number of patterns are evident from the data in Table 1. Firstly, the nature of the counter ion exerts a major influence (compare **1b**, **1b'** and **1b''**), although it is not readily apparent why this should be so. Secondly, a benzyl group is much better than a methyl group as the quaternising moiety on the thiazolium nitrogen - a benzyl group here should lead to more facile formation of the thiazolium-derived ylide (compare **1b** with **1d**, and **1c** with **1e**). Finally, there is some advantage in having a non-polar group as R_3 (compare **1b** with **1c**, and **1d** with **1e**).[8]

We speculate that the main binding interaction of the thiazolium salt with the calcium silicate is one of ion exchange (Figure 1). The nature of the counter ion may play an important role here; however, it is not apparent why the iodide salt should show no catalytic activity. Also shown in Figure 1 is an acid-base interaction between a silicate anion and the thiazolium C-2 hydrogen. The possibility exists that water, present in calcium silicate, and polarised by silicate anions, acts as the base species. Generation of hydroxide ion, however, can be excluded since it is well established that addition of hydroxide ion to the thiazolium ring causes it to cleave with loss of all catalytic activity.[9] Lowering the water content of the calcium silicate (from the equilibrium value of 6% to

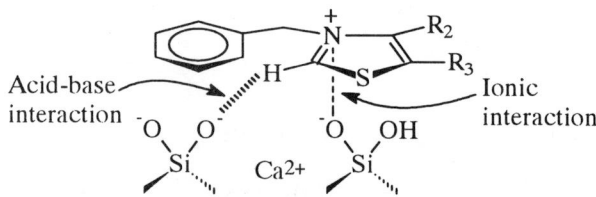

Figure 1 Schematic view of the interaction of a thiazolium salt with the surface of calcium silicate

less than 1%) reduced the activity of the supported thiazolium salts (the yield of benzoin fell by a factor of four). Water is known to play a complex role in heterogeneous catalysis.[3] An alternative role for the water of hydration present in the calcium silicate is to disrupt the microcrystalline structure of the thiazolium salts. A set of XRD patterns is shown in Figure 2 corresponding to those of calcium silicate, **1b**, and **1b** supported on calcium silicate (10% w/w). Although not very diagnostic as to the nature of the interaction between **1b** and CaSiO₃, these XRD patterns indicate that the thiazolium salt is not present in its original crystalline form when supported on CaSiO₃. Attenuated total reflectance (ATR) IR spectra of **1b** and supported **1b** showed some differences, but the extra peaks in the spectra of supported **1b** were not readily assignable.

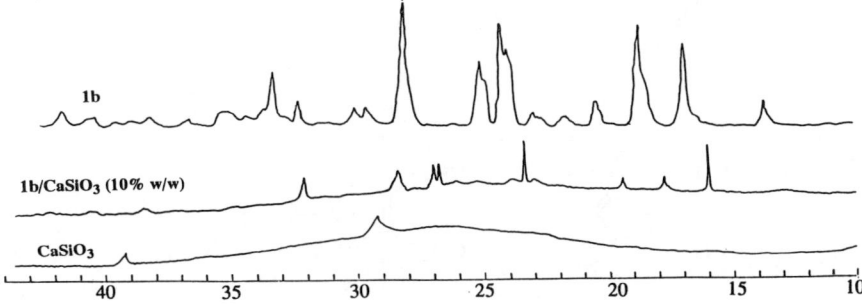

Figure 2 XRD patterns of calcium silicate, **1b**, and supported **1b**

In simple terms, it appeared that benzaldehyde was the mobile component in the system and reacted with thiazolium-derived ylide, generated by interaction of the thiazolium ion with some base species present in calcium silicate. We turned our attention to studying the nature of the interaction between calcium silicate and the thiazolium ion more closely. In particular, we wished to establish if a thiazolium-derived ylide, or ylide complex, was generated in a sequestered form on the calcium silicate surface. Given that the supported thiazolium salts had proved so effective we felt that there was a reasonable possibility that the amount of ylide species occurring in this system would be sufficient to allow its direct observation.

1.2 Search for a Sequestered Thiazolium-derived Ylide or Ylide Complex

Thiazolium-derived ylides have proved to be very elusive species. The situation as it stood when we initiated this work was that there had been no direct observation of any thiazolium ylide despite detailed efforts.[4] Ylides, generated from thiazolium ions known to be good catalysts for the benzoin condensation, proved to be highly reactive. When formed in solution, in the absence of any carbonyl substrate, the ylide derived from an N-benzyl thiazolium salt was found to react rapidly with the parent thiazolium ion giving rise to the symmetrical dimer **2** initially, which then rearranged to **3**.[4] One way of limiting this reaction path is to form the ylide in an environment where it does not have translational mobility, such as on a solid surface; a catalytically active ylide sequestered in this way will, most likely, occur as some sort of ylide complex. An alternative approach is to modify the structure of the thiazolium ion such that the derived ylide is restricted in its reactivity patterns, either by steric or electronic effects, or both; such an ylide may not be highly active as a catalytic entity but will have characteristics generic of this structure type. This latter approach has been successfully used by Arduengo in the preparation of isolable imidazolium derived ylides.[10] He has, very recently, reported the first example of a stable, isolable, thiazolium derived ylide **4**, and the symmetrical dimer **5**.[11]

$R_4 = $ 1-(2,6-diisopropylphenyl)

We decided to try solid state ^{13}C NMR spectroscopy, using the cross polarisation magic angle spinning (CPMAS) technique, to study one of the supported thiazolium chloride salts. As ^{13}C NMR data were available for a variety of imidazolium-derived ylides and ylide complexes at that time, this approach was deemed the most appropriate. We were also aware that Kheir and Haw had successfully observed the *aci* anion of nitromethane using solid state ^{13}C NMR, by supporting nitromethane on magnesium oxide.[12] A solution phase ^{13}C NMR spectrum of **1b** is shown in Figure 3. In Figure 4 the solid state spectra of **1b** and supported **1b** (10% w/w) are shown. The spectrum of solid **1b** clearly shows all the required resonances. The spectrum obtained from supported **1b** shows two sets of peaks most of which are only slightly shifted from those of **1b** itself,

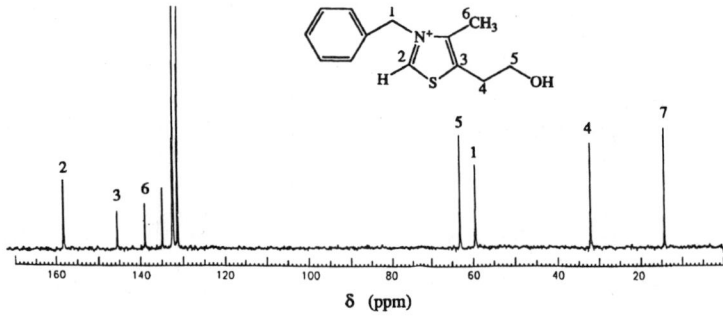

Figure 3 ^{13}C NMR (D_2O) of **1b**

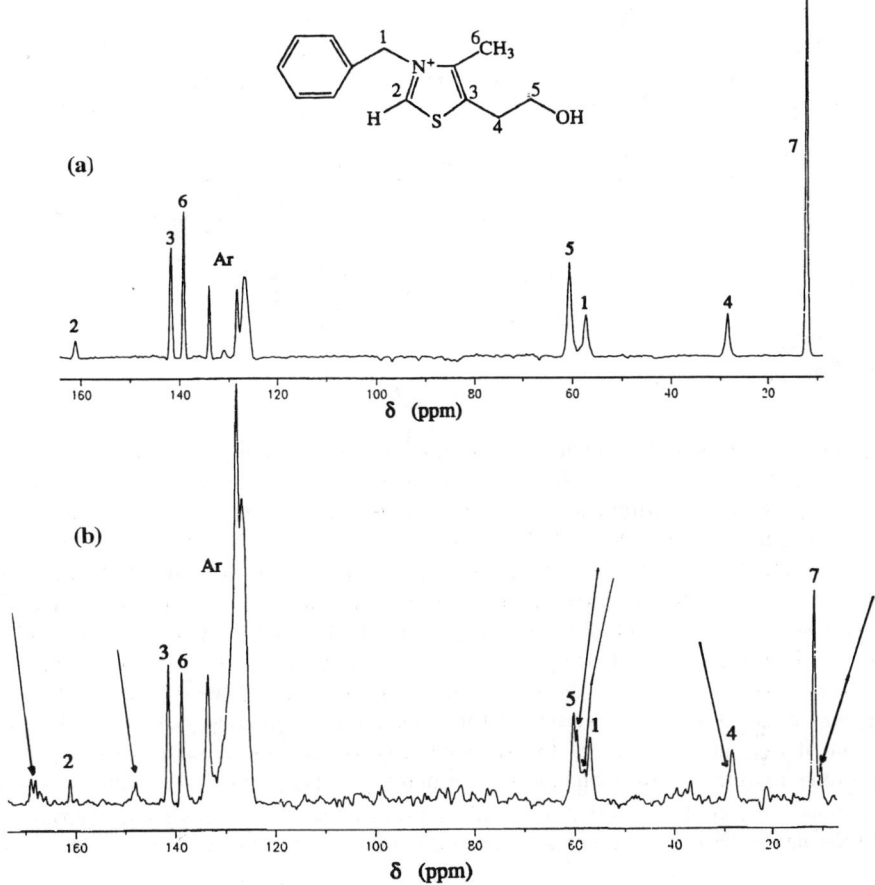

Figure 4 ^{13}C NMR CPMAS spectrum of, (a) **1b**, and (b) **1b** on calcium silicate (10% w/w)

but one of these is more significantly displaced. A peak at 81 ppm, which is characteristic of the rearranged dimer - observed as the end product in solution[4] - is not seen. Thus, one advantage of the solid-supported thiazolium system is established. The new resonance at 168 ppm ought to correspond to C-2 of some species derived from **1b**. It is not, however, attributable to the free ylide species. By analogy with the imidazolium-derived ylides the chemical shift for C-2 of the thiazolium-derived ylide was expected to occur somewhere between 210 - 230 ppm (Arduengo recently reported a value 254.3 ppm for C-2 of **4**). Given that the most significantly shifted peak (at 168 ppm) for supported **1b** is reasonably attributable to C-2, this is indicative of a strong interaction between the C-2 hydrogen of **1b** and calcium silicate. Depending on the strength of this interaction, and therefore the extent of bond breaking of the C-2 carbon-hydrogen bond, one could describe the resultant species either as a complex of the thiazolium ion or of the derived ylide.

(very dynamic system)

175 ppm (^{13}C NMR)

$X^- = PF_6, CF_3SO_3$

$R_1 = 1$-(2,4,6-trimethylphenyl)

6

The complex **6** is a well characterised species:[10d] it is essentially a complex of an ylide and the parent imidazolium ion. It is relevant in that it shows the potential for complex formation between the parent azolium type species and a basic partner: a variety of complexes of imidazolium-derived ylides have been characterised by Arduengo *et al.*[10c] The ^{13}C NMR chemical shift for C-2 of **6** is at 175 ppm and is shifted significantly less downfield than that of the corresponding ylide. On that basis it is probable that the species observed here, with C-2 at 168 ppm, is a complex involving a strong interaction of the C-2 hydrogen with a basic partner on the calcium silicate surface. No complexes have been reported to date for the thiazolium ylide species **4**; however, it is likely that complexes analogous to those observed for imidazolium structures, can occur. That the extra set of resonances observed for **1b**, supported on calcium silicate, corresponds to some such structure is reasonable, but must remain speculative for the present.

2 CONCLUSIONS

Our present work clearly shows that thiazolium salts retain their catalytic activity to an exceptional degree when supported on calcium silicate. More work is required to define fully the nature of the interaction of these thiazolium salts with calcium silicate. Arduengo has shown that it is possible to form a stable thiazolium-derived ylide when suitable substituents are used. It is not yet clear, however, that this species is catalytically active: it does not dimerise very readily.[11] The challenge still remains to directly observe a catalytically active thiazolium-derived ylide. In this context, our results demonstrate the potential for the formation and detection of catalytically active ylide species on a simple basic solid: the formation of dimeric structures is avoided which is a problem in solution. Enhancing the basicity of the calcium silicate, by impregnation with potassium t-butoxide[5] or some other base, may be one way of increasing the concentration of an ylide or a derived complex in this system.

References

1. R. Kluger, *Chem. Rev.* 1987, **87**, 863 and references therein.

2. R. Breslow, *J. Am. Chem. Soc.* 1958, *80*, 3719.
3. J. H. Clark, A. P. Kybett and D. J. Macquarrie, *Supported Reagents-Preparation, Analysis and Applications*, VCH, NY, 1992.
4. Y-T. Chen and F. Jordan, *J. Org. Chem.* 1991, *56*, 5029 and references therein.
5. S. Chalais, P. Laszlo and A. Mathy, *Tetrahedron Lett.*, 1985, **26**, 4453.
6. '*The Merck Index*', 12th Edit., ed. S. Budavari, Merck and Co., Inc, NJ, 1996, 1754.
7. O. Kennedy and T. Smyth, *J. Chem. Research (S)*, 1993, 188.
8. K. Karimian, F. Mohanazadeh and S. Rezai, *J. Heterocycl. Chem.* 1983, **20**, 1119.
9. R. Kluger, J. Chin and T. Smyth, *J. Am. Chem. Soc.* 1981, **103**, 884 and references therein.
10. (a) A. J. Arduengo, III, H. V. R. Dias, R. L. Harlow and M. Kline, *J. Am. Chem. Soc.*, 1992, **114**, 5530. (b) A. J. Arduengo, III, H. V. R. Dias, J. C. Calabrese and F. Davidson, *Organometallics*, 1993, **12**, 3405. (c) A. J. Arduengo, III, M. Tamm and J. C. Calabrese, *J. Am. Chem. Soc.*, 1994, **116**, 3625. (d) A. J. Arduengo, III, S. F. Gamper, M. Tamm, J. C. Calabrese, F. Davidson and H. A. Craig, *J. Am. Chem. Soc.*, 1995, **117**, 572.
11. A. J. Arduengo, III, J. R. Goerlich and W. J. Marshall, *Liebigs Ann./Recueil*, 1997, 365.
12. A. A. Kheir and J. F. Haw, *J. Am. Chem. Soc.*, 1994, **116**, 817.

SURFACE ORGANOMETALLIC CHEMISTRY: THE SILICA SURFACE AS AN UNUSUAL REACTION MEDIUM FOR HIGH-YIELD SYNTHESES OF CARBONYL CLUSTERS AND RELATED METAL CARBONYL COMPOUNDS

E. Cariati, E. Lucenti, D. Roberto and R. Ugo

Dipartimento di Chimica Inorganica, Metallorganica e Analitica and Centro CNR, Università di Milano, Via G. Venezian 21, 20133 Milano, Italy

1 INTRODUCTION

In recent years there has been considerable interest in the use of the surfaces of inorganic oxides as a non conventional reaction medium for the preparation of various organometallic complexes and clusters.[1-9] The first surface-mediated organometallic synthesis was reported by Fischer et al. 35 years ago; they observed that $MCl_3 \cdot nH_2O$ (M=Ir,[10] Rh[11]) adsorbed on silica gel yielded, respectively, $[Ir(CO)_3Cl]_n$ and $[Rh(CO)_2Cl]_2$, after successive thermal treatments with a stream of Cl_2 and then of CO. Nevertheless, it was a long time before the potential of this innovative synthetic method was realized.

In surface-mediated organometallic syntheses, the surface of a solid such as an inorganic oxide plays the role of the solvent in conventional syntheses. The synthesis variables include the nature and loading of the metal salt or organometallic precursor adsorbed on the solid, the physical and chemical properties of the surface, the nature and composition of the gaseous phase, temperature, pressure and reaction time. The products synthesized on the surface are recovered simply by extraction or sometimes by sublimation.

A large amount of work has been devoted particularly to the synthesis of metal carbonyl complexes and clusters.[1-9] As a general trend, strongly basic surfaces such as MgO favor the formation of anionic metal carbonyl clusters[1-3] while uncharged carbonyl complexes and clusters are generated on the surface of rather neutral supports such as SiO_2.[4-6] However, we found recently that both neutral and anionic osmium carbonyl clusters can be synthesized by controlled reduction of silica-supported $[M(CO)_3Cl_2]_2$ or MCl_3 (M=Ru, Os) in the presence of alkali carbonates.[7-9] Addition of a relatively weak base, such as an alkali carbonate, may change the properties of the surface, thus increasing its basicity and allowing the synthesis of anionic metal carbonyl clusters even on silica. This observation is of great synthetic interest because anionic metal carbonyl clusters prepared on MgO are partially entrapped by the MgO matrix and therefore cannot be quantitatively recovered[1] whereas they can be easily extracted from the silica surface even if added with an alkali carbonate.[7-9]

Therefore the silica-mediated syntheses of metal carbonyl complexes and clusters conducted on Areosil, a nonporous silica, can be divided in two kinds: (1) syntheses on the surface of silica (2) syntheses on the surface of silica added with alkali carbonates.

2 SYNTHESES ON THE SURFACE OF SILICA

2.1 Syntheses of $[Ir(CO)_3Cl]_n$, $[Rh(CO)_2Cl]_2$, $[Ru(CO)_3Cl_2]_2$ and $[Os(CO)_3Cl_2]_2$ from the Corresponding Metal Chloride

Using the silica surface as reaction medium, $[Ir(CO)_3Cl]_n$, $[Rh(CO)_2Cl]_2$, $[Ru(CO)_3Cl_2]_2$ and $[Os(CO)_3Cl_2]_2$ can be easily prepared in high yields by reductive carbonylation of metal chlorides, working under rather mild conditions (low temperatures and CO at atmospheric pressure). In fact, when $IrCl_3 \cdot nH_2O$ supported on silica (3-5 wt% Ir with respect to silica) is heated at 150°C under CO in a closed vessel, $[Ir(CO)_3Cl]_n$ sublimes on the walls of the vessel outside the oven. The reaction is complete after 24h, affording 76-83% yields of $[Ir(CO)_3Cl]_n$.[5] Besides, reaction of $RhCl_3 \cdot nH_2O$ supported on silica (1.5-5 wt% Rh with respect to silica) with CO in a closed vessel and at room temperature affords $[Rh(CO)_2Cl]_2$ which can be easily separated from the support by extraction with dichloromethane (80-84% yields).[5] These silica-mediated syntheses of $[Ir(CO)_3Cl]_n$ and $[Rh(CO)_2Cl]_2$ are much more convenient than conventional synthetic methods in solutions and than those on silica which require successive thermal treatments with a stream of Cl_2 and then of CO.[10-11]

Treatment of $RuCl_3 \cdot nH_2O$ supported on silica (2-5 wt% Ru with respect to silica) with CO (1 atm) at 100°C for 2 days affords $[Ru(CO)_3Cl_2]_2$ which can be easily extracted with acetone (88-93% yields).[5] The osmium analogue, $[Os(CO)_3Cl_2]_2$, can be prepared in a similar way but working at a higher temperature (180°C). Also in this case, yields are excellent (90%).[5]

2.2 Syntheses of Metal Carbonyl Clusters

2.2.1 Syntheses of $[Ir_4(CO)_{12}]$, $[Rh_4(CO)_{12}]$ and $[Rh_6(CO)_{16}]$ from the Corresponding Metal Chlorides. The reductive carbonylation of $IrCl_3 \cdot nH_2O$ supported on silica (3 wt% Ir with respect to silica) can be made to proceed after the formation of $[Ir(CO)_3Cl]_n$ by controlling the temperature (70°C) and by increasing the amount of water (10 wt% H_2O with respect to silica) to give selectively $[Ir_4(CO)_{12}]$ (68% yield after 4 days). This synthesis is comparable to conventional syntheses of $[Ir_4(CO)_{12}]$ in solution.[5]

On the contrary, silica-supported $RhCl_3 \cdot nH_2O$ cannot be converted directly to $[Rh_4(CO)_{12}]$ or $[Rh_6(CO)_{16}]$. When the reductive carbonylation is carried out at room temperature, the reaction stops at $[Rh(CO)_2Cl]_2$ even in the presence of water. At higher temperatures, for example 70°C, complete sublimation of the dimer occurs, preventing further reduction to zerovalent rhodium clusters. However, when silica-supported $[Rh(CO)_2Cl]_2$ (1.5 wt% Rh with respect to silica) is treated at room temperature with CO and water (89 wt% water with respect to silica), a reduction to $[Rh_6(CO)_{16}]$ (21%) and $[Rh_4(CO)_{12}]$ (47%) occurs in 24h. However yields are not competitive with those of the traditional solution methods to prepare $[Rh_4(CO)_{12}]$ and $[Rh_6(CO)_{16}]$.[5]

2.2.2 Syntheses of $[H_4M_4(CO)_{12}]$ from $[M_3(CO)_{12}]$ (M=Ru, Os). When silica-supported $[Os_3(CO)_{12}]$ (2 wt% of Os with respect to silica) is treated in a closed vessel with H_2 (1 atm) at 100°C for 5 days, it is converted in 68% yield into $[H_4Os_4(CO)_{12}]$, which is recovered from the surface with dichloromethane.[4] The particularly mild pressure required by this synthesis compared to the high pressure (120 atm H_2) required in inert solvents like octane in order to obtain similar yields,[12] can be explained by the activation of physisorbed $[Os_3(CO)_{12}]$ by interaction with surface silanol groups to give the silica-anchored species $[HOs_3(CO)_{10}OSi\equiv]$ which acts as a reactive and labile intermediate.[13]

Similarly, [H$_4$Ru$_4$(CO)$_{12}$] is synthesized in 78% yield on the silica surface from [Ru$_3$(CO)$_{12}$] by treatment with H$_2$ (1 atm) working at 50°C for 4 days.[4]

2.2.3 Syntheses of [HOs$_3$(CO)$_{10}$Y] (Y=OH, OR, Cl, Br, O$_2$CR) from [Os$_3$(CO)$_{12}$]. The facile and quantitative activation of [Os$_3$(CO)$_{12}$] by the silica surface, *via* reaction with surface silanol groups to give [HOs$_3$(CO)$_{10}$OSi≡], provides a new convenient general route to the selective, high-yield synthesis under mild conditions of various related osmium carbonyl clusters such as [HOs$_3$(CO)$_{10}$Y] (Y=OH, OMe, OBu, OPh, Cl, Br, O$_2$CCF$_3$, O$_2$CCH$_3$). These silica-mediated syntheses compare favorably with the more conventional syntheses in solution because (i) differently from other related intermediates used in solution, the reactive intermediate [HOs$_3$(CO)$_{10}$OSi≡] is easily obtained in one step and in nearly quantitative yield starting from [Os$_3$(CO)$_{12}$][15] and (ii) the reaction can be carried out in *one pot* (Table 1).[14]

In summary, [HOs$_3$(CO)$_{10}$OSi≡] provides a new simple direct route to the preparation of various [HOs$_3$(CO)$_{10}$Y] clusters which were previously prepared in solution by a series of steps and with much lower total yields from [Os$_3$(CO)$_{12}$].[14]

Table 1 *Syntheses of various clusters starting from [Os$_3$(CO)$_{12}$] via [HOs$_3$(CO)$_{10}$OSi≡]*

Product	Reaction conditions	T (°C)	t(h)	Total yield from [Os$_3$(CO)$_{12}$](%)
[H$_4$Os$_4$(CO)$_{12}$]	H$_2$, 1 atm	150	24	94
[HOs$_3$(CO)$_{10}$(OH)]	H$_2$O/toluene	95	5	91
[HOs$_3$(CO)$_{10}$(OBu^n)]	*n*-Butanol	118	20	84
[HOs$_3$(CO)$_{10}$(OMe)]	Methanol and HBF$_4$.Et$_2$O	65	24	52
[HOs$_3$(CO)$_{10}$(OPh)]	PhOH/heptane	98	5	64
[HOs$_3$(CO)$_{10}$Cl]	37% HCl (aq)/CH$_2$Cl$_2$	40	6	84
[HOs$_3$(CO)$_{10}$Br]	48% HBr (aq)/CH$_2$Cl$_2$	40	7	86
[HOs$_3$(CO)$_{10}$(O$_2$CCF$_3$)]	excess CF$_3$CO$_2$H/toluene	90	6	70
[HOs$_3$(CO)$_{10}$(O$_2$CCH$_3$)]	excess CH$_3$CO$_2$H/toluene	90	10	56

3 SYNTHESES ON THE SURFACE OF SILICA ADDED WITH ALKALI CARBONATES

3.1 Syntheses of Neutral and Anionic Osmium Carbonyl Clusters from $[Os(CO)_3Cl_2]_2$ or $OsCl_3$ in the Presence of Na_2CO_3 or K_2CO_3

The reductive carbonylation of silica-supported $[Os(CO)_3Cl_2]_2$ to $[Os_3(CO)_{12}]$ does not occur readily due to the easy sublimation of chlorocarbonyl osmium complexes and clusters at relatively high temperatures and to the difficulty of removing chloro ligands from osmium at lower temperatures.[16] For this reason, the addition of weak bases to silica was investigated in order to favor removal of the chloro ligands.[7-8] We found that when silica-supported $[Os(CO)_3Cl_2]_2$ (2-15 wt% Os with respect to silica) is stirred with a slurry of Na_2CO_3 in CH_2Cl_2 (molar ratio Na_2CO_3:Os=2:1), dried, and heated with CO (1 atm) at 200°C, $[Os_3(CO)_{12}]$ is obtained in 76-82% yields. By working at 150°C under 1atm of H_2, the major product is $[H_4Os_4(CO)_{12}]$ (70-83% yields). By increasing the amount of added Na_2CO_3 or K_2CO_3 and by working under well defined conditions, various anionic clusters ($[H_3Os_4(CO)_{12}]^-$, $[H_2Os_4(CO)_{12}]^{2-}$, $[Os_5C(CO)_{14}]^{2-}$ and $[Os_{10}C(CO)_{24}]^{2-}$) can be also prepared in high yields (Scheme 1). The selectivity of the synthesis is controlled by the choice of the (i) nature and amount of the alkali carbonate (Na_2CO_3 or K_2CO_3) added to the silica surface; (ii) temperature; (iii) reaction time; and (iv) gas-phase composition (CO, H_2 or mixtures CO/H_2). Interestingly the nature of the added alkali carbonate governs the selectivity towards the synthesis of different osmium clusters; in particular silica-supported K_2CO_3 behaves as a stronger base than silica-supported Na_2CO_3.[7-8]

The various osmium clusters can also be prepared in excellent yields from silica-supported $OsCl_3$ in a two-step route:[7-8]
(i) formation of the silica-bound species $[Os(CO)_3Cl_2HOSi\equiv]$ by carbonylation of silica-supported $OsCl_3$[5] and (ii) addition of the alkali carbonate followed by further reduction with H_2 or CO under specific conditions.

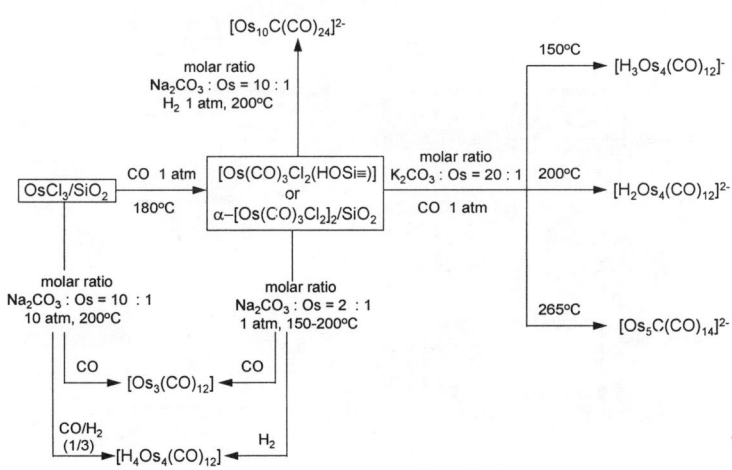

Scheme 1

3.2 Syntheses of Neutral and Anionic Ruthenium Carbonyl Clusters from RuCl₃ in the Presence of Na₂CO₃ or K₂CO₃

Like in the case of osmium, addition of an alkali carbonate to the silica surface favors removal of chloro ligands from the ruthenium coordination sphere.

When [Ru(CO)₃Cl₂HOSi≡] (2-5 wt% Ru with respect to silica), prepared by carbonylation of silica-supported RuCl₃,[5] is treated with a slurry of Na₂CO₃ in CH₂Cl₂ (molar ratio Na₂CO₃:Ru=3:1), followed by evaporation of the solvent and further reaction with 1 atm of CO at 110°C for 48h, [Ru₃(CO)₁₂] is obtained in 93% yield. Interestingly, by increasing the ruthenium loading (15 wt% Ru with respect to silica) [Ru₃(CO)₁₀Cl₂] is the major product (Scheme 2). This different selectivity can be explained by a non-homogeneous dispersion of Na₂CO₃ on the silica surface when a slurry in CH₂Cl₂ is used for its deposition. This low homogeneity of the Na₂CO₃ surface dispersion at high loading leads to a lower surface basicity than that expected for a 3:1 molar ratio of Na₂CO₃:Ru, and therefore to a more difficult removal of chloro ligands from the ruthenium coordination sphere with formation of [Ru₃(CO)₁₀Cl₂] instead of [Ru₃(CO)₁₂]. In agreement with this hypothesis, when Na₂CO₃ is deposited on the silica surface by using a water solution instead of a CH₂Cl₂ slurry, a better dispersion of the base is reached and only [Ru₃(CO)₁₂] is formed. By working under 1 atm of CO + H₂ (molar ratio 1:3) instead of pure CO, [H₄Ru₄(CO)₁₂] is obtained in high yields (86-94%) with metal loadings of either 5 or 15 wt% of Ru with respect to silica. By adding K₂CO₃ in order to increase the surface basicity, various anionic clusters ([HRu₆(CO)₁₈]⁻, [H₃Ru₄(CO)₁₂]⁻, [Ru₆C(CO)₁₆]²⁻) can be also prepared in high yields (Scheme 2).

The unusual selectivity of these silica-mediated syntheses of various ruthenium and osmium carbonyl clusters in the presence of alkali carbonates can be explained by the initial formation on the surface of reactive species such as [M(CO)ₓ(OR)₂]ₙ (M=Ru, Os; R=H, Si≡; x=2, 3) which, under CO, aggregates first to the key intermediates [HM₃(CO)₁₀OR] (R=H, Si≡) and then to [M₃(CO)₁₂].[9,17] The latter, due to the unexpected high basicity of the surface added with an alkali carbonate such as K₂CO₃, evolves into anionic species.[17]

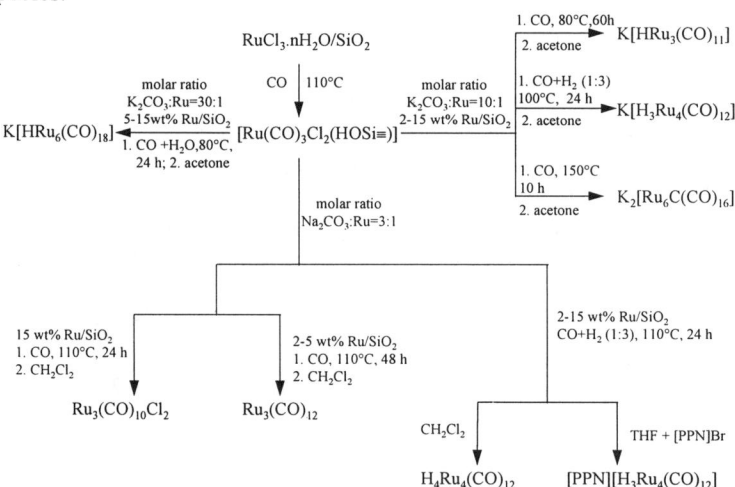

Scheme 2

4 CONCLUSION

The silica surface could show a great potential as an efficient, although unusual, reaction medium for organometallic synthesis since it allows selective and high yield syntheses of metal carbonyl complexes and clusters. Obviously the use of a solid as a reaction medium is not limited as in solution by boiling points; therefore it is possible to work at atmospheric pressure even at relatively high temperatures.

The isolation of the final products is generally easy: neutral and anionic species can be totally extracted with a suitable solvent.

From a synthetic point of view yields and reaction conditions are usually better and milder than those of the related syntheses in solution. In addition, the possibility of achieving, in the majority of cases, high yields and selectivities even working with high metal loadings of the silica surfaces gives to this synthetic method the potential of being used for the preparation of good amounts of metal carbonyl complexes or clusters by using only a few grams of silica.

References

(1) Lamb, H. H.; Fung, A. S.; Tooley, P. A.; Puga, J.; Krause, R.; Kelley, M. J.; Gates, B. C. *J. Am. Chem. Soc.* 1989, **111**, 8367.
(2) Gates, B. C. *J. Mol. Catal.* 1994, **86**, 95.
(3) Kawi, S.; Xu, Z.; Gates, B. C. *Inorg. Chem.* 1994, **33**, 503.
(4) Dossi, C.; Psaro, R.; Roberto, D.; Ugo, R.; Zanderighi, G. M. *Inorg. Chem..* 1990, **29**, 4368.
(5) Roberto, D.; Psaro, R.; Ugo, R. *Organometallics.* 1993, **12**, 2292.
(6) Roberto, D.; Psaro, R.; Ugo, R. *J. Organomet. Chem.* 1993, **451**, 123.
(7) Roberto, D.; Cariati, E.; Psaro, R.; Ugo, R. *Organometallics.* 1994, **13**, 734.
(8) Roberto, D.; Cariati, E.; Ugo, R.; Psaro, R. *Inorg. Chem.* 1996, **35**, 2311.
(9) Roberto, D.; Cariati, E.; Lucenti, E.; Respini, M.; Ugo, R. *Organometallics.* 1997, vol. 16, p. 4531.
(10) Fischer, E. O. Brenner, K. S. *Z. Naturforsh B* 1962, **17**, 774.
(11) Brenner, K. S.; Fischer, E. O.; Fritz, H. P.; Kreiter, G. G. *Chem. Ber.* 1963, **96**, 2632.
(12) Johnson, B. F. G.; Lewis, I.; Raithby, P. R.; Sheldric, G. M.; Wong, K.; McPartlin, M. *J. Chem. Soc., Dalton Trans.* 1978, 673.
(13) Roberto, D.; Pizzotti, M.; Ugo, R. *Gazz. Chim. Ital.* 1995, **125**, 133.
(14) Roberto, D.; Lucenti, E.; Roveda, C.; Ugo, R. Submitted to *Organometallics.*
(15) Barth, R.; Gates, B. C.; Knozinger, H.; Hulse, J. *J. Catal.* 1983, **82**, 147.
(16) Roberto, D.; Psaro, R.; Ugo, R. *J. Mol. Catal.* 1994, **86**, 109.
(17) Cariati, E.; Recanati, P.; Roberto, D.; Ugo, R. Submitted to *Organometallics.*

POLYMER-SUPPORTED TRANSITION METAL COMPLEX ALKENE EPOXIDATION CATALYSTS

D C Sherrington

Department of Pure and Applied Chemistry
University of Strathclyde
295 Cathedral Street
Glasgow
G1 1XL

1   INTRODUCTION

Though the last decade or so has seen a rapid development of a wide range of novel homogeneous transition metal complex catalysts, few of these have found technological application on anything other than a very small scale. This is surprising since many of these species show extremely high activity and selectivity (chemo-, stereo-, regio- and enantio-) and at the same time often operate under very mild conditions (low temperature and pressure). One of the reasons for the lack of exploitation is that often the metal involved is costly (e.g. Rh, Ru), secondly the catalysts sometimes have rather low stability and thirdly, the engineering of plants using soluble metal complexes is difficult and costly. For some time now it has been recognised that a potential solution to these problems is to replace one (or more) of the catalyst metal complex ligands with an organic polymeric ligand, where the latter is, for example, a crosslinked polymer resin particle or "bead".[1,2] Such heterogenised species then offer all the engineering and manipulative advantages of classic heterogeneous catalysts, while retaining the same metal ion co-ordination sphere and hence, in principle, the same catalytic properties as those of the soluble metal complex analogue.

Research and development in this area peaked in the late 1980 s when, in the case of alkene hydrogenation, hydroformylation, hydrosilylation and isomerisation catalysts, it was found that catalyst leaching was a major problem.[3] Gradually emphasis has shifted to include oxidation catalysts, and with the drive towards more environmentally acceptable chemistry, the 1990 s has seen a resurgence of interest in polymer-supported catalysts with a focus not only on catalytic activity and turnover, and catalytic selectivity, but also on catalyst stability and re-cycling, even in academic laboratories. The acceptability of polymer-supported species in large scale industrial processes is now high, as a result of the widespread adoption of polymeric sulphonic acid resin catalysts; likewise in small scale processes, and in the laboratory, the advent of resin-supported combinatorial synthesis has removed much of the previous prejudice against this methodology.

1.1 Polymer-supported Mo(VI) alkene epoxidation catalysts

The industrial production of propylene oxide by direct oxidation of propylene with alkyl hydroperoxides uses a soluble Mo(VI) catalyst (Halcon or Arco Process)[4] and although the Shell Company have reported a heterogeneous Ti(IV)-silica catalyst,[5] it is not clear if this process is actually operated. The development of a heterogeneous Mo(VI)-based catalytic system is therefore both a strategically and scientifically important target.

Our first attempts[6,7] with polymer-supported species involve the chelating resins shown in Figure 1. Immobilised Mo(VI) proved to be very active and more so than the

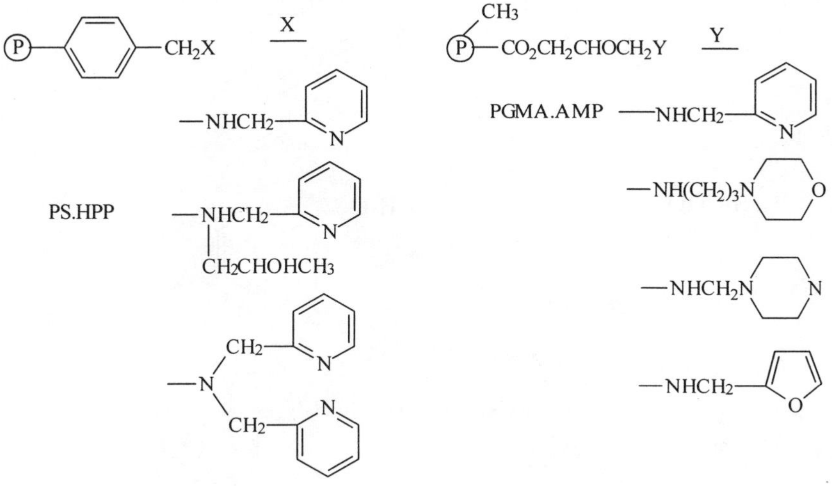

Figure 1   *Structures of poly(chloromethylstyrene)-based (PCMS) and poly(glycidyl-methacrylate)-based (PGMA) chelating resins for Mo(VI) immobilisation.*

analogous V(V) species in the epoxidation of cyclohexene using *t*-butylhydroperoxide as the mono-oxygen source. Unfortunately all the polymeric catalysts displayed a significant induction period, which could however be eliminated by pre-oxidising the catalyst with the hydroperoxide prior to a catalytic reaction. Perhaps a more important negative however, again characteristic of all the resins studied, was the tendency to leach Mo(VI) during repeated use, although some of the polymers did show evidence of stabilisation with ageing. A crude measure for the likely technological potential of polymeric ligands in oxidation reactions is afforded by their behaviour in thermogravimetric analysis (TGA). Data for the polymethacrylate based resin, PGMA.AMP, and its Mo(VI) complex in oxygen show that at ~200°C both the non-loaded and Mo(VI)-loaded resins start to undergo rapid oxidative decay with progressive loss of mass.[8] Polystyrene-based systems display similar instability.

We reported the use of a very thermo-oxidatively stable porous resin, polybenzimidazole (PBI) as a support for Mo(VI) (PBI.Mo) in the epoxidation of cyclohexene in 1995.[9] This polymer catalyst proved to be highly active, but also required no pre-oxidation to generate the active species. Subsequently a range of imidazole-containing resins were synthesised, loaded with Mo(VI), and shown to be similarly active without any activation step (Figure 2). Furthermore, the supported catalysts could be

PBI

Ps.PyIm

Ps5BzCOO

PGMA.PyIm

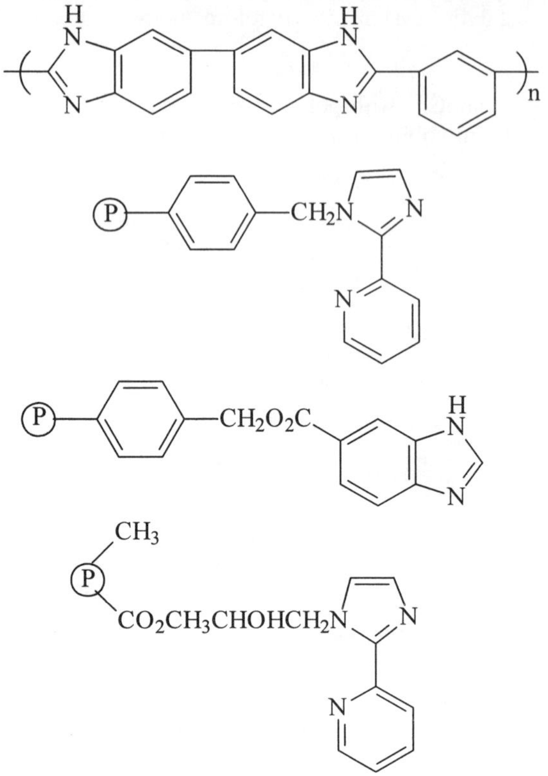

Figure 2   *Imidazole-containing resins for Mo(VI) immobilisation*
(P) = vinyl polymer backbone)

recycled in batch with essentially no leaching of Mo (detection limited <0.2%), apart from a small loss (2-5%) in the first use (weakly bound or physically trapped component?). Despite this stability, however, the activity of PBI.Mo was found to drop on recycling in cyclohexene epoxidations,[10] and while the loss of activity could be reduced by oxidising in hydroperoxide, this treatment seemed simply to delay the deactivation of the catalyst. On the optimistic side, however, TGA data for PBI and PBI.Mo in oxygen showed, as expected, that these species are stable beyond ~400°C and so offer a real prospect for technological application,[8] given a maintenance of catalytic activity.

The first epoxidations of propylene using PBI.Mo were in 1994[11] with a detailed description of the behaviour of a full range of resins following later.[8] Table 1 shows the first impressive data for PBI.Mo using conditions which represent a realistic technological

Table 1   *Epoxidation of propene using homogeneous and PBI-supported heterogeneous Mo^{VI} catalysts^a*

	Propylene oxide yield (%)^{b,c}	
t/h	MoO$_2$(acac)$_2$	PBI.MoO$_2$acac
1	47.6	89.4
2	77.1	90.6
3	97.7	83.4
4	100.0	92.6

^{a} *Reaction conditions:* [catalyst] mass equiv. to 0.12 mmol Mo, [TBHP] (anhydrous solution in toluene) vol. equiv. to 10 mmol, propene 10 g, 238 mmol, 1.0 ml chlorobenzene (GLC int. std), 22 ml 1,2-dichloroethane, 400 psi He at 80°C. ^{b} Products analysed by high resolution capillary GLC. ^{c} GLC yield based on conversion of TBHP *i.e.* 10 mmol TBHP = 10 mmol propylene oxide = 100% yield.

Table 2   *Epoxidation of propene using TBHP Recycling experiments with catalysts PBI.Mo, Ps.HPP.Mo and PGMA.AMP.Mo*

	PBI.Mo^a		Ps.HPP.Mo^b		PGMA.AMP.Mo^b	
Run	Yield propylene^c oxide (%)	Mo leached^d (%)	Yield propylene oxide (%)	Mo leached (%)	Yield propylene oxide (%)	Mo leached (%)
1	59.0	2.9	98.7	2.9	84.5	0.7
2	68.4	0^f	85.5	2.2	82.8	0.7
3	74.8	0	90.1	2.2	94.8	0.7
4	74.1	0	79.9	3.6	88.6	0.7
5	80.4	0	91.3	4.4	96.6	0.7
6	84.6	0	99.0	2.2	93.6	0.7
7	87.4	0	65.2	2.9	99.0	2.1
8	89.7	0	85.9	0.7	83.9	0.7
9	94.8	0	87.0	0.7	82.0	0
10	99.8	0	-	-	81.7	0

^{a} Non-activated catalyst. ^{b} Catalyst activated for 4 h. ^{c} Yield after 1 h based on conversion of TBHP; *i.e.* 10 mmol TBHP = 10 mmol propylene oxide ≡ 100% yield. ^{d} Expressed as a % of metal originally loaded on polymer sample. Detection limit indicates <0.2%.

test. Table 2 shows the remarkable ageing behaviour of PBI.Mo over 10 recycles, along with comparative data for a polystyrene-based species (PsHPP.Mo) and a polymethacrylate-based system PGMA.AMP.Mo. Though under the conditions used the activity of PBI.Mo is the poorest initially, progressive use leads to a very potent and highly selective catalyst. Furthermore, apart from the first catalytic run there is negligible leaching of Mo from PBI.Mo, whereas the other two species, while maintaining reasonable activity over this time scale, show a steady leakage of Mo. The recycling behaviour of PBI.Mo in propylene epoxidation contrasts markedly with its re-use in cyclohexene epoxidation, and this substrate (and potentially product) dependence of recyclability is well known with traditional inorganic-based heterogeneous catalysts. In the latter instance the behaviour is often associated with contamination and physical blocking of the catalyst by side-products, or products arising from subsequent reaction of the main product (e.g. oligomerisation and tar formation). This seems to be the most likely explanation for the deactivation of catalyst in the case of cyclohexene, and the process responsible may well be ring-opening of the product cyclohexene oxide. The contrasting increase in catalytic activity and selectivity in the case of propylene epoxidation most likely arises from microenvironmental changes in the polymer support allowing a larger percentage of the bound Mo(VI) to become accessible for catalysis. However, our recent studies[12] of the structure of the complex on the support (see below) also suggest that changes in the nature of the complex or the complexes present may also occur on prolonged oxidation and so the overall picture is by no means clear-cut at the moment.

In addition to cyclohexene, propylene and oct-1-ene, PBI.Mo is also an efficient catalyst for epoxidation of styrene, methylenecyclohexane, 4-vinyl cyclohexene and *cis/trans* 1,3-pentadiene where in each case behaviour very similar to that of a soluble complex, $MoO_2(acac)_2$, is observed.[13] The results using allyl alcohol, bromide and chloride are much poorer and side reactions seem to be a major problem with these substrates. Attempts to oxidise vinyl acetate yielded only acetic acid.

1.2   Structure of immobilised Mo(VI) species

We have already reported[8] that polymer resins with chelating ligands supporting Mo display Mo=O bands ($\sim 916, 933$ cm^{-1}) and a Mo-O-Mo band (~ 725 cm^{-1}) in their FTIR spectra, whereas PBI.Mo and the other imidazole-containing polymers show only the Mo=O characteristic stretches. PBI.Mo also shows some evidence for residual acetylacetonate ligand. This led us to propose structures (**1**) and (**2**) as the species present on chelating and imidazole-containing resins respectively (Figure 3), with the slow oxidative break-up of (**1**) being responsible for the observed induction periods with these resins. We now know that the picture is more complicated than this[12] and we will report in more detail on this in due course. However, exhaustive structural investigations of four polymer-supported Mo(VI) species have now been performed – these are PBI.Mo and PBI.Mo activated by oxidation, and PSHPP.Mo and its activated analogue. XPS data show two Mo centres in the non-activated species, collapsing to one in their activated counterparts. The electronic spectra (200-1280 nm) indicate a much broader absorption for PSHPP.Mo extending down to 920 nm, suggesting the presence of a range of structural species. Finally, EXAFS data show very significant changes when PSHPP.Mo is activated but hardly any change when PBI.Mo is activated. The latter data was still in process of further refinement and fitting and will be published in due course. These recent studies have therefore confirmed a clear structural difference between Mo immobilised on a chelating resin and on PBI.

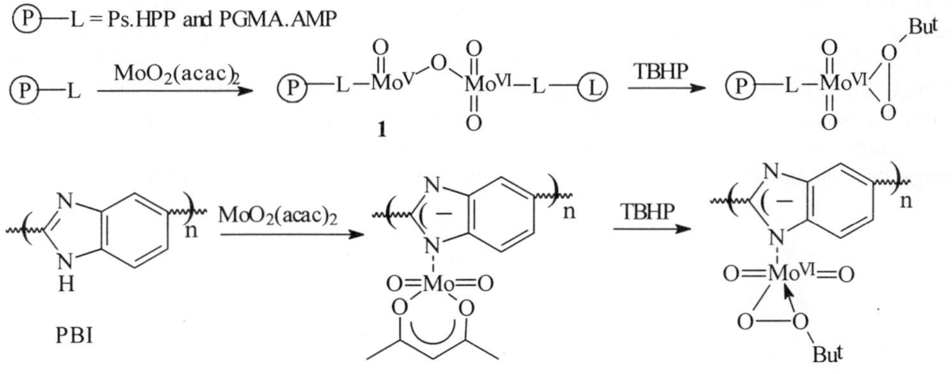

Figure 3   *Proposed structures of catalyst centres on PSHPP.Mo; PGMA.AMP.Mo; and PBI.Mo.*

1.3   Polyimide–supported Mo(VI)

Although polybenzimidazole-supported Mo(VI) displays many characteristics of technological importance in an epoxidation catalyst, the polymer itself is rather expensive. The polyimide range of polycondensation species are well developed high performance polymers[14] and display thermo-oxidative stability at least as good as PBI. We have previously shown that these materials can be synthesised in the form of porous spherical particulates by suspension polycondensation in paraffin oil,[15] using traditional chemistry involving a bis-anhydride and bis-aromatic amine. In this instance the precursors are relatively cheap and so in principle supports based on these species are more likely to be cost effective. By using a functional bis-amine we have also been able to synthesise polyimide particulates containing ligands and other exploitable functional groups,[16] and some of these prove capable of binding Mo(VI). Thus a polyimide species with a 1,2,4-triazole unit in the backbone (**3**) not only binds Mo(VI) but is an active catalyst in cyclohexene epoxidation. Although this shows some loss of activity on recycling the prospects for further development look good.

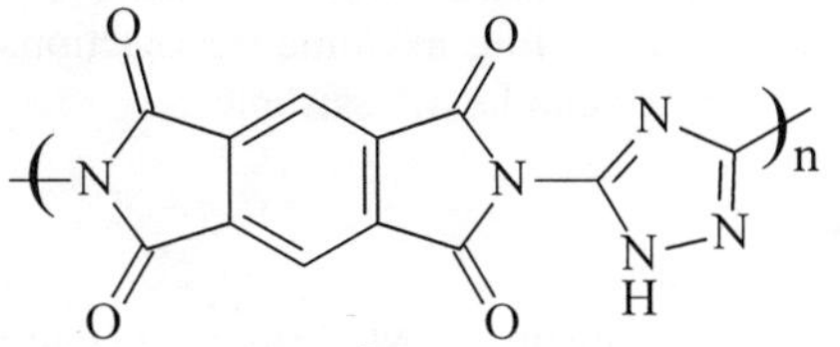

3

1.4   Poly(tartrate ester)-supported Ti(IV) asymmetric epoxidation catalysts

A key alkene asymmetric epoxidation catalyst developed and exploited in recent years is the Sharpless Ti(IV)/tartrate species.[17] This is a homogeneous species and would be experimentally more convenient if available in a heterogeneous form. An early attempt by Farrall *et al*[18] described a single tartrate ester unit bound to a polystyrene resin, and though the Ti(IV) complex with this was an active catalyst the level of asymmetric induction achieved (ee%) was only modest 50-60%. Nevertheless, this was encouraging and somewhat surprisingly the work was not followed up. We have now synthesised a range of poly(tartrate ester)s (**4**) with the optically active ligand as a component of the

$$\text{HO} \quad CO_2 + $$

$$\text{HO}^{\prime\prime\prime\prime} \quad CO_2 - R \frac{}{}_n$$

R = $(CH_2)_2$; $(CH_2)_6$; $(CH_2)_8$; $(CH_2)_{12}$; $p\text{-}CH_2C_6H_4CH_2$; $m\text{-}CH_2C_6H_4CH_2$

4   a   b   c   d   e   f

polymer backbone.[19] Use of these with Ti(OPr^i)$_4$ and *t*-butylhydroperoxide as the oxidant under typical Sharpless conditions gives good yields of epoxide product from *trans*-hex-2-en-1-ol with ee% up to 80% (Table 3). Interestingly there is a significant dependence on the structure of the polymer ligand and at first it seemed that the poorly soluble polymeric complexes performed worst of all. Initially the best result was with the species derived from 1,8-octanediol, and interestingly though the parent polyester is insoluble in the methylene chloride CH_2Cl_2 solvent, it does form a soluble complex when the Ti(OPr^i)$_4$ is added. This led us to make the corresponding ester from 1,12-dodecandiol which itself is soluble in CH_2Cl_2 and forms a soluble complex with Ti(OPr^i)$_4$. Significantly the results with this catalyst prove to be poorer! Thus, solubility alone is not the only controlling factor. Since these initial studies we have been able to optimise the use of the polyester from octanediol to obtain even higher ee%, and to do so with fully heterogeneous catalyst. Full details of this will be reported in due course.[20]

CONCLUDING REMARKS

There is little doubt now that polymer-supported metal complex catalysts are going to be exploited in synthesis both in the laboratory and in the plant. There are significant possibilities both with large-scale commodity processes and with smaller scale specialities. Much remains to be done on the academic side both in developing new possibilities and in understanding currently available systems. Our own intention is to pursue the strategically important area of oxidations, including asymmetric reactions, and to help our industrial colleagues wherever they feel we can be of assistance.

ACKNOWLEDGEMENT

The author is grateful to S. Simpson, M.M. Miller, G. Olason, J.K. Karjalainen and L. Canali for their dedicated experimental work, to Professor O. Hormi for useful discussions, and to B.P. Chemicals and Neste for their financial support.

Table 3   *Epoxidation of <u>trans</u>-hex-2-en-1-ol with tBHP catalysed by L-(+)-polyester 4 and Ti(OPr^1)$_4$^{a}*

Ligand	Molar ratio 4; Ti:tartrate	T/°C	Reaction time/h	Epoxide yield (%)	Isolated yield (%)	Work-up	ee (%)
DMT^b	100:5:6	-30	3	91	44	A	≥98
4	100:17:20	-20	3	51	50	B	8
4	100:17:20	-20	3	63	63	B	55
4	100:5:10	-20	3	22	59	B	79
4	100:17:20	-20	7	92	58	B	79
4	100:12:12	-20	3	65	60	B	41
4	100:5:10	-20	3	21	46	B	52
4	100:10:30	-20	3	80	61	B	77
4	100:17:20	-20	3	75	61	B	64
4	100:17:20	-20	3	73	42	B	47
4	100:17:20	-20	3	74	80	B	68

^{a} see ref. 19 for conditions; ^{b} dimethyltartrate-L-(+)

REFERENCES

1.   F.R. Hartley, "Supported Metal Complexes, a New Generation of Catalysts", Reidel Pub. Co., Dordrecht, Holland, 1985.
2.   D.C. Sherrington, *Pure and Appl. Chem.*, 1998, **60**, 401.
3.   P.E. Garrou and B.C. Gates in "Syntheses and Separations Using Functional Polymers", Eds., D.C. Sherrington and P. Hodge, J. Wiley and Sons, Chichester, U.K., 1988, Chap. 3, p123.
4.   J. Kollar, U.S. Pat., 3 350 422, 1967; 3 357 635, 1967; 3 507 809, 1970; 3 625 981, 1971 to Halcon International.
5.   H.P. Wulff, Br. Pat., 1 249 079, 1971; U.S. Pat., 3 923 843, 1975 to Shell Oil.
6.   D.C. Sherrington and S. Simpson, *J. Catalysis*, 1991, **131**, 115.
7.   D.C. Sherrington and S. Simpson, *Reactive Polymers*, 1993, **19**, 13.
8.   M.M. Miller, D.C. Sherrington and S. Simpson, *J. Chem. Soc., Perkin Trans. 2*, 1994, 2091.
9.   M.M. Miller and D.C. Sherrington, *J. Catalysis*, 1995, **152**, 368.
10.   M. Miller and D.C. Sherrington, *J. Catalysis*, 1995, **152**, 377.
11.   M.M. Miller and D.C. Sherrington, *J. Chem. Soc., Chem. Comm.*, 1994, 55.
12.   F. Morazzoni, C. Canevali, J. Reedijk, G. van Albada, J. Corker, S. Leinonen and D.C. Sherrington, unpublished results.
13.   G. Olason, PhD Thesis, University of Strathclyde. To be published in *Macromol. Symp.*
14.   D. Wilson, H.D. Stenzenberger and P.M. Hergenrother, *Polyimides*, Chapman and Hall, New York, 1990.
15.   T. Brock, D.C. Sherrington and J. Swindell, *J. Mater. Chem.*, 1994, **4**, 229.
16.   J.H. Ahn and D.C. Sherrington, *J. Chem. Soc., Chem. Comm.*, 1996, 643.

17.   T. Katsuki and K.B. Sharpless, *J. Amer. Chem. Soc.*, 1980, **102**, 5974.
18.   M.J. Farrall, M. Alexis and M. Trecarten, *Nouv. J. Chim.*, 1983, 7, 449.
19.   L. Canali, J.K. Karjalainen, D.C. Sherrington and O. Hormi, *J. Chem. Soc., Chem. Comm.*, 1997, 123.
20.   J.K. Karjalainen, D.C. Sherrington and O. Hormi, to be published in *J. Chem. Soc., Perkin Trans 2*.

ENHANCED SELECTIVITY WITHOUT LOSS OF ACTIVITY IN THE ALKYLATION OF BENZENE USING HEXAGONAL MESOPOROUS SILICA (HMS) - SUPPORTED ALUMINIUM CHLORIDE[1]

Peter M. Price,[a] James H. Clark,[a] Keith Martin,[b] Duncan J. Macquarrie,[a] and Tony W. Bastock[b]

[a] Department of Chemistry, University of York, Heslington, York, YO1 5DD, UK
[b] Contract Chemicals Ltd., Knowsley Business Park, Prescot, Merseyside, L34 9HY, UK

The selective *mono*alkylation of aromatics using alkenes is an important goal notably in the production of linear alkylbenzenes for the manufacture of detergents[2]. Traditional homogeneous catalysts such as $AlCl_3$ and HF give good rates of reaction but poor product selectivities[2,3]. Their use also leads to health and safety, plant corrosion and waste disposal problems and is being phased out. Zeolites are solid acids that are safe to handle and can lead to much improved product selectivities but their activity especially in reactions involving larger alkenes, is poor - the reaction of benzene with 1-dodecene catalysed by HZSM-4 for example, is carried out at high temperatures and pressures[2].

Recently we reported the first example of an immobilised form of aluminium chloride which exhibited activity in alkylations comparable to that of $AlCl_3$ itself but offers the advantages of a solid acid catalyst and is reusable[4]. These supported reagents were based on broad pore size distribution silicas and acid-activated clays but still led to a small but significant improvement in selectivity towards monoalkylation. Further improvements in product selectivity are however extremely important especially in these environmentally conscious days when waste minimisation is becoming a priority in many industries[5].

We have now discovered that by using a narrow pore size distribution hexagonal mesoporous silica (HMS)[6] as a support results in substantial improvements in selectivity towards *mono* alkyation. The 24Å HMS support was prepared as described elseware[6]. The supported reagent was prepared by refluxing a well-stirred slurry of anhydrous $AlCl_3$, the support (pre-dried for 18h at 300°C) and benzene for 1h under dry nitrogen. The mixture is then cooled to room temperature and the alkene (0.5 mole equivalent with respect to benzene) added over a period of 0.5-1h. To further improve the selectivity of the catalyst any contribution from external catalytic sites was eliminated; this was achieved in two ways. The first involved poisoning the external catalytic sites with a bulky amine, triphenylamine. This was added to the catalyst in a solution of benzene, and the mixture stirred for 30 min at room temperature. The best results were observed at an optimum $AlCl_3$:Ph_3N ratio of 5:1. The second method involved treatment of the dried HMS support with a bulky silane, triphenylchlorosilane ($0.03molg^{-1}$ in toluene), prior to catalyst preparation so as to remove the external hydroxyl groups. All reactions gave improved selectivity towards *mono*alkylation compared with both $AlCl_3$ and aluminium chloride immobilised on acid-activated clays; remarkably the catalyst activity remained high (with a substantial loss only occurring for the poisoned catalyst when using 1-hexadecene) and comparable to that of the best homogeneous catalyst. The enhancement in *mono*alkylation selectivity increased with increasing size of the alkene: 1-hexene< 1-octene< 1-dodecene< 1-tetradecene< 1-hexadecene, consistent with increasing molecular sieve effect (figure 1).

Figure 1   Comparison of catalysts selectivities over a range of alkenes

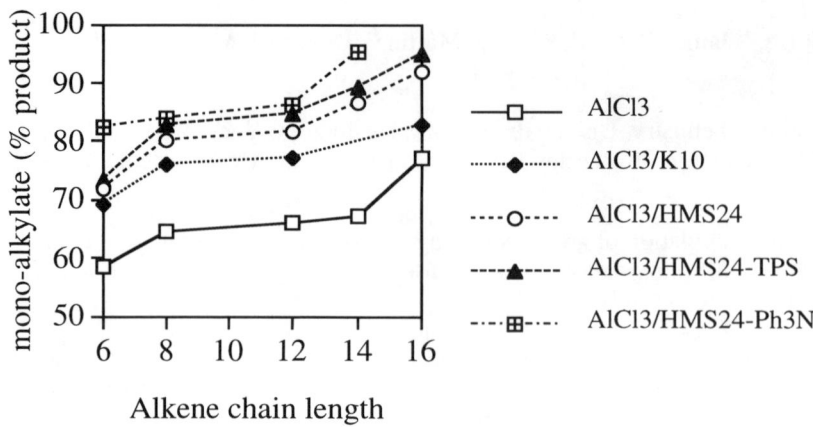

Small, but significant improvements in selectivity were also observed towards the 2-phenyl isomer (table 1); this is the preferred isomer for the detergent industry due to its better water solubility, emulsibility and biodegradability.

Table 1   Comparison of catalyst selectivities towards the 2-phenyl isomer

	Alkene (% 2-phenyl isomer)		
Catalyst[a]	**1-dodecene**	**1-tetradecene**	**1-hexadecene**
AlCl$_3$	29.6	29.9	35.2
AlCl$_3$/HMS$_{24}$	34.0 (38.0)[c]	35.5	35.2
AlCl$_3$/HMS$_{24}$-TPS	34.9	35.0	38.3
AlCl$_3$/HMS$_{24}$-Ph$_3$N	34.5	36.5	-
HZSM-4[b]	37.5	-	-

All reactions carried out with a PhH:alkene ratio of 2:1. [a] Aluminium chloride loading = 1.5 mmolg^{-1}, equivalent amount used for unsupported AlCl$_3$. [b] PhH:1-dodecene ratio was 4:1, reaction carried out at 205°C and 14 bar pressure. [c] Reaction carried out at a PhH:1-dodecene ratio of 4:1.

References
1. *UK Pat.* Appl. P96302GB.
2. J.L.G. de Almeida, M. Dufaux, Y. Ben Taarit and C. Naccache, J. Am. Oil Chem. Soc., 1 **71**, 675.
3. S.Sivasanker and A.Thangaraj, *J.Catal.,* 1992, **138**, 386.
4. J.H.Clark, K.Martin, A.J.Teasdale and S.J.Barlow, *Chem.Commun.,* 1995, 2037.
5. J.H.Clark ed., *Chemistry of Waste Minimisation,* Chapman and Hall, London, 1995.
6. P.T.Tanev and T.J.Pinnavaia, *Science,* 1995, **267**, 865.

α-PINENE OXIDE ISOMERIZATION PROMOTED BY MIXED OXIDES

NICOLETTA RAVASIO, MICHELA FINIGUERRA, MICHELE GARGANO
Centro C.N.R. MISO, Dipartimento di Chimica dell'Università, via Amendola 173,
I-70126 Bari (Italy)

We recently carried out a study on the activity of some mixed cogels, in the presence and in the absence of copper, in some acid catalyzed reaction, like the addition of an hydroxy group to a C=C double bond,[1] double bond isomerization[2] and the ene reaction of citronellale to give isopulegol.[3] The target of this investigation was both to propose an heterogeneous alternative to the use of mineral or homogeneous Lewis acids and to have a deeper insight on the role of supported copper on the surface acidity of the mixed oxide.

Here we wish to report our results on the activity of SiO_2-Al_2O_3, SiO_2-TiO_2 and SiO_2-ZrO_2 mixed cogels in the isomerization of α-pinene oxide **1** to campholenic aldehyde **2**, an intermediate in the synthesis of α- and β-santalol.

| 1 | 2 | 3 | 4 | 5 | 6 |

Also this reaction is catalyzed by $ZnBr_2$ in the homogeneous phase with selectivities of about 85% to **2**,[4] and only recently an heterogeneous system giving up to 75% of campholenic aldehyde **2** at total conversion has been reported.[5]

In Table 1 some representative results are reported: best activity and selectivity were obtained in the presence of SiO_2-Al_2O_3 containing 1.2% of Al_2O_3 (SiAl) and SiO_2-ZrO_2 (4,7% ZrO_2, SiZr), activated at 450°C in toluene. These two solids allow to reach a 72% yield in aldehyde **2**. The product distribution found is in agreement with the Lewis acid character of these mixed oxides.

In this reaction the presence of copper lowers the support activity and modifies the product distribution in a significant way. Thus, the formation of pinocamphone **4** is much reduced, whereas the yield of carveol **6** encreases. According to the relevant literature[6], this is consistent with a more pronounced Broensted acidity.

Table 1[a]

Isomerization of α-pinene oxide in the presence of different solid acids

cat	T (°C)	t (min)	2	3	4	5	6
Si-Ti	90	60	57	9	7	2	14
Cu/SiTi	90	60	56	5	3	6	10
Si-Al	90	5	63	6	13	2	9
Si-Al*	90	5	64	7	11	4	2
Si-Al*	25	5	72	7	14	2	3
Cu/SiAl *	25	15	58	7	6	5	8
Si-Zr	90	15	64	6	11	1	5
Si-Zr	60	40	66	6	11	1	7
Si-Zr*	60	20	72	6	12	-	4
Cu/SiZr	90	15	61	5	6	5	9

[a] = 100% conversion, catalyst dehydrated at 270°C, toluene as a solvent, GC analysis on a 60 m Supelcowax 10, injection T 140°C
* = catalyst dehydrated at 450°C

1 N.Ravasio, M.Antenori, F.Babudri, M.Gargano, *Tetrahedron Letters*, accepted for publication
2 N.Ravasio, M.Antenori, M.Gargano, 'Catalysis of Organic Reactions', R.Malz ed., Marcel Dekker, New York, 1996, p. 413
3 N.Ravasio, M.Antenori, F.Babudri, M.Gargano, 'Heterogeneous Catalysis and Fine Chemicals IV', H.U.Blaser, A.Baiker, R.Prins eds., Elsevier, Amsterdam, 1997, p.625
4 B.Arbusow, *Chem. Ber.* , 1935, **68**, 1430
5 A.T.Liebens, C.Mahaim, W.F.Holderich, 'Heterogeneous Catalysis and Fine Chemicals IV', H.U.Blaser, A.Baiker, R.Prins eds., Elsevier, Amsterdam, 1997, p.587
6 J.Kaminska, M.A.Schwegler, A.J.Hoefnagel, H. van Bekkum, *Recl. Trav. Chim. Pays-Bas*, 1992, **111**, 432

CONTRAST BETWEEN ISOPROPANOL OLIGOMERISATION ACTIVITY OF $Cu^{\delta+}$ SPECIES UPON TiO_2 AT CA. 473K AND PREDOMINANTLY DEHYDROGENATION AND DEHYDRATION ACTIVITIES OVER Cu/CeO_2 AND Cu/MgO

Z. Wang. J. Cunningham and J. McCarthy
Department of Chemistry,
University College Cork,
Ireland

This work investigated the catalytic conversion of isopropanol over copper supported on a number of metal oxides. In order to achieve and maintain a high dispersion of Cu upon TiO_2, CeO_2 or MgO a reduction of Cu(II) precursors dispersed upon the oxides was made with hydrazine. After exposure to atmosphere, the first catalytic activity was very low, but increased greatly after each catalyst was reduced in-situ for two hours in a flow of hydrogen at 523K before again establishing a flow of isopropanol vapour in an argon carrier gas. An on-line GC was used to monitor the exit gases for acetone, propene, unreacted isopropanol and any higher molecular weight products across a reaction temperature range from room temperature to 723K.

Over the Cu/CeO_2 and Cu/MgO catalysts the predominant conversion of isopropanol at a reaction temperature adequate to achieve 50% conversion (460 and 400K, respectively) was highly selective dehydrogenation to acetone. On increasing the reaction temperature for those systems, dehydration to yield propene, progressively replaced the dehydrogenation and became the dominant catalysed elimination reaction at 513K. However over similarly prepared and prereduced Cu/TiO_2 materials, onset of a significant competition against dehydrogenation by a catalysed oligomerisation process occurred. This process which yielded oxygenated oligomeric products, up to C_{21}, became evident over in-situ prereduced Cu/TiO_2 at a reaction temperature of 440K and rose to a maximum at ca. 520K before its progressive replacement by dehydration at higher temperatures. The extent to which a deficit existed between the isopropanol converted and the sum of the detected amounts of acetone and/or propene served as an approximate measure of the oligomerisation activity at temperatures in that range.

Progress to the complex mixture of oligomers, only identified in part by GC-MS, implied a complex sequence of conversions upon the Cu/TiO_2 surfaces. These include isopropanol dehydrogenation; acetone condensation to 4 hydroxy-4 methylpentane 2-one followed by its dehydration to mesityl oxide; subsequent aldol condensations with acetone to yield phorone and its isomers, and hydrogenations of α,β unsaturated ketones. Only the Cu/TiO_2 catalyst displayed the multifunctional catalytic action required to promote this sequence of conversions.

Investigations into the effect of preparation of the Cu/TiO_2 catalyst showed that a suspension-adhesion technique, in which the TiO_2 support was added to an aqueous solution of copper acetate that had been reduced by hydrazine, yielded a superior catalyst than a traditional wet impregnation technique. Catalysts, which were pre-treated in-situ in a flow of hydrogen at 473K, proved to be slightly better in the conversion of isopropanol than those, which were pre-treated in a flow of argon at 473K.

Further investigations into the effects of introducing co-reactants into the isopropanol gas stream upon the selectivity for High Molecular Weight Products (HMWP) yielded some insights into the nature of the active sites involved in the oligomerisation process. The addition of N_2O which could be expected to oxidise all surface copper atoms from Cu^0 to Cu^I and also to remove many oxygen anion vacancies at the created copper-titania interface in the prereduction, caused a drastic reduction in the selectivity towards the HMWP. A high temperature (973K) reduction in hydrogen, which would ensure a complete phase transition of the titania from anatase to rutile and also cause an agglomeration of the copper particles, totally quenched the production of the oligomers. Additions of acetone or hydrogen to the reactant gas stream also decreased oligomer production but only to extents consistent with some displacement of the $(CH_3)_2CHOH$ / $(CH_3)_2CO + H_2$ equilibrium to the left.

In an attempt to investigate possible involvement of hydride species in the multifunctional catalytic activity of Cu/TiO_2, 3% Palladium was added to give a $(Pd + Cu)/TiO_2$ catalyst likely to have the ability of creating more hydride species than the Cu/TiO_2 catalyst alone. However, in as much as the composition of the complex oligomeric product could be assessed from the incomplete GC-MS analyses, this remained very similar over both catalysts, the Cu/TiO_2 yielding up to C_{21} and the $(Cu + Pd)/TiO_2$ yielding up to C_{18} oligomers.

Our results indicate that sites featuring copper, in a partially oxidised state, $Cu^{\delta+}$, stabilised by the titania surface sites having adjacent oxygen anion vacancies within TiO_2 (e.g. $Cu^{n+}.....V_{ox}.....Ti^{4+}$) were crucially important for the maintenance of oligomeric activity. Such interpretation is consistent with recent literature reports of some $Cu^{\delta+}$ persistence in pre-reduced Cu/TiO_2[1,2] and for production of many anion vacancies in regions of semiconducting oxides adjacent to small metallic particles[3].

References

1. F. Boccuzzi, M. Baricco, and E. Gaglielminotti, *Appl. Surf. Sci*; 1993, 70 147.
2. H.W. Chen. J.M. White and J.G. Ekhart. *J. Catal.*, 1986, 99 293.
3. J.C. Frost, *Nature*, 1988, 334, 577.

CATALYSTS FOR BIOMASS REFORMING

D. Sutton and J.R.H. Ross

Centre For Environmental Research
University of Limerick
Limerick
Ireland

1 INTRODUCTION

Much research in biomass gasification is being carried out world-wide with several books [1,2] and papers [3,4] published. The gasification of biomass yields synthesis gas (H_2, CO, CO_2) which has several industrial uses. Research into the role of catalysts for biomass gasification is of growing interest to several research groups around the world. The development of such catalysts can further the optimisation of the gasification process, making the process more competitive with conventional methods.

1.1 Gasification Process

The gasification process yields several products among which is tar, an unwanted by-product of all gasification processes. The tar contains a range of up to 300 compounds of varying carbon chain length and form [5]. Tar causes corrosion and blockages of equipment. Ideally a catalyst placed downstream of the gasifier can reform the tar to synthesis gas.

In this work thermogravimetric analysis has shown that peat decomposes to tars and gases between 230°C and 500°C. The volatiles are passed over a catalyst which is located downstream of the gasifier and which has separate temperature control. The gas stream exiting the catalyst reactor is analysed by on-line GC for H_2, CO, CH_4, CO_2 in order to examine the effectiveness of each catalyst tested.

2 CATALYSTS INVESTIGATED

Several catalysts were investigated the results for three of which are reported here. 15 wt% nickel on alumina , 1 wt% platinum on zirconia and 2 wt% nickel on zirconia. Catalysts were pre-reduced in a stream of hydrogen and nitrogen before volatiles from the gasification were introduced. All results reported are for the catalysts at 800°C.

2.1 Conversion of Tar to Carbon Gases

The conversion of tar to carbon gases (CO, CO_2, CH_4) using the various catalysts was up to 80%. In the absence of catalyst the thermal cracking of the tar to carbon gases at 800°C was of the order of 12%.

Figure 1 shows the mole percentage composition of product gases exiting the catalytic reactor. The "blank" data represent a typical composition of the gas stream exiting the gasifier prior to entering the second catalytic reactor. The absence of catalyst at 800°C ("thermal effect" data) does not increase the selectivity towards the desired synthesis gas composition.   The 15 wt% nickel on alumina completely reformed the methane present and reduced the carbon dioxide content. The hydrogen to carbon monoxide ratio was of the order of 2:1. The 1 wt% platinum on zirconia had a similar effect with a 2:1 ratio of hydrogen to carbon monoxide and almost complete reforming of the methane. However, the amount of carbon dioxide was 5% greater than that for the 15% Ni/Al$_2$O$_3$. The 2 wt% nickel on zirconia also gave a 2:1 ratio of hydrogen to carbon monoxide. However, 5% methane which is undesirable in a synthesis gas stream for industrial usage was present.

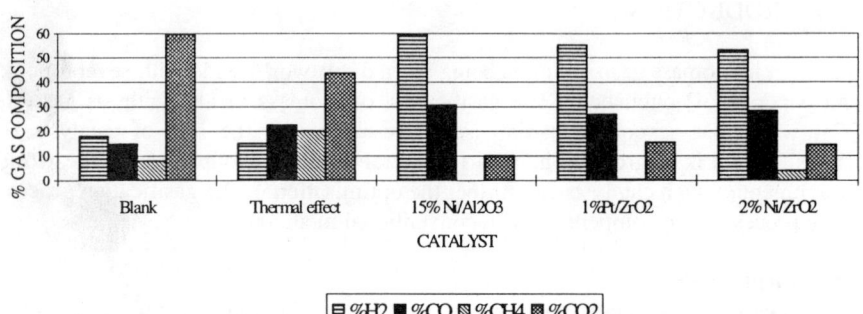

Figure 1 *Mole percentage of product gases exiting the catalytic reactor (800°C)*

3 CONCLUSIONS
Work carried out has shown nickel on alumina and platinum on zirconia catalysts to be effective in reforming tar to synthesis gas composition in a separate reactor downstream from the gasification process. Further work will centre on the lifetime of the catalysts and the nature of their deactivation.

References
1. A.V. Bridgwater and J.L. Kuester, 'Research in Thermochemical Biomass Conversion', Elsevier Applied Science, London, 1989.
2.   R.P. Overend, T.A. Milne and L.K. Mudge, 'Fundamentals of Thermochemical Biomass Conversion', Elsevier Applied Science, London, 1985.
3. J. Corella and M.P. Aznar, *Ind. Eng. Chem. Res.*, 1991, **30**, 2252.
4. V.A. Frolov and R. Lyons, Proc. 10th World Hydrogen Energy Conference, Cocoa Beach, Florida, USA, June 1994, **1**, 157.
5. R. Cristian and E. Chornet, *J. Anal. Appl. Pyrolysis*, 1983, **5**, 261.

THE SELECTIVITY OF ACTIVE SITES ON OXIDE CATALYSTS

F E Cassidy, C Batiot and B K Hodnett,

Dept. of Chemical and Environmental Sciences,
University of Limerick, Limerick, Ireland

The selectivity of active sites on oxide catalysts has been evaluated through the use of selectivity-conversion plots constructed from literature data. These selectivity-conversion plots were generated for a variety of catalysts for each reaction over a range of temperatures and space velocities.

Figure 1 shows two examples of these plots for the oxidation of ethylbenzene to styrene and propane oxidation to acrolein. In all cases studied an upper limit could be identified beyond which experimental studies have not yet progressed. Data points which fall below this limit are assumed to arise from operation with poor catalysts or in non-optimised conditions.

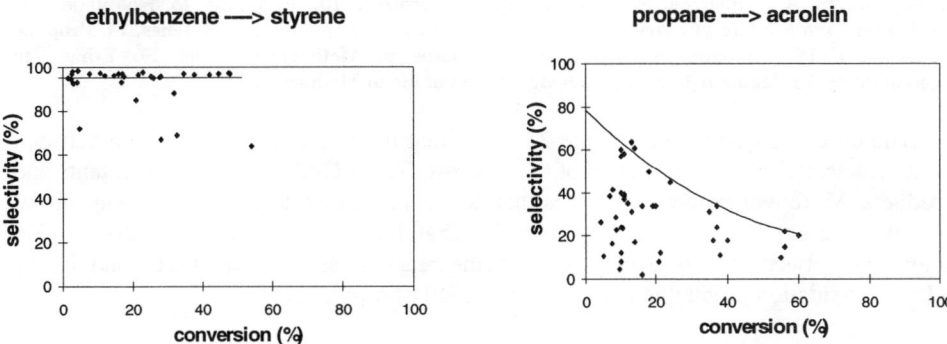

Figure 1 *Selectivity-conversion plots for the oxidation of ethylbenzene to styrene*[1,2] *and propane to acrolein* [3-5]

Generally selectivity in oxidation catalysis involves activation of the reactant through rupture of a C-H bond, whereas diminishing selectivity is associated with rupture of any bond in the selective oxidation product. As a means of validating this hypothesis the upper selectivity limit, attained at a fixed conversion, in all reactions used in this study was plotted against the function:

$$\mathbf{D^{\circ}HC\text{-}H \text{ (reactant)} - D^{\circ}HC\text{-}H \text{ or } C\text{-}C \text{ (product)}}$$

where $D°H_{C-H}$ (reactant) is the bond dissociation enthalpy of the weakest C-H bond in the substrate and $D°H_{C-H}$ or C-C (product) is the bond dissociation enthalpy of the weakest bond in the selective oxidation product.

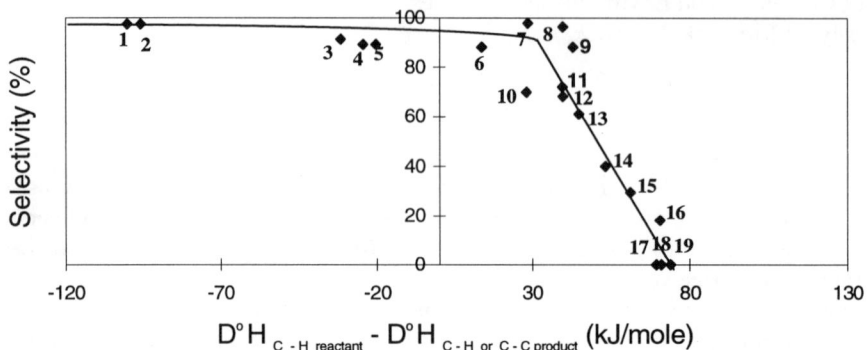

$$D°H_{C-H\ reactant} - D°H_{C-H\ or\ C-C\ product}\ \text{(kJ/mole)}$$

Figure 2 : *Selectivity in product versus $D°H$ C-H reactant - $D°H$ C-H or C-C product at 30% conversion.*
[1. Ethylbenzene to Styrene, 2. 1-Butene to Butadiene, 3. Acrolein to Acrylic Acid, 4. Ethane to Ethylene, 5.n-Butane to Maleic Anhydride, 6. Propene to Acrolein, 7. Methanol to Formaldehyde, 8. Ethanol to Acetaldehyde, 9.Isobutene to Methacrolein, 10. Isobutane to Isobutene, 11. Methacrolein to Methacrylic Acid, 12. Propane to Propene, 13. n-Butane to Butenes, 14. Propane to Acrolein, 15. Methane to Ethane, 16. Isobutane to Methacrylic Acid, 17. Ethane to Acetaldehyde, 18. Methane to Formaldehyde, 19. Isobutane to Methacrolein][7-9]

The observed correlation (Figure 2) shows that there is a clear relationship between limiting selectivities and the nature of the weakest C-H or C-C bonds in the reactants and products. Moreover, Figure 2 indicates that active sites in oxide catalysts are capable of activating target bonds selectively provided that the difference in bond dissociation enthalpies between the weakest bond in the reactant and the weakest bond in the selective oxidation product is not more than 30-40 kJ mol^{-1}.

REFERENCES

1. M. Turco, G. Bagnasko, P. Ciambi, A. La Ginestra and G. Russo, *Stud. Surf. Sci. Catal.,* 1990, **55**, 327
2. F.M. Baustida, J.M. Campelo, A. Garcia, D. Luna, J.M. Marinas and R.A. Quiros, *Stud. Surf. Sci. Catal.,* 1994, **82**, 759
3. Y.C. Kim, W. Ueda and Y. Moro-Oka, *Stud. Surf. Sci. Catal.,* 1990, **55**, 491
4. I. Matsuura and N. Kimura, *Stud. Surf. Sci. Catal.,* 1994, **82**, 271
5. Y.C. Kim, W. Ueda and Y. Moro-Oka, *Appl. Catal.,* 1991, **70**, 175
6. C. Batiot and B.K. Hodnett, *Appl. Catal.* A, 1996, **137**, 179
7. N. Mizuno, M. Tateishi and I. Iwamoto, *Appl. Catal.* A, 1994, **118**, L1
8. I. Matsuura, H. Oda and K. Oshida, *Catal. Today,* 1993, **16**, 547
9. W.D. Zhang, D.L. Tang, X.P. Zhou, H.L. Wan, K.R. Tsai, *J. Chem. Soc. Chem. Comm.,* 1994, 771

Subject Index